Global Biodiversity
Volume 1
Selected Countries in Asia

Global Biodiversity

Volume 1

Selected Countries in Asia

Edited By
T. Pullaiah, PhD

Apple Academic Press Inc.   Apple Academic Press Inc.
3333 Mistwell Crescent   9 Spinnaker Way
Oakville, ON L6L 0A2   Waretown, NJ 08758
Canada   USA

© 2019 by Apple Academic Press, Inc.

First issued in paperback 2021

Exclusive worldwide distribution by CRC Press, a member of Taylor & Francis Group

No claim to original U.S. Government works

Global Biodiversity, Volume 1: Selected Countries in Asia
ISBN 13: 978-1-77463-131-7 (pbk)
ISBN 13: 978-1-77188-707-6 (hbk)

Global Biodiversity, 4-volume set
ISBN 13: 978-1-77188-751-9 (hbk)

Library and Archives Canada Cataloguing in Publication

Global biodiversity (Oakville, Ont.) Global biodiversity / edited by T. Pullaiah, PhD.

Includes bibliographical references and indexes.
Contents: Volume 1. Selected countries in Asia.
Issued in print and electronic formats.
ISBN 978-1-77188-707-6 (v. 1 : hardcover).--ISBN 978-0-42948-774-3 (v. 1 : PDF)

1. Biodiversity--Asia. 2. Biodiversity--Europe. 3. Biodiversity--Africa. I. Pullaiah, T., editor II. Title.

QH541.15.B56G66 2018   578.7   C2018-905091-8   C2018-905092-6

CIP data on file with US Library of Congress

Apple Academic Press also publishes its books in a variety of electronic formats. Some content that appears in print may not be available in electronic format. For information about Apple Academic Press products, visit our website at **www.appleacademicpress.com** and the CRC Press website at **www.crcpress.com**

Contents

About the Editor

T. Pullaiah, PhD
Former Professor, Department of Botany,
Sri Krishnadevaraya University, Anantapur, Andhra Pradesh, India,
E-mail: pullaiah.thammineni@gmail.com

T. Pullaiah, PhD, is a former Professor at the Department of Botany at Sri Krishnadevaraya University in Andhra Pradesh, India, where he has taught for more than 35 years. He has held several positions at the university, including Dean, Faculty of Biosciences, Head of the Department of Botany, Head of the Department of Biotechnology, and Member of Academic Senate. He was President of Indian Botanical Society (2014), President of the Indian Association for Angiosperm Taxonomy (2013) and Fellow of Andhra Pradesh Akademi of Sciences. He was awarded the Panchanan Maheshwari Gold Medal, the Dr. G. Panigrahi Memorial Lecture award of the Indian Botanical Society and Prof. Y.D. Tyagi Gold Medal of the Indian Association for Angiosperm Taxonomy, and a Best Teacher Award from Government of Andhra Pradesh. Under his guidance 54 students obtained their doctoral degrees. He has authored 46 books, edited 17 books, and published over 330 research papers, including reviews and book chapters. His books include *Ethnobotany of India* (5 volumes published by Apple Academic Press), *Flora of Andhra Pradesh* (5 volumes), *Flora of Eastern Ghats* (4 volumes), *Flora of Telangana* (3 volumes), *Encyclopaedia of World Medicinal Plants* (5 volumes, 2nd edition), and *Encyclopaedia of Herbal Antioxidants* (3 volumes). He was also a member of Species Survival Commission of the International Union for Conservation of Nature (IUCN). Professor Pullaiah received his PhD from Andhra University, India, attended Moscow State University, Russia, and worked as a Postdoctoral Fellow during 1976–78.

Contributors

M. Al-Zein
Nature Conservation Center, American University of Beirut, Lebanon

Leonid Averyanov
Komarov Botanical Institute, Russian Academy of Sciences, Russia, E-mail: av_leonid@mail.ru

Siddhartha B. Bajracharya
National Trust for Nature Conservation (NTNC), Khumaltar, Lalitpur, Nepal,
E-mail: sid.bajracharya@gmail.com

Kjetil Bevanger
Scientific Advisor and Senior Research Scientist, the Norwegian Institute for Nature Research (NINA),
Pb. 5685 Torgaden, NO-7485 Trondheim, Norway, E-mail: Kjetil.bevanger@nina.no

Shekhar R. Biswas
Department of Biology, Lakehead University, Thunder Bay, ON P7B 5E1, Canada

S.-W. Breckle
Department of Ecology, Wasserfuhr 24-26, D-33619 Bielefeld, Germany, E-mail: sbreckle@gmx.de

Gantigmaa Chuluunbaatar
Institute of General and Experimental Biology, Mongolian Academy of Sciences, Mongolia

Nathan Conaboy
Zoological Society of London, UK

M. M. Dehshiri
Department of Biology, Boroujerd Branch, Islamic Azad University, Boroujerd, Iran
E-mail: dehsiri2005@yahoo.com

I. C. Hedge
Royal Botanic Garden, Edinburgh, EH3 5LR, Scotland, UK, E-mail: I.Hedge@rbge.ac.uk

Naing Zaw Htun
Deputy Director, Nature and Wildlife Conservation Division, Forest Department,
Ministry of Natural Resources and Environmental Conservation,
the Republic of the Union of Myanmar, E-mail: nzhtun@gmail.com

Nature Iraq
Pak City Apartments, Block A2, Floor 1, Flat No.8, Near City Center Mall, Kurdistan Region, Iraq,
E-mail: info@natureiraq.org

Irawati
Center for Plant Conservation, Botanic Gardens – Indonesian Institute of Sciences, Indonesia, E-mail:
nfn.irawati@gmail.com

M. Itani
Nature Conservation Center, American University of Beirut, Lebanon

Terbish Khayankhirvaa
Department of Biology, National University of Mongolia

A. Latiff
School of Environmental and Natural Resource Science, Universiti Kebangsaan Malaysia, 43600
Bangi, Selangor, Malaysia, E-mail: pakteh48@yahoo.com

P. K. Loc
Department of Botany, Faculty of Biology, Hanoi University of Science, Vietnam,
E-mail: pkeloc@yahoo.com

Urgamal Magsar
Institute of General and Experimental Biology, Mongolian Academy of Sciences, Mongolia

Sharif A. Mukul
Department of Environmental Management, School of Environmental Science and Management,
Independent University, Bangladesh, Dhaka 1229, Bangladesh / Tropical Forests and People Research
Centre, University of the Sunshine Coast, Maroochydore DC, Qld 4558, Australia /
School of Geography, Planning and Environmental Management, The University of Queensland,
Brisbane, Qld 4072, Australia, E-mail: sharif_a_mukul@yahoo.com, smukul@iub.edu.bd

Fumiko Nakao
Nature Conservation Bureau, Ministry of the Environment, Japan, 1-2-2, Kasumigaseki, Chiyoda-ku,
Tokyo, 100-8975, Japan

Tohru Nakashizuka
Research Institute for Humanity and Nature, 457-4 Kamigamo-Motoyama, Kitaku, Kyoto, 603-8047,
Japan

Takafumi Ohsawa
Nature Conservation Bureau, Ministry of the Environment, Japan, 1-2-2, Kasumigaseki, Chiyoda-ku,
Tokyo, 100-8975, Japan, E-mail: takafumi_osawa@env.go.jp

T. Pullaiah
Department of Botany, Sri Krishnadevaraya University, Anantapur 515003, A.P., India,
E-mail: pullaiah.thammineni@gmail.com

M. D. Rafiqpoor
Nees - Institute for Biodiversity of Plants, University of Bonn, North Rhine-Westphalia, Germany,
E-mail: d.rafiqpoor@uni-bonn.de, www.lotus-salvinia.de

Ramakrishna
Former Director of Zoological Survey of India, Kolkata, India, E-mail: ramakrishna.zsi@gmail.com

A. Z. M. Manzoor Rashid
Department of Forestry and Environmental Science, School of Agriculture and Mineral Sciences,
Shahjalal University of Science and Technology, Sylhet 3114, Bangladesh

Krishna K. Shrestha
Professor and Former Head, Central Department of Botany, Tribhuvan University, Kathmandu, Nepal
E-mail: kkshrestha123@gmail.com

Gombobaatar Sundev
Department of Biology, National University of Mongolia, Mongolian Ornithological Society, Mongolia,
E-mail: gomboo@num.edu.mn; info@mos.mn

Salma N. Talhouk
Department of Landscape Design and Ecosystem Management, Faculty of Agricultural and Food
Sciences, Nature Conservation Center, American University of Beirut, Lebanon,
E-mail: ntsalma@aub.edu.lb

Didik Widyatmoko
Center for Plant Conservation, Botanic Gardens – Indonesian Institute of Sciences, Indonesia, E-mail:
nfn.irawati@gmail.com

M. D. Yen
Department of Zoology, Faculty of Biology, Hanoi University of Science, Vietnam

Abbreviations and Acronyms

CAS	California Academy of Sciences
CBD	Convention on Biological Diversity
CBNRM	Community Based Natural Resources Management
CHAL	Chitwan Annapurna Landscape
CITES	Convention on International Trade in Endangered Species of Wild Fauna and Flora
CMC	carboxymethyl cellulose
CR	Critically Endangered
DPR	Department of Plant Resources
ECOSOC	United Nations Economic and Social Council
EN	Endangered
EW	Extinct in the Wild
FD	Forest Department
FECOFUN	Federation of Community Forestry Users Nepal
FFI	Fauna & Flora International
GDP	gross domestic product
GRA	genetic resource area
IAS	Invasive Alien Species
IBA	Important Bird Areas
IBCAS	Institute of Botany, Chinese Academy of Sciences
ICIMOD	International Centre for Integrated Mountain Development
ILUC	indirect land use change
IPA	Important Plant Areas
IPBES	Intergovernmental Science-Policy Platform on Biodiversity and Ecosystem Services
IPCC	Intergovernmental Panel on Climate Change
ITCZ	Inner Tropical Convergence Zone
IUCN	International Union for Conservation of Nature
KBA	Key Biodiversity Areas
KL	Kangchenjunga Landscape
KRG	Kurdish Regional Government
KSL	Kailash Sacred Landscape
LARI	Lebanese Agricultural Research Institute
LGM	last glacial maximum
LPI	living planet index

LRMP	Land Resource Mapping Project
LSGC	Laboratory for Seed Germination and Conservation
MA	millennium ecosystem assessment
MAB	Man and Biosphere
MOAI	Ministry of Agriculture and Irrigation
MSY	maximum sustained yield
NARC	Nepal Agricultural Research Council
NAST	Nepal Academy of Science & Technology
NBWL	National Board for Wildlife
NCNPP	Nature Conservation and National Parks Project
NEPA	National Environmental Protection Agency of Afghanistan
NS	national strategies
NT	Near Threatened
NTCA	National Tiger Conservation Authority
NWCD	Nature and Wildlife Conservation Division
PA	Protected Area
PAS	Protected Area System
PGR	Plant Genetic Resources
PMR	plant micro-reserves
POES	percentages of endemic species
SHL	Sacred Himalayan Landscape
SNP	Sagarmatha National Park
SPA	seed production area
TAL	Tarai Arc Landscape
TFM	tree flora of Malaya
TFSS	tree flora of Sabah and Sarawak
TU	Tribhuvan University
UNCED	UN Conference on Environment and Development
UNEP	United Nations Environment Programme
USF	Unclassed State Forest
VU	Vulnerable
WCCB	Wildlife Crime Control Bureau
WDPA	World Database on Protected Areas
WHC	World Heritage Convention
XTBG	Xishuangbanna Tropical Botanical Garden

Preface

The term 'biodiversity' came into common usage in the conservation community after the 1986 National Forum on BioDiversity, held in Washington, DC, and publication of selected papers from that event, titled *Biodiversity*, edited by Wilson (1988). Wilson credits Walter G. Rosen for coining the term. Biodiversity and conservation came into prominence after the Earth Summit, held at Rio de Janeiro in 1992. Most of the nations passed biodiversity and conservation acts in their countries. Biodiversity is now the buzzword of everyone from parliamentarians to laymen, professors, and scientists to amateurs. There is a need to take stock on biodiversity of each nation. The present attempt is in this direction.

The main aim of the book is to provide data on biodiversity of each nation. It summarizes all the available data on plants, animals, cultivated plants, domesticated animals, their wild relatives, and microbes of different nations. Another aim of the book series is to educate people about the wealth of biodiversity of different countries. It also aims to project the gaps in knowledge and conservation. The ultimate aim of the book is for the conservation of biodiversity and its sustainable utilization.

The present series of the four edited volumes is a humble attempt to summarize the biodiversity of different nations. Volume 1 covers *Biodiversity of Selected Countries in Asia*, Volume 2 presents *Biodiversity of Selected Countries in Europe,* Volume 3 looks at *Biodiversity of Selected Countries in Africa*, and Volume 4 contains *Biodiversity of Selected Countriess in the Americas and Australia*. In these four volumes, each chapter discusses the biodiversity of one country. Competent authors have been selected to summarize information on the various aspects of biodiversity. This includes brief details of the country, ecosystem diversity/vegetation/biomes, and species diversity, which include plants, animals and microbes. The chapters give statistical data on plants, animals, and microbes of that country, and supported by relevant tables and figures. They also give accounts on genetic diversity with emphasis on crop plants or cultivated plants, domesticated animals, and their wild relatives. Also mentioned are the endangered plants and animals and their protected areas. The book is profusely illustrated. We hope it will be a desktop reference book for years to come.

Biodiversity of some countries could not be presented in this book. This needs explanation. I tried to contact as many specialists as possible from these countries but was unable to convince these experts to write chapter on biodiversity of their country.

The book will be useful to professors, biology teachers, researchers, scientists, students of biology, foresters, agricultural scientists, wild life managers, botanical gardens, zoos, and aquaria. Outside the scientific field it will be useful for lawmakers (parliamentarians), local administrators, nature lovers, trekkers, economists, and even sociologists.

Since it is a voluminous subject, we might have not covered the entire gamut; however, we tried to put together as much information as possible. Readers are requested to give their suggestions for improvement for future editions.

I would like to express my grateful thanks to all the authors who contributed on the biodiversity of their countries. I thank them for their cooperation and erudition.

I wish to express my appreciation and help rendered by Ms. Sandra Sickels, Rakesh Kumar, and the staff of Apple Academic Press. Their patience and perseverance has made this book a reality and is greatly appreciated.

—T. Pullaiah, PhD

CHAPTER 1

Global Biodiversity Threats and Development Trends

KJETIL BEVANGER

Scientific Advisor and Senior Research Scientist, The Norwegian Institute for Nature Research (NINA), Pb. 5685 Torgarden, NO-7485 Trondheim, Norway, E-mail: Kjetil.bevanger@nina.no

1.1 What Is Biodiversity?

Taking a global perspective on biodiversity, it is easy to get depressed, partly because the media prefers to focus on disasters and catastrophic events, but also because we are everyday eyewitnesses to how species, landscapes, and ecosystems worldwide are dwindling and destroyed. However, there are many good and dedicated forces out there when it comes to the nature conservancy and biodiversity protection. Fortunately, an increased understanding of the fact that biodiversity and ecosystem services are crucial for our own and grandchildren's survival and quality of life has become more and more decisive for political decisions worldwide.

The concept of biodiversity is quite young. In 1968, Raymond Fredric Dasmann used the term biological diversity in his book *A Different Kind of Country*. Biodiversity is simply a contraction of these two words. Over the years, however, its logic has been defined and explained in several ways. But biodiversity is simply a concept that encompasses the variety and range of life forms on planet Earth, and as such a measure of *"the variety of life at every hierarchical level and spatial scale of biological organizations: genes within populations, populations within species, species within communities, communities within landscapes, landscapes within biomes, and biomes within the biosphere"* (Wilson, 1988). The UN Convention on Biological Diversity (CBD, 1992) uses the following definition: *"Biological diversity means the variability among living organisms from all sources including, inter alia, terrestrial, marine and other aquatic ecosystems and the ecological complexes of which they are a part; this includes diversity within species, between species and of ecosystems."*

1.2 Biogeographic Classification

Lifeforms and number of organisms are not evenly distributed (e.g., Myers and Giller, 1988). Huge areas on the planet have few or no organisms at all. This relates particularly to polar and desert regions as well as high mountainous landscapes. There are also huge areas with reduced biodiversity, like most cities and densely populated districts, cultivated land, and industrial forests. In fact, more than 50% of the species in the world are (probably) found in tropical rainforests; although these are ecosystems covering only about 7% of the earth land surface (Wilson, 1988).

Because of land area similarities and differences when it comes to, for example, species number, distribution, and abundance, botanists and zoologist have debated how to make useful biogeographic classification systems. Concepts like life zones, biotic provinces, faunal and floral regions, realms, biomes, etc., have been heavily discussed for more than 150 years (Udvardy, 1969). More recently, ecoregion has become a concept commonly used to describe ecosystem characteristics on a finer scale (Olson et al., 2001; Bailey, 2014).

1.2.1 Biomes

Biomes are broad habitat and vegetation types, spanning across biogeographic realms, being useful units for assessing global biodiversity and ecosystem services because they stratify the globe into ecologically meaningful and contrasting classes. The biome concept has survived in modern textbooks of ecology, and are described within chapters discussing community structure patterns. It is frequently underlined that ecologists and biogeographers operate with differing numbers of biomes based on differences and similarities in flora and fauna as well as environmental factors. There are of course no sharp boundaries when it comes to communities, ecoregions or biomes; however, it is definitely convenient from a biological point of view to categorize the landmasses based on dominating ecological characteristics. Classical terrestrial biomes are tundra, taiga, temperate forest, grassland, chaparral, desert, and tropical rainforests (Udvardy, 1975; Cox et al., 1976; Begon et al., 1996, 2006).

Tundra is characterized by permafrost and mainly found north of the Arctic Circle and the tree line. However, small areas of alpine tundra can be found even at high altitudes in some tropical mountains. Water is only available for some short periods during summer, and low growing bushes, lichen, mosses, and grasses dominate the vegetation. The number of vascular plants decreases significantly from the Low to the High Arctic areas. Greenland and Ellesmere Island are left with about 100 species, while in Antarctica only two native higher plants exist (Begon et al., 2006) (Figure 1.1).

Both the Nearctic and Palearctic have a broad belt of mainly coniferous forest, frequently known as the *Taiga* region (Figure 1.2). It is definitely the

Figure 1.1   Arctic, ice covered habitats within the tundra biome. East coast of Greenland in April.

largest biome in the world, with 27% of the world's forest cover, occupying 11% of the Northern Hemisphere land area (http://www.wilds.mb.ca/taiga/tbsfaq.html). The number of tree species is low, particularly in the northern parts, and a mixture of pines (*Pinus* spp.), larch (*Larix* spp.), birch (*Betula* spp.), and aspen (*Populus* spp.) is common. To the north, vast areas with spruce (*Picea* spp.) dominates. In this part of the taiga, the permafrost limits the vegetation growth, and only during some few summer months, the upper layer of the soils is thawed and able to support the root system with water.

Figure 1.2 Taiga; close to the city of Arkhangelsk, Russia. Huge areas covered
with forests dominated by coniferous tree species with clear signs of
logging activity.

Temperate forests comprise several forest mixtures in the temperate zone,
particularly at the northern hemisphere, like the temperate deciduous forest,
temperate coniferous forests, temperate rainforest, and temperate broadleaf
and mixed forest (Begon et al., 2006). Although these forests dominate in
North America and the central and northern parts of Europe, low latitude
evergreen and dripping forests like those found in New Zealand is categorized within this biome as well. However, deciduous trees losing their leaves
during the fall dominate most temperate forests because they are distributed
in the areas with winter temperatures below zero. Due to this seasonality,
there is also a diverse flora of vascular plants on the forest floor, particularly
during the spring, before the leaves on the trees have developed. Insects,
mammals, and birds are also numerous, but the majority of the birds are
migrants spending the breeding period in the northern parts before migrating
to lower latitudes during winter (Figure 1.3).

Grassland is used to recognize another biome having a variety of
grass-dominated ecosystems found both in temperate and tropical regions.
Temperate grasslands could be prairie (North America), steppes (Asia),
pampas (South America) and veldt (South Africa), while savanna is a typical

Figure 1.3 Moist, temperate forests. At Haast River, New Zealand.

tropical grassland (Figure 1.4). Most grasslands experience drought periods, though grazing animals are more important in the ecosystem dynamics. One of the best-known examples is the greater Serengeti-Mara Ecosystem with its famous wildebeest (*Connochaetes gnou*) and zebra (*Equus zebra*) migrations. Another important driving force to maintaining grasslands is fire. Due to seasonal as well as long- and short-term shifts in precipitation, the availability of food for mammals, birds, or insects varies and may result in mortality and/or specific migrating patterns (Figures 1.5 and 1.6).

The global distribution of grasslands correlates, not surprisingly, with the most densely human populated areas. Grasslands have been transformed into cultivated areas with monocultures and only a few dominating grass species like maize (*Zea mays*), wheat (*Triticum aestivum*), barley (*Hordeum vulgare*), rye (*Secale cereale*), and oats (*Avena sativa*) in the temperate regions and rice (*Oryza sativa*) in the tropics. These few species represent the staple food sources for people worldwide, however, at the cost of a significant decrease in grassland biodiversity. Huge areas are also used for milk and meat production, which is being dominated mainly by three species: cow (*Bos taurus*), goat (*Capra aegagrus hircus*), and sheep (*Ovis aries*) at the cost of a variety of naturally occurring mammals.

Most continents have areas with dry summers and wet, mild winters, creating so-called Mediterranean landscapes and climate types (Noble and

Figure 1.4   Tropical grassland. Savanna. Serengeti Plains, Tanzania.

Figure 1.5   Grassland at Cooma, New South Wales, Australia.

Bradstock, 1989). Here the chaparral (*maquis*) biome occurs (Gjaerevoll, 1973) (Figure 1.7). Chaparral ecosystems experience less precipitation than grasslands and the vegetation is dominated by drought-resistant plant species (scrub and brushes), though periods with rain during late winter

Figure 1.6   Grassland, Prairie, Colorado, USA.

Figure 1.7   Chaparral vegetation, Mallorca, Spain.

and spring open up for several annual vascular plant species. Periodic fires are an important part of this ecosystem dynamics as several plant species are adapted to a fire regime and will in fact not germinate without being burned.

Deserts are found on all continents and are characterized by very low precipitation over the year; normally less than 250 mm. Such ecosystems are found in extreme cold areas like in Antarctica and Mongolia (Gobi). Hot deserts, like Sahara, Sonoran, and Atacama, are just the same, although they have some fire resistant plant and animal species with specific adaptations. When an unpredictable rainfall occurs, it is an overwhelming sight to see the normally bare sand dunes converted into blossoming meadows. Huge amounts of seeds are produced within a short period surviving inactive in the dry sand until the next rains. The plants of the Sonoran desert, like the saguaro cactus (*Carnegiea gigantea*), have developed another strategy, that is, to store water to stand long periods of droughts (Figure 1.8).

The most productive, and species-rich biome is the *tropical rainforests* (Longman and Jeník, 1987). Its main distribution follows the Equator but may extend up to Tropics of Cancer and the Tropics of Capricorn. Thus, it is not a surprise that several ecosystems are encompassed by this designation. Among the best-known types are the *lowland equatorial evergreen rainforests* in the Amazon and Congo basins in South America and Central

Figure 1.8 Sonoran desert with Saguaro cactus, Arizona, USA.

Africa, respectively. In Asia, it is found in the Philippines, Indonesia, and New Guinea. Frequently these ecosystems receive more than 2000 mm rain per year. The Amazonian rainforests also include several types of *flooded rainforests*. Another type is the *moist tropical seasonal forest*. Here, the overall precipitation is high, but restricted to a warm summer season, while the cooler winter season is drier. During winter, some of the tree species may even lose some of their leaves. These forests are particularly distributed in Indonesia, the Indian subcontinent, Central America and West Africa. The *cloud forests* or the mountain rainforests, appear in somewhat cooler climate between 1500 and 2500 meters above sea level, but may be found up to 3300 meters (Bruijnzeel and Veneklaas, 1998).

Rainforests are very old ecosystems that have developed within environmental conditions of regular precipitation patterns and high solar radiation. They are the most productive and diverse ecosystems on the planet, but also of those disappearing at a very high speed. Some few hundred years back, rainforest covered up to 20% of the land surface; now only 5–7% is left. The broadleaved evergreen trees in the rainforests have a tight canopy being the main biomass-producing engine. Typically, however, below the upper canopy, there are two or more vegetation layers. Compared to the canopy diversity the forest floor is a big contrast as very little sunlight penetrates down. The sparse vegetation allows animals and humans to move easily around in most places (Figures 1.9–1.12).

1.2.2 Ecoregions

Revising the old classification biogeographic systems, Olson et al. (2001) came up with a global map of 867 terrestrial ecoregions categorized within 14 biomes and 8 biogeographic realms (Palearctic, Nearctic, Afrotropical, Indomalayan, Oceanian, Australian, Antarctic, and Neotropical). These modifications are a significant improvement from a conservation and management perspective. The identification of 867 ecoregions has provided politicians and environmental managers at different levels with a new and strong tool for long-term planning. As such, earlier maps have not been effective due to the coarse biodiversity units. The average size of the 867 ecoregions is 150,000 km², compared to 740,000 km² of the Udvardy (1975) biotic provinces. This new ecoregion map has come into use in several ways, for example, when looking at possible biodiversity losses and threats from different encroachments. A detailed list of the ecoregions is found on the WWF website http://wwf.panda.org/about_our_earth/ecoregions/ecoregion_list/.

Figure 1.9   Interior equatorial rainforest, Madre dos Dias, Amazon Basin, Peru.

1.3   Biodiversity Patterns and Species Number

1.3.1   How Many Species?

Biodiversity focuses mainly on species, and questions on the number of species on the planet have triggered both scientist and policy makers for a

Figure 1.10 Scarlet macaw (*Ara macao*) licking clay at Madre dos Dias river, Peru.

Figure 1.11 Equatorial rainforest tree dimension in Amazon, Peru.

Figure 1.12 Mountain rainforest, Salak Mountain, Java Indonesia.

very long time. The estimates have differed; however, they have steadily been adjusted upwards, particularly due to research documenting the huge species numbers recorded in rainforest areas. In 1988, Wilson (1988) stated that "the absolute number falls somewhere between 5 and 30 million." In 2011, Mora and his coworkers compiled existing information on species number and predicted 8.7 million (± 1.3 million SE) eucaryotic species globally, of which 2.2 million (± 0.18 SE) are marine (Mora et al., 2011) (see, Table 1.1). Due to their findings, they assumed that 86% and 91% of existing species on Earth and in the ocean, respectively, remain unknown.

Last year, Locey and Lennon (2016) published a study including microbial diversity. They used scaling laws to predict global, microbial diversity. Scaling laws expose how physiological, ecological, and evolutionary limitations hold across genomes, cells, organisms, and communities of very varying dimension. A classic example is how species number (S) scale with an area (A); $S = cA^z$ (z is the slope of the species–area relationship in log–log space) (MacArthur and Wilson, 1967).

To test whether scaling laws apply to microorganisms or not, Locey and Lennon (2016) compiled approximately 5.6×10^6 species from ~35,000 sites of both macro- and microorganisms around the world, and documented comparable rates of scaling in commonness and uncommonness across

Table 1.1 Cataloged and Predicted Species Number on Earth and in the Ocean.

Species	Earth			Ocean		
	Cataloged	Predicted	±SE	Cataloged	Predicted	±SE
Eucaryotes						
Animalia	953,434	7,770,000	958,000	171,082	2,150,000	145,000
Chromista	13,033	27,500	30,500	4,859	7,400	9,640
Fungi	43,271	611,000	297,000	1,097	5,320	11,100
Plantae	215,644	298,000	8,200	8,600	16,600	9,130
Protozoa	8,118	36,400	6,690	8,118	36,400	6,690
Total	1,233,500	8,740,000	1,300,000	193,756	2,210,000	182,000
Procaryotes						
Archaea	502	455	160	1	1	0
Bacteria	10,358	9,680	3,470	652	1,320	436
Total	10,860	10,100	3,630	653	1320	436
Grand total	**1,244,360**	**8,750,000**	**1,300,000**	**194,409**	**2,210,000**	**182,000**

Source: Mora et al., 2011

microorganisms and higher plants and animals. They found a universal scaling law predicting the abundance of dominant species across 30 orders of magnitude to the scale of all microorganisms. Combining this scaling law with the lognormal model of biodiversity, they estimated that Earth has about one trillion (10^{12}) microbial species, which is far beyond earlier approximations.

1.3.2 Biodiversity Patterns

Issues on biodiversity distribution are of fundamental importance in many aspects when it comes to how biodiversity should be managed to the best for future generations. Present threats to biodiversity, being land use change, invasive species, climate change, or human population growth, cannot be handled successfully without such knowledge. Fortunately, the last decades have provided tools, such as remote sensing technology, to both large-scale data sampling and huge dataset analyzes (Gaston, 2000).

The decisive processes when it comes to how biodiversity patterns and species diversity are regulated is not always obvious as both unknown historical, as well as accidental events, are involved. The reason why a certain bird or group of birds exist on a remote island could be that an ancestor arrived by chance due to extremely windy conditions. However, many other factors need to be present; for example, the "immigrants" must be of different sex, and the living conditions should be good with few competitors, predators, and suitable food and nesting conditions. Isolation can result in a remarkable speciation and adaptive evolution, of which Darwin's finches is a classic example (Lamichhaney et al., 2015; Almén et al., 2015).

The pattern of species diversity is regulated in space and time by many aspects (Brown, 1988), particularly temperature, precipitation, soils, altitude, and inter- and intra-specific factors. The topic has been heavily discussed, not at least how terrestrial and marine ecosystem diversity is influenced by latitudinal and altitudinal gradients (Dobzhansky, 1950; MacArthur, 1965; Terborgh, 1985; Stevens 1989, 1992; Rex et al., 1993; Rosenzweig, 1995). The fact that tropical ecosystems are the most diverse in numerous taxa compared to higher latitude ecosystems is definitely connected to history. Equatorial ecosystems are very old, having developed for probably millions of years. Moreover, a mild climate with high solar influx is an efficient biomass producing system allowing more species to survive per unit area. At higher latitudes, periods with glaciation have eradicated most species within relatively short intervals.

Marine ecosystem patterns are less known (Irigoien et al., 2004). However, in some areas, a parallel to terrestrial ecosystem patterns are found. In the Great Barrier Reef, soft corrals expose a "hotspot" in taxonomic richness on the mid-shelf between 13° and 16°, and decreases with increasing latitude, and is low south of 21° (Fabricius and De'ath, 2001). It is, however, important to keep in mind that some of the most productive ecosystems are found in Antarctica and the Southern Ocean because of the upwelling dynamics (Anderson et al., 2009). The nutrient water brought up to the surface triggers a huge phytoplankton growth, creating food surplus for small crustaceans which in turn form the basis for a diverse marine fauna of whales, seals, birds, and fish.

1.3.3 Biodiversity Hotspots

"Biodiversity hotspots" is commonly used among conservationists, ecologists, and politicians. The concept was originally introduced by Myers (1988), and defined as "*tropical forest areas that feature exceptional concentrations of species with exceptional levels of endemism, and that face exceptional degrees of threat.*" Thus, the basic idea was that it is crucial to prioritize when it comes to area conservation, also because of the economy involved; that is, identify those areas where protection is critically needed, and which will also yield a high biodiversity return when saving numerous species.

The system has been revised several times (Myers, 1990; Myers et al., 2000; Mittermaier et al., 2011). However, being adopted by Conservation International already in 1989, it has become an important management tool. Myers et al. (2000) identified 25 hotspots meeting the criteria of containing at least 0.5% or 1500 vascular plants as endemics, and suffered a 70% loss of primary vegetation. For the time being, 36 biodiversity hotspots have been identified, supporting close to 60% of the plant, bird, mammal, reptile, and amphibian species in the world. Majority of these hotspots are located in tropical forests, representing less than 2.5% of the land surface of the planet. It is estimated that the hotspots have lost close to 90% of their original land areas (https://en.wikipedia.org/wiki/Biodiversity_hotspot). The biodiversity hotspot concept has been criticized because its criteria mainly are based on vascular plants, not considering ecosystem services, as well as not addressing the concept of cost (Kareiva and Marvier, 2003; Possingham and Wilson, 2005; Daru et al., 2014).

1.4 Biodiversity Conservation

1.4.1 Global Protected Areas

All countries have their own history when it comes to protected areas (PAs). Looking into this issue, we will see that many countries have had ideas on areas that should be managed to remain pristine several hundred years back. It is, however, only over the last 50 years or so that policies on PAs have become a common global issue. The main reason is, of course, the human population growth and the senseless resource outtake in several areas, as well as the growing awareness of the long-term consequences. Overexploitation and unsustainable harvest of resources, like the whaling by the Norwegians and other countries in the Southern Ocean in the eighteenth century, is a terrifying example. Starting to see that species like the blue whale disappeared and became close to exterminated, became a warning and wakeup call for consorted actions. There are numerous other examples worldwide, both when it comes to species and habitats that have been lost due to an increasing human population growth and the need for resources to stay alive. Unfortunately, as humans, we also have an inherited greediness and lack of long-term planning abilities that frequently overrule rational actions. The globalization process has luckily made countries come together also when it comes to environmental issues, making the basis for increased understanding, cooperation, and international agreements.

Today PAs are regarded as decisive tools for maintaining ecosystems and species in most countries (Geldman et al., 2013). It is estimated that PAs globally cover about 12.7% of the land surface (Bertzky et al., 2012). There is no doubt, however, that in several countries, PAs are insufficient to take care of those ecosystems and species that need both long- and short-term protection (Joppa and Pfaff, 2011). Some countries advertising with a high percentage of their land area protected, typically have avoided including the most productive parts, which of course are the most controversial ones when it comes to political actions. Reviewing 2599 publications, Geldman et al. (2013) concluded that PAs have contributed to forest habitat conservation, but the evidence was inconclusive regarding maintenance of species populations. However, the overall conclusion was that PAs deliver positive outcomes *"but there remains a limited evidence base, and weak understanding of the conditions under which PAs succeed or fail to deliver conservation outcomes."*

1.4.2 International Commitments and Cooperation

Today, most countries have a dedicated ministry taking care of environmental issues. A historical view reveals that very few multilateral or bilateral agreements on international conservation existed prior to 1900. The focus was typically connected to economic interests and resource exploitation like fishery boundaries and fishing rights, without addressing environmental or ecological issues as a topic for cooperation. This is also reflected in the treaties agreed upon before World War II, such as "Convention to Protect Birds Useful to Agriculture" (USA–Canada 1902) and "International Convention on the Regulation on Whaling" (1931).

Before 1972, no country had a specific cabinet minister for conservation management. In 1972, the Norwegian botanist Olav Gjaerevoll, from the Labour Party, became minister of environment in Norway, the same year as the United Nations Conference on the Human Environment was convened ("The Stockholm Conference"). As early as 1968, Sweden suggested to the United Nations Economic and Social Council (ECOSOC) to have a UN conference focusing on human/environment interactions (https://en.wikipedia.org/wiki/United_Nations_Conference_on_the_Human_Environment). Another important event in 1972 was the UN General Assembly decision on United Nations Environment Programme (UNEP) to act as a principal point for environmental action.

The Stockholm Conference and its declarations represent a paradigm shift regarding awareness of trans-boundary environmental challenges and cooperation needs. Fortunately, it had become evident that developing adaptive and correcting mechanisms is a critical task to undertake. Another important event was the World Commission of Environment and Development, also known as the "Brundtland Commission." The UN General Assembly formed the commission in 1983 to develop a long-term environmental strategy for sustainable development, finalized in the Brundtland report "Our Common Future," in 1987. An important concept introduced in the report was "sustainable development." At the UN Conference on Environment and Development (UNCED), known as the Earth Summit, or Rio Summit, in 1992, the concept was revised, as principles relating to national accountabilities and international cooperation on environmental protection were included.

The Rio Summit resulted in the Rio Declaration, consisting of 27 principles intended to guide countries in their work on future sustainable development. It also resulted in Agenda 21, which is a UN non-binding action plan regarding sustainable development. Not least, it created "The Convention on Biological Diversity" (CBD), as it was opened for signature at the Rio Summit. There is no doubt that this has become a key document when it comes to developing national conservation strategies and sustainable biodiversity development.

One year before the Stockholm Conference, that is, in 1971, an important convention aiming at biodiversity protection was signed in the city of Ramsar in Iran, the so-called Ramsar Convention. It is also called the Convention on Wetlands as it aims to protect wetlands of international importance, particularly waterfowl habitats. Thus, it became the first environmental convention for the protection of specific habitats and not species. A resolution on international trade in endangered species of wild fauna and flora (later known as CITES) was adopted as early as 1963 among the World Conservation Union (IUCN) members, and 12 years later CITES entered force. In 1979, an important convention to protect migratory species was signed in Bad Godesburg, a suburb of the German city Bonn. It entered into force in 1983 and is known as the Convention on the Conservation of Migratory Species of Wild Animals, or the Bonn Convention.

Another important event regarding focus on biodiversity and human well-being took place in 2001 when UN initiated the Millennium Ecosystem Assessment (MA), governed by a multi-stakeholder board including representatives of international institutions, governments, ethnic minorities, NGOs, and trade. The secretariat was coordinated by UNEP, and more than 1360 scientists from 95 countries contributed (Millennium Ecosystem Assessment, 2005). In short, the MA aimed to *"assess the consequences of ecosystem change for human well-being and establish the scientific basis for actions needed to enhance the conservation and sustainable use of ecosystems and their contributions to human well-being."* Based on CBD and other international conventions, the MA initiative was supported by several governments recognizing that biodiversity plays a serious role in supporting ecosystem services.

In 2012, an Intergovernmental Science-Policy Platform on Biodiversity and Ecosystem Services (IPBES) was established in Panama as a mechanism to arrange for scientific information in answer to demands from policymakers. The IPBES secretariat is held by the German government and sited on the UN campus in Bonn, placed under the umbrellas of four UN entities: UNEP, UESCO, FAO, and UNDP; and administered by UNEP. The current membership includes 126 governments.

1.4.3 Ecosystem Services

Humans have probably always recognized our dependence on nature. Even Plato understood that deforestation could cause soil erosion. However, it was not until the late 1940s that scientists (e.g., Leopold, 1949) started to write about the human dependence of the environment. In 1970, the term "environmental services" was introduced (SCEP 1970), but despite that scientists and naturalist have discussed ecosystem services indirectly for many years, it was not until the Millennium Ecosystem Assessment published its reports in 2005 the concept was thoroughly defined and popularized.

The Millennium Ecosystem Assessment, or MA report (2005), focuses on the associations between human well-being and ecosystems, and define ecosystem services as the benefits we obtain from ecosystems. These services are grouped into four categories, where the so-called supporting services are considered the basis for the services of the other three categories. The supporting services include nutrient recycling, primary production, and soil formation, as these are the ecosystem elements that are necessary for the regulations like food, clean water, and flood; that is, the provisioning services.

The provisioning services are products we can obtain from the ecosystems, like game meat, fishes, crops, fruit, spices, etc. It also includes all types of raw materials, like firewood, timber for house building and hides. Not least is access to clean water vital. Plants and animals also contribute with pharmaceuticals and genetic resources like those that can improve crop yield. Rivers can be used for hydropower production and forest material for biomass fuel. Ornamental resources also derive from ecosystems, be it for handicraft, decoration and souvenirs, aquarium fishes, shells, etc.

Regulating services are those obtained from the regulation of ecosystem processes, such as climate regulation and control, waste decomposition and detoxification, water and air purification, and pest and disease control. The fourth category is about cultural services. These are nonmaterial benefits we, as humans, may obtain through spiritual enrichment, cognitive development, reflection, recreation, aesthetic experiences, etc.

In particular, The MA report focuses on direct and indirect drivers of change in ecosystems, the present conditions of those services, and how changed ecosystem services will, and have affected human well-being. How can ecosystem changes impact income and material needs, good social relationships, health and choice independence?

1.5 Biodiversity Threats and Drivers of Change

The term "tipping point" has become very common over the last years, not least in the scientific literature (van Nes et al., 2016). Previously, it was broadly used to describe something that enters and penetrates a threshold, causing changes in a system and forcing it to operate at a new state. Although it originally seems to have been used about social conflicts, it has now become an extremely popular term in ecology and environmental science. This is probably partly due to a book by Malcolm Gladwell, issued in 2000 (Gladwell, 2000).

Self-propelled accelerating change in ecological systems is well known. In tropical regions, there are examples of tree mortality caused by storms opening up for the invasion of grasses and wildfires, which in turn can kill more trees and finally turn the tropical forest into an open savanna (Hoffmann et al., 2012). Van Nes et al. (2016) propose that "tipping point" simply should be used to describe a situation where accelerating change caused by a positive feedback drives the system to a new state. Thus, it is a very relevant term when it comes to biodiversity threats like consequences of species extinction, climate change, invasive species, etc.

1.5.1 Species Extinction

Wilson (1988) predicted that the species extinction rate due to current human interventions, and particularly rainforest destruction, is 1,000 to 10,000 times background rates typical over Earth's history. Several researchers have tested this estimate later, and so far, none has reached another result.

The overall conclusion is that the human population increases over the last few hundred years have changed the diversity of life on all continents and caused a tremendous biodiversity loss. The negative changes regarding biodiversity have been more rapid over the past 50 years than ever in the history of man, and the most disturbing thing is that the projections and future scenarios points in the same direction; the negative trends will continue or even accelerate.

The MA reports summarize the negative, declining trends in several taxa, pinpointing among other facts that globally amphibians, African mammals, agricultural land birds, British butterflies, Caribbean and IndoPacific corals, and commonly harvested fish species expose declining trends. About 100 extinctions of birds, mammals, and amphibians have been documented over the last 100 years; that is, 100 times the background rates. Of well-studied higher taxonomic groups (mammals, birds, amphibians, conifers, and cycads), between 10% and 50% are currently threatened with extinction, according to IUCN criteria. About 12, 23, and 25% of birds, mammals and conifers respectively, currently face extinction, as well as 32% of amphibians, however, that could be an underestimate. Among evergreen palm-like plants (cycads) a threat of 52% has been recorded. In addition, the genetic diversity has declined globally, not at least among domesticated species. Over the last 50 years, or since the "Green Revolution" (http://www.agbioworld.org/biotech-info/topics/borlaug/borlaug-green.html), it has been a complete change in farming organizations due to the intensification of agricultural systems and specialization by plant breeders etc., resulting in a considerable reduction in genetic diversity of both domesticated plants and animals.

1.5.2 *Entering the Sixth Mass Extinction?*

In general, planet Earth is supposed to have experienced five mass extinction episodes (Wake and Vredenburg, 2008). The first two are questioned because some analyses indicate that the extinction scale was not significantly higher than in other events (Alroy, 2008). The oldest mass extinction, however, is supposed to have occurred approximately 439 million years ago (Mya) (Ordovician-Silurian period). About 25% of the families and 60% of the genera of marine organisms seem to have disappeared due to variations in sea levels resulting from widespread glaciation followed by a period of global warming (Jablonski, 1995; Erwin, 2001).

Some 364 Mya, in the late Devonian period, 22 and 57% of marine families and genera, including close to all jawless fishes disappeared (Jablonski,

1995; Erwin, 2001). The upbringing for this event is supposed to be global cooling as a response to a meteor colliding with our planet; this is supported by the fact that taxa associated with warm water, in particular, were affected.

The worst of the five mass extinctions took place some 252 Mya, during the Permian-Triassic period. Approximately 95% of all taxa, marine as well as terrestrial, disappeared, including 53% of marine families, 84% of marine genera and 70% of terrestrial plants, insects, and vertebrates (Jablonski, 1995; Erwin, 2001). The background for this disaster is debated, but it is generally agreed upon that volcanic activity leading to severe climatic change, including oxygen loss in the sea, was the main reason.

Towards the end of the Triassic period, some 199–214 Mya, another extinction took place. Supposedly, massive volcanic activity and lava floods in the Atlantic Ocean caused a substantial global warming. The marine taxa were particularly affected, and some 22 and 53% of marine families and genera were lost. Several terrestrial taxa vanished as well (Jablonski, 1995; Erwin, 2001).

The best-documented and most recent mass extinction took place some 65 Mya, between the Cretaceous-Tertiary periods when 16 and 47% of families and genera among marine organisms were lost. Among vertebrate families, 18% disappeared, including non-avian dinosaurs. The reason for this catastrophic event is also debated; however, several scientists believe an asteroid hit the planet Earth in the Gulf of Mexico (Jablonski, 1995; Erwin, 2001).

Being in the Anthropocene period (Zalasiewicz et al., 2008), the question is whether this will be the period for a new mass extinction event. Some scientists even think we already are "in the midst of the sixth mass extinction," as there are extensive indications on an ongoing extinction episode (Wake and Vredenburg, 2008). Hundreds of anthropogenic vertebrate eradications are documented in prehistoric and historic times (Ceballos and Ehrlich, 2009; Ceballos et al., 2015). On the islands of tropical Oceania, about 1800 bird species, described from subfossil remains, have vanished during the approximately last 2000 years; that is, after humans colonized the area (Steadman, 2006). Over the last 400–500 years, there are many written records about animal extinctions like the dodo (*Raphus cucullatus*), being eradicated in the seventeenth century, Steller's sea cow (*Hydrodamalis gigas*), eradicated in the eighteenth century and the Rodrigues giant tortoise (*Cylindraspis peltastes*), eradicated in the nineteenth century. Several extinction records of reptiles, amphibians, freshwater fishes and other animals have been made since the beginning of the twentieth century (Ceballos et al., 2015).

Recently a background rate of two mammal extinctions per 10,000 species per 100 years has been estimated, which is the double of the previous rough estimates (Barnosky et al., 2011). Using this as the background rate, Ceballos et al. (2015) found that modern rates of vertebrate extinction were much higher. According to vertebrate taxa evaluations by IUCN, 338 extinctions have been documented since 1500. Another 279 species have either become "extinct in the wild" or listed as "possibly extinct." A majority of the extinctions have occurred during the last 100 or so years, that is, since 1900. The Ceballos et al. (2015) "highly conservative" and "conservative" modern extinction rates for vertebrates were 8 to 100 times higher than the background rate. The estimates can be discussed, however, there is no doubt that if the current extinction speed continues, we will soon be deprived of several biodiversity paybacks.

1.5.3 Human Population Growth

Compared to other species, *Homo sapiens* are quite new to planet Earth, appearing some 130,000 to 160,000 years ago. The human population growth rate has been slow most of the time, and it was not until 1804 that the estimated human population reached one billion. Since then, however, the growth curve has more or less followed an exponential pattern. In 1999, we reached 6 billion, and for the time being, we are 7.5 billion (http://www.worldometers.info/world-population/), although demographers have estimated that the growth rate will slow down and that we will not reach 9 billion until 2050.

From now on, human population growth will mainly take place in developing countries, and that is part of the problem as the majority of rainforests and biodiversity are located within the borders of these countries. To support the people with adequate resources it is inevitable that pristine rainforest and other ecosystems rich in biodiversity become targets for unsustainable use of ecosystem services, for example, legal and illegal logging, bushmeat consumption and general land use change. Indirect consequences are for example, air and water pollution from toxic materials, greenhouse gases and excess of nutrients causing algal blooming, oxygen depletion and fish mortality.

Thus, it is unquestionable that human population growth is the largest environmental problem and challenge we are facing today. Unfortunately, it is a very complex question to deal with. Intervening people's traditional lifestyle is something politicians do not want to do. The only exception seen so far is China with its one-child policy (1978–2015). The success of this is a significant population growth reduction; however, the long-term consequences remain to be seen. Another strategy that most politicians agree on is education. The Sub-Saharan Africa has the highest birthrate in the world; frequently called a demographic-economic paradox, as most countries in that region are very poor. The number one factor shown to impact fertility, in general, is education and age of marriage (Shirahase, 2000). The number of years at the school for girls is closely associated with the number of children born and time of first birth. So far, this is the best and least controversial medication to reduce population growth also in developing countries.

1.5.4 Climate Change

The Fifth Assessment Report (AR5), the latest report from the Intergovernmental Panel on Climate Change (IPPC), was issued in 2014. The overall conclusions are clear; humans influence the global climate system. The recently observed emission of greenhouse gases has never been higher, and the climate changes have had widespread impacts on our ecosystems (IPCC, 2014).

A long range of human activities, including rainforest destruction, influence the greenhouse gas concentrations in the atmosphere. The contributing estimates vary, however, deforestation and land conversion are supposed to contribute approximately 30% of global greenhouse emission. Diversity loss is supposed to reduce ecosystem resilience to climate change and other disturbances.

Depending on what scenarios regarding global warming we will see, there will be significant impacts on biodiversity. Climate change and global warming are supposed to be disastrous to many areas, as frequencies of floods, droughts, erosion etc. will threaten large-scale infrastructure investments and densely populated areas. Scenarios including sea level increase will hit several large cities forcing people to settle elsewhere.

Warmer regional temperatures have already affected species distribution, population sizes, the timing of reproduction or migration patterns as well as a rise in outbreaks of pests and diseases. Coral reefs have been bleached due to warmer seawater. By the end of the twenty-first century, climate change could be the dominant driver of biodiversity loss and changes in ecosystem services on a global scale (MA, 2005).

Arctic habitats are supposed to respond intensively to global warming. One of the species that have been predicted to suffer is the polar bear (*Ursus maritimus*). If the species loses its sea-ice habitats and experience-reduced access to the primary prey (seals), it could be disastrous in several ways. The polar bear is hunting seals from solid sea ice stands. Higher temperatures will contribute to an earlier melt down of the sea ice, which will force the animals to mainland areas before they have accumulated sufficient fat reserves to survive the period of limited food in the late summer and early fall.

1.5.5 Land Use Change

The MA (2005) predicts a persistent fast conversion of ecosystems until 2050. The predictions include a 10–20% change of current grassland and forested areas to other uses, particularly because of an increased need of agricultural lands, but also due to the expansion of densely populated areas and infrastructure development. Consequently, global extinctions will continue until the number of species reaches equilibrium with remnant habitats. Over the period 1970–2050, the equilibrium number of plant species is estimated to be reduced by 10–15% due to habitat loss. However, this is probably an underestimate as stressors like climate change and pollution are not included. In the same manner, modifications of rivers and watercourses will generate extinctions of fish species.

A typical example of consequences of land use change is connected to biofuel production. Biofuels are important for several countries to meet the international commitments on CO_2 emission. The biofuel production normally takes place on cropland previously used for food and fodder, which is still a needed production. Thus, it may be displaced to previously non-cropland areas like forests and grassland. This is typically known as indirect land use change (ILUC). However, atmospheric greenhouse gas concentrations may increase due to biofuel production because forests and grasslands absorb huge amounts of CO_2 (European Commission, 2012).

Another example of ecosystem change is mangroves being transformed for aquaculture purposes. It is estimated (where data is available) that approximately 35% of the mangroves worldwide have been lost over the last two decades. This covers about 50% of the total mangrove areas in the world. In several Asian developing countries like Vietnam, the development of aquaculture in mangrove ecosystems is significant (Tong et al, 2004; Arizumi et al., 2015). Using SPOT-satellite images, it was possible to learn how shrimp aquaculture affected the mangrove ecosystem in the Mekong Delta. Knowledge on the spatial distribution of shrimp ponds and mangroves, and the transformation of mangroves is very important to understand mangrove conservation and sustainable use. In many places, shrimp aquaculture is supposed to be the main threat to mangroves.

1.5.6 Alien Species

Ecosystems are dynamic, that is, they are in a sort of equilibrium, which means they are in a relatively stable state and resilient towards changing impacts. However, within a given period the abundance and distribution of different organisms will change, depending on the length of the period and environmental change extent (Begon et al., 2006). Ecosystems rarely have an option to remain long enough to form "stable" systems. Humans have always had a qualitative and quantitative impact on vegetation and animals.

Human population growth, together with an increased globalization, has led to a significant rise in a number of alien species in several regions over the past two hundred years (di Castri, 1989). Alien species are species that have come to a new area because of human activity, whether through intentional introduction or unintentionally (Vitousek et al., 1996; McKinney and Lockwood, 1999). One impact from alien species is that species distribution becomes more homogeneous, that is, differences between species composition from one location to another on average are shrinking. Another factor intensifying this development is that species unique to specific regions are experiencing higher rates of extinction. In recent decades, exotic invasive species have gained increased attention as a threat to indigenous biodiversity, and many countries have worked out their own "black lists" (Bevanger et al., 2007).

The same alien species causing loss of biodiversity frequently also bring financial losses (Pimentel 2002). It has, for instance, become evident that agriculture, forestry, and fisheries are suffering very high costs, and our own health is greatly compromised as well. Estimates show that alien species inflict an annual loss close to 120 billion dollars in the USA, and that about 42% of Red Listed species are particularly endangered (Pimentel et al., 2004). The topic of alien species is large and complex, and can be understood in many ways just as the term exotic species can be. Climate change will for instance trigger changes in species distribution and allow new species to establish. Thus, it is not always easy to separate natural species responses from responses caused by humans.

1.5.7 Specific Biome Threats

The MA (2005) report assessed 14 biomes and more than 50% of these have experienced a 20–50% transformation to human use. Temperate and Mediterranean forests and temperate grasslands are particularly affected as about 75% of the biome's native habitat has been changed into cultivated lands.

Annually human-induced fires destroy on average one percent of the Mediterranean forests. Over the previous 50 years, the conversion rates have been highest in tropical and subtropical dry forests, that is, mainly in developing countries. Industrial countries, however, have historically experienced the same type of changes. There is a terrific logging pressure in temperate rainforests, being the number one threat. Even Canada has been criticized for their logging practices. The Taiga Biome is seriously impacted by logging as well due to e.g., paper production need, infrastructure development, air pollution, expansion of human settlements, etc. In tundra regions, mining and gas and oil extraction are encroachments influencing the biodiversity.

Temperate grasslands are heavily grazed by livestock throughout the world, and have been overexploited by humans by agricultural production and infrastructure development. Some of the famous grassland areas are the savannas, being home to huge numbers of indigenous species like in the Serengeti-Masai Mara in Tanzania and Kenya. When people invade non-protected savannas, bringing livestock with them, this frequently forces the native animals to leave, or they will die due to overgrazing and lack of food.

Large marine areas have been polluted by oil spills from crude oil carriers, and it is an increasing concern about the plastic pollution. Birds, fishes, sea mammals and other organisms are suffering due to small and large plastic fragments accumulating in marine environments all over the world. In freshwater hydropower plants and water reservoirs are blocking migrating routes for several fish species. Runoff from agricultural areas contributes to eutrophication and oxygen depletion. Estuaries, frequently created where the big rivers meet the sea, are commonly destroyed because of agricultural development and city expansions, as most cities are established close to such areas. Diking and filling in estuaries to create farmland normally destroy these ecosystems. It is very unfortunate as many of these areas are vital to e.g., migratory birds as they are used for stopovers and as feeding places.

1.6 Economics of Biodiversity

To assist politicians and decision makers many ecosystem services are assigned economic values (Jones-Walters and Mulder, 2009). Loss of biodiversity has a tremendous economic impact as species, being on land or in water, are the basement for functional ecosystems and ecosystem services. About 78% of the world's poorest people depend on what they can get from healthy ecosystems (MA 2005), including clean water. World Bank estimates on crimes affecting ecosystem services negatively in developing countries amount to more than US$70 billion a year. The high rate of deforestation, estimated to approximately 13 million hectares is a critical loss, of not only animal and plant habitats but also an unknown number of microorganisms that could be of importance to human health in the future.

The Amazon area, together with countries like Peru, Brazil, Indonesia, Myanmar and the Democratic Republic of Congo, are among the regions where the extent of illegal logging is significant. The demand for human and animal food, timber and paper, fuel and minerals are the main reasons to the rainforest depletion, however, road building and hydropower plant construction contribute substantially as well. Several countries contribute with huge money to governments in rainforest regions to compensate for conservation. However, these have also created problems, as corrupt politicians may get involved. In one of the best-known cases ("The Borneo Case"), billions of dollars have become white-washed all over the world (http://theborneocase.com/).

Over the last 30–40 years, the world has lost approximately 40% of warm water coral reefs, with dramatic consequences, not at least on local economies for about 350 million people by reducing the coastal protection and fish habitats. Approximately 9% of the world's coral reefs are located in the Caribbean, within 38 countries. They are very important to the region's economy supporting more than 43 million people and generating more than US$3 billion from ecotourism, fisheries, etc. Unfortunately, less than 20% of the original coral cover is intact today. The Australian Great Barrier Reef is facing a bad situation as well. It is estimated that as much as half of the coral in the northern third of the 2000 km^2 reef is dead due to too high ocean temperatures, and in 2016 scientists estimated that 93% of the whole reef is affected by bleaching (https://www.theguardian.com/environment/2016/apr/19/great-barrier-reef-93-of-reefs-hit-by-coral-bleaching). Additionally, mining and energy companies want to construct a shipping lane through the reef to get a more direct link with export markets.

Today less than 1% of known species are used for living. According to Myers (1984), people have utilized about 7,000 plants species for food. There are, however, at least another 75,000 species that can be used; several being superior to the common crop species, and many others are potential medicinal species. Among the insects, large numbers are vital as crop pollinators and as parasites and predators for pest control. Bacteria, fungi and a long range of microorganisms have definitely a potential for pharmaceuticals, food, etc. (Wilson, 1988). Many industries depend on plants and animals to produce their products, being fabric and clothes (e.g., silk), oils, lubricants, perfumes, dyes, waxes, latexes, resins, etc. Thus, the economic importance of biodiversity cannot be questioned.

Another aspect is how ecotourism generates money. Countries like Tanzania and Kenya have their main income due to the large animal herds still living in famous protected ecosystems and National Parks like the Serengeti Masai–Mara. Worldwide several hundred million people are occupied within the industry of ecotourism. Definitely, this industry will become increasingly important for many countries and people in the future, if we are able to maintain the protected areas.

The Living Planet Index (LPI) (originally developed by WWF and UNEP-WCMC) is presenting its results biannually in the WWF Living Planet Report. The biodiversity is measured by collecting population data of vertebrate species by estimating an average abundance change over time (http://wwf.panda.org/about_our_earth/all_publications/lpr_2016/). The global LPI is based on controlled information on 14,152 monitored populations of 3,706 mammals, birds, fishes, amphibians and reptiles from around the world, being an important indicator of the planet's ecological state. From 1970 to 2012 the index shows a 58% overall decline in vertebrate population abundance. On average the vertebrate species population size decreased by more than half in about 40 years, that is, an annual 2% decline. This type of indexes are very important, not only as an indicator of ecological condition, but largely also for future human wellbeing and economy.

1.7 Conclusion

Humans are vitally dependent on biodiversity and ecosystem services for economic and social welfare and short- and long-term survival. Unfortunately, there is no doubt that the present threats to species and ecosystems have dimensions never experienced before due to a planet having too many people. Species extinction and ecosystem degradations because of human activities continue at an alarming rate, and we are now supposed to reach an era, frequently called the sixth mass extinction period.

Over the last decades, we have mapped how many species there are on our planet, and how they are distributed. We have managed to create a network of protected areas and established international regulations and adopted international commitments. We have increased our understanding of how significant ecosystem services are to our health and economy and what the biodiversity threats and drivers of change are. Thus, our toolbox should be sufficiently equipped to enable actions for the survival of future generation. It is for ourselves to decide.

Keywords

- biogeographic classification
- conservation
- patterns
- species number

References

Almén, M. S., Lamichhaney, S., Berglund, J., Grant, B. R., Grant, P. R., Matthew, T., Webster, M. T., & Andersson, L., (2015). Adaptive radiation of Darwin's finches revisited using whole genome sequencing. *Bioessays, 38*, 14–20.

Alroy, J., (2008). Dynamics of origination and extinction in the marine fossil record, *Proc. National Acad, Sci. United States of America (PNAS), 105*(suppl.), 11436–11542.

Anderson, R. F., Ali, S., Bradtmiller, L. I., Nielsen, S. H. H., Fleisher, M. Q., Anderson, B. E., & Burckle, L., (2009). Wind-Driven upwelling in the Southern Ocean and the deglacial rise in atmospheric CO_2. *Science, 323*, 1443–1448.

Arizumi, Y., Fuki, H., & Hirose, K., (2015). Land use change in mangrove areas of southern Vietnam using satellite images. https://www.researchgate.net/publication/266224788.

Bailey, R. G., (2014). Ecoregions. *The Ecosystem Geography of the Oceans and Continents,* 2nd edn., Springer, New York.

Barnosky, A. D., Matzke, N., Toliya, S., Wogan, G. O., Swartz, B., Quental, T. B., Marshall, C., McGuire, J. L., Lindsey, E. L., Maguire, K. C., Mersey, B., & Ferrer, E. A., (2011). Has the Earth's sixth mass extinction already arrived? *Nature, 471*, 51–57.

Begon, M., Harper, J. L., & Townsend, C. R., (1996). *Ecology. Individuals, Populations and Communities,* 3rd edn., Blackwell Science Ltd.

Begon, M., Townsend, C. R., & Harper, J. L., (2006). *Ecology. From Individuals to Ecosystems,* 4th edn., Blackwell Science Ltd.

Bertzky, B., Corrigan, C., Kemsey, J., Kenney, S., Ravilious, C., Besancon, C., & Burgess, N. D., (2012). *Protected planet report: tracking progress towards global targets for protected areas.* IUCN and UNEP-WCMC, Gland, Switzerland and Cambridge, UK.

Bevanger, K., Fremstad, E., & Ødegaard, F., (2007). Dispersal and effect of alien species. In: *Norwegian Black List. Ecological Risk Analysis of Alien Species*, Gederaas, L., Salvesen, I., & Viken, Å., (eds.). The Norwegian Biodiversity Information Centre, Norway, pp. 19–50.

Brown, J. H., (1988). Species diversity. In: *Analytical Biogeography. An Integrated Approach to the Study of Animal and Plant Distribution*, Myers, A. A., & Giller, P. S., (eds.). Chapman and Hall, London, pp. 57–89.

Bruijnzeel, L. A., & Veneklaas, E. J., (1998). *Climatic conditions and tropical Montane forest productivity: The fog has not lifted yet. Ecology, 79*, 3–9.

Ceballos, G., & Ehrlich, P. R., (2009). Mammal population losses and extinction crisis. *Proc. National Acad. Sci. United States of America (PNAS), 106*, 3841–3846.

Ceballos, G., Ehrlich, P. E., Barnosky, A. D., García, A., Pringle, R. M., & Palmer, T. M., (2015). Accelerated modern human-induced species losses: Entering the sixth mass extinction. *Science Advances, 1*, DOI: 10.1126/sciadv.1400253.

Cox, C. B., Healey, I. N., & Moore, P. D., (1976). *Biogeography,* 2nd edn. Blackwell Scientific Publications, Oxford, UK.

Daru, B. H., Van der Bank, M., & Davies, J. T., (2014). Spatial incongruence among hotspots and complementary areas of tree diversity in southern Africa. *Diversity and Distributions*, 1–12. DOI: 10.1111/ddi.12290.

Di Castri, F., (1989). History of biological invasions with emphasis on the Old World. In: *Biological Invasions: A Global Perspective*, Drake, J., Di Castri, F., Groves, R., Kruger, F., Mooney, H. A., Rejmanek, M., & Williamson, M., (eds.). Wiley, New York, pp. 1–30.

Dobzhansky, T., (1950). Evolution in the tropics. *American Scientist*, *38*, 209–221.

Erwin, D. H., (2001). Lessons from the past: Biotic recovery from mass extinction. *Proc. National Acad. Sci. United States of America (PNAS)*, *98*, 1399–1403.

European Commission, (2012). Commission staff working document, Impact assessment, Accompanying the document proposal for a Directive of the European Parliament and of the Council amending directive 98/70/EC relating to the quality of petrol and diesel fuels and amending directive 2009/28/EC on the promotion of the use of energy from renewable sources. Brussels, *17*, p. 10.

Fabricius, K. E., & De'ath, G., (2001). Biodiversity on the Great Barrier Reef: Large-scale patterns and turbidity-related local loss of soft coral taxa. In: *Oceanographic Processes of Coral Reefs: Physical and Biological Links in the Great Barrier Reef*, Wolanski, E., (eds.). CRC Press, London, pp. 127–144.

Gaston, K. J., (2000). Global patterns in biodiversity. *Nature*, *405*, 220–227.

Geldman, J., Barnes, M., Coad, L., Craigie, I. D., Hockings, M., & Burgess, N. D., (2013). Effectiveness of terrestrial protected areas in reducing habitat loss and population decline. *Biological Conservation*, *161*, 230–238.

Gjaerevoll, O., (1973). *Plant Geography.* Scandinavian University Press, Oslo. (In Norwegian).

Gladwell, M., (2000). *The Tipping Point.* Little, Brown and Company.

Hoffmann, W. A., Geiger, E. L., Gotsch, S. G., et al., (2012). Ecological thresholds at the savanna – forest boundary: How plant traits, resources and fire govern the distribution of tropical biomes. *Ecol. Lett.*, *15*, 759–768.

IPCC, (2014). Climate change 2014: Synthesis Report. Contribution of Working Groups I, II and III to the Fifth Assessment Report of the Intergovernmental Panel on Climate Change, Core Writing Team, Pachauri, R. K., & Meyer, L. A., (eds.). IPCC, Geneva, Switzerland.

Irigoien, X., Huisman, J., & Harris, R. P., (2004). Global biodiversity patterns of marine phytoplankton and zooplankton. *Nature*, *429*, 864–867.

IUCN, (1994). Guidelines for Protected Areas Management Categories, IUCN: Cambridge, UK and Gland, Switzerland.

Jablonski, D., (1995). *Extinction Rates.* Oxford University Press, pp. 25–44.

Jones-Walters, L., & Mulder, I., (2009). Valuing nature: The economics of biodiversity. *J. Nature Conservation*, *17*, 245–247.

Joppa, L. N., & Pfaff, A., (2011). Global protected area impacts. *Proc. R. Soc. B. – Biol. Sci.*, *278*, 1633–1638.

Kareiva, P., & Marvier, M., (2003). Conserving biodiversity coldspots. *American Scientist*, *91*, 344–351.

Lamichhaney, S., Berglund, J., Almén, M. S., Maqbool, K., Grabherr, M., Martinez-Barrio, A., Promerová, M., Rubin, C.J., Wang, C., Zamani, N., Grant, B. R., Grant, P. R., Webster, M. T., & Andersson, L., (2015). Evolution of Darwin's finches and their beaks revealed by genome sequencing. *Nature*, *518*, 371–375.

Leopold, A. A., (1949). *Sand County Almanac and Sketches from Here and There.* Oxford University Press, New York.

Locey, K. J., & Lennon, J. T., (2016). Scaling laws predict global microbial diversity. *Proc. National Acad. Sci. United States of America (PNAS), 113,* 5970–5975.

Longman, K. A., & Jeník, J., (1987). *Tropical Forest and its Environment,* 2nd edn. Longman Scientific & Technical, Wiley & Sons, New York.

MacArthur, R. H., & Wilson, E. O., (1967). *The Theory of Island Biogeography.* Princeton University Press, Princeton, New Jersey, USA.

MacArthur, R. H., (1965). Patterns of species diversity. *Biological Review, 40,* 510–533.

McKinney, M. L., & Lockwood, J. L., (1999). Biotic homogenization: A few winners replacing many losers in the next mass extinction. *TREE, 14,* 450–453.

Millennium Ecosystem Assessment, (2005). *Ecosystems and Human Well-Being.* Current state and trends, findings of the condition and trends working group. Island Press, Washington, Covelo, London.

Mittermeier, R. A., Turner, W. R., Larsen, F. W., Brooks, T. M., & Gascon, C., (2011). Global biodiversity conservation: The critical role of hotspots. In: *Biodiversity Hotspots: Distribution and Protection of Conservation Priority Areas,* Zachos, F. E., & Habel, J. C. (eds.). Springer, Berlin, Heidelberg.

Mora, C., Tittensor, D. P., Adl, S., Simpson, A. G. B., & Worm, B., (2011). How many species are there on Earth and in the ocean? *PLoS Biol., 9*(8), e1001127. doi:10.1371/journal.pbio.1001127.

Myers, A. A., & Giller, P. S., (1988). *Analytical Biogeography.* An integrated approach to the study of animal and plant distribution, Chapman and Hall, London.

Myers, N., (1984). *The Primary Source: Tropical Forests and Our Future.* Norton, W. W., New York.

Myers, N., (1988). Threatened biotas. "Hot spots" in tropical forests. *The Environmentalist, 8,* 187–208.

Myers, N., (1990). The biodiversity challenge: Expanded hot-spots analysis. *The Environmentalist, 10,* 243–256.

Myers, N., Mittermeier, R. A., Mittermeier, C. G., Da Fonseca, G. A. B., & Kent, J., (2000). Biodiversity hotspots for conservation priorities. *Nature, 403,* 853–858.

Noble, J. C., & Bradstock, R. A., (1989). *Mediterranean Landscapes in Australia.* CSIRO: Australia.

Olson, D. M., Dinerstein, E., Wikramanayake, E. D., Burgess, N. D., Powell, G. V. N., Underwood, E. C., D'Amico, J. A., Itoua, I., Strand, H. E., Morrison, J. C., Loucks, C. J., Allnut, T. F., Ricketts, T. H., Kura, Y., Lamoreux, J. F., Wettengel, W. W., Hedao, P., & Kassem, K., (2001). Terrestrial ecoregions of the world: A new map of life on earth. *BioScience, 51,* 933–938.

Pimentel, D., (2002). *Biological Invasions,* Economic and biodiversity management. Kluwer Academic Publishers, Dordrecht.

Pimentel, D., Zuniga, R., Morrison, D. (2005). Update on the environmental and economic costs associated with alien-invasive species in the United States. *Ecological Economics 52*(3), 273–288.

Possingham, H., & Wilson, K., (2005). Turning up the heat on hotspots. *Nature, 436,* 919–920.

Rex, M. A., Stuart, C. T., Hessler, R. R., Allen, J. A., Sanders, H. L., & Wilson, G. D. F., (1993). Global-scale latitudinal patterns of species diversity in the deep-sea benthos. *Nature*, *365*, 636–639.

Rosenzweig, M. L., (1995). *Species Diversity in Space and Time*. Cambridge University Press.

SCEP (Study of Critical Environmental Problems), (1970). *Man's Impact on the Global Environment*. MIT Press, Cambridge.

Shirahase, S., (2000). Women's increased higher education and the declining fertility rate in Japan. *Review of Population and Social Policy*, *9*, 47–63.

Steadman, D. W., (2006). *Extinction and Biogeography of Tropical Pacific Birds*. Chicago University Press, Chicago.

Stevens, G. C., (1989). The latitudinal gradient in geographical range: How so many species coexist in the tropics. *American Naturalist*, *133*, 240–256.

Stevens, G. C., (1992). The elevational gradient in altitudinal range: An extension of Rapoport's latitudinal rule to altitude. *American Naturalist*, *140*, 893–911.

Terborgh, J., (1985). The vertical component of plant species diversity in temperate and tropical forests. *American Naturalist*, *126*, 760–766.

Tong, P. H. S., Auda, Y., Populus, J., Aizpuru, M., Habshi, A. A., & Blasco, F., (2004). Assessment from space of mangroves evolution in the Mekong Delta, in relation to extensive shrimp farming. *Int. J. Remote Sensing*, *10*, 4795–4812.

Udvardy, M. D. F., (1969). *Dynamic Zoogeography, With Special Reference to Land Animals*. Van Nostrand Reinhold Company, New York, USA.

Udvardy, M. D. F., (1975). *A Classification of the Biogeographical Provinces of the World*. IUCN Occasional paper no. 18. International Union for Conservation of Nature and Natural Resources, Morges, Switzerland.

Van Nes, E. H., Arani, B. M. S., Staal, A., Van der Bolt, B., Flores, B. M., Bathiany, S., & Scheffer, M., (2016). What do you mean, "Tipping point"? *Trends in Ecology & Evolution*, *31*, 902–904.

Vitousek, P. M., D'Antonio, C., Loope, L., & Westbrooks, R., (1996). Biological invasions as global environmental change. *American Scientist*, *84*, 468–478.

Wake, D. B., & Vredenburg, V. T., (2008). Are we in the midst of the sixth mass extinction? A view from the world of amphibians. *Proc. National Acad. Sci. United States of America (PNAS)*, *105*, 11466–11473.

Wilson, E. O., (1988). *Biodiversity*, National Academy Press, Washington, DC.

Zalasiewicz, J., Williams, M., Smith, A., et al., (2008). Are we now living in the Anthropocene? *GSA Today*, *18*, 4–8.

CHAPTER 2

Biodiversity in Afghanistan

S.-W. BRECKLE,[1] I. C. HEDGE,[2] and M. D. RAFIQPOOR[3]

[1]Department of Ecology, Wasserfuhr 24-26, D-33619 Bielefeld/Germany,
E-mail: sbreckle@gmx.de
[2]Royal Botanic Garden, Edinburgh, EH3 5LR, Scotland, UK,
E-mail: I.Hedge@rbge.ac.uk
[3]Nees-Institute for Biodiversity of Plants, University of Bonn, Germany,
E-mail: d.rafiqpoor@uni-bonn.de

2.1 Introduction

The term biodiversity has different facets. It can be defined as the existence of a wide variety of plant and animal species in their natural environment, but it can also cover the biodiversity of ecosystems and also genetic diversity. As a general rule, when assessing the biodiversity of a region, the number of species in a particular group of organisms has to be taken into account; flowering plants, non-flowering plants, fungi, birds, mammals, reptiles, etc. There are two main preconditions: (i) knowledge of an accurate number of species, where often there are problems of delimitations; and (ii) an awareness of all the species in the relevant region. The situation in Afghanistan is fairly good. There has been much research on its plants and some animal groups for a very long time; often jointly by the foreign scientists with Afghan counterparts. Flowering plants are certainly the best-studied and below we mostly refer to them; the relevant literature is very substantial and up-to-date, though presently few research is possible. Other groups are less well or poorly known, but in the references section, we have tried to cover the most important literature for all biota.

2.2 Nature Geography as a Precondition for Understanding Biodiversity

With an area of 652,089 km^2 (Shank, 2006) Afghanistan lies on the Asian continent between 29°30'–38°30' north latitude and 60°30'–74°50' east longitude. It is a very mountainous land with snowy peaks up to c. 7,000 m (its highest is Nowshaq: 7,485 m) in the eastern Hindu Kush. High and lesser elevated mountains, deeply eroded valleys, high plateaus, inter-montane basins (e.g., Kabul- and Koh-Daman basin) and wide pediments characterize the general topography and the mountainous nature of the country. The high mountain range of Hindu Kush, the mountainous region of central Afghanistan including Kohre Baba (its summit Shah Fuladi: 5,143 m), Ferozkoh in the west, and Tirbande Turkistan (Paropamesus 4,500–1,250 m) in the north, the Safed Koh (its summit Sikaram: 4,745 m) in east Afghanistan, the semi-desert and steppe regions of the southwest and those of north Afghanistan (1,250 to less than 500 m), and finally the arid desert-lowlands of the southwest (1,000 to less than 500 m) are the major natural landscape units in Afghanistan (Figure 2.1).

2.3 Ecosystem Diversity

For the altitudinal arrangement of the landforms and the actual morphodynamic processes, the geological basement, the seasonal temperature fluctuations, the quantity and distribution of rainfall, and the difference between the various aspects of the mountain slopes are very essential. They determine the level of the permanent snow line, as well as the vertical limits of solifluction and patterned grounds, block glaciers, the permafrost, of vegetation formations, and the settlements and land-use forms in the mountainous country.

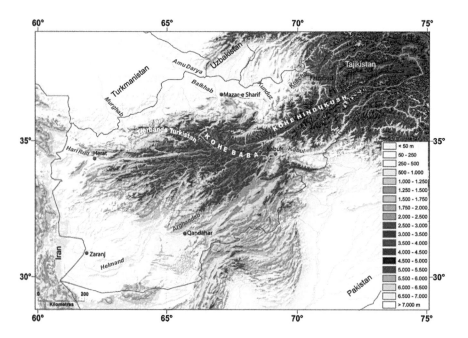

Figure 2.1 The physical nature of Afghanistan (prepared in GIS: Rafiqpoor).

Geologically, there are two clearly defined major tectonic lineaments in Afghanistan (*Kabul-Herat* and *Moqor-Chaman* Line) that divide the country into three main blocks (micro-continents = terranes):

1. *North Afghan-Tajikistan* block,
2. *East Iran-central Afghan* block, and
3. *East and southeast Afghan-Pakistan* block.

These three great blocks are fragmented by tectonic faults into various mosaics and smaller units. In particular, the two tectonic lineaments are striking features by causing altitude leaps of several hundred meters in the landscape. Along these two lineaments, ophiolites can be found as indicators for fossil continental margins (Krumsiek, 1980) (for more information see, Breckle and Rafiqpoor, 2010).

In Afghanistan, we can generally recognize six major geological and tectonic units with very different petrographic characteristics:

1. The crystalline basement of the mountainous region of central Afghanistan and the Hindu Kush, as well as the metamorphites of eastern Afghanistan.
2. The Mesozoic block between the Kabul-Herat- and Chaman-Moqor tectonic lineaments.
3. The upper Cretaceous-Tertiary block in the north.
4. The Tertiary of Baluchistan-geosyncline in the south.
5. The sedimentary basins of early Quaternary in the southwest.
6. The loess and sand-covered areas of late Quaternary in north and south.

On these geological fundaments, varied soil types are developed each depending on the specific regional climatic conditions and altitudinal belts (Rafiqpoor and Breckle, 2010).

2.4 Climate

For a country with a high-continental climate and high-mountain nature, the temperatures, its annual variability, and altitudinal zonation determine the living conditions of the people, the rhythm in land-use, and the life cycle of natural vegetation. The term *continental* refers in this context to the climatologically defined thermal continentality describing the thermal properties of an area, in which the amplitude between the absolute maximum temperature in summer and the absolute minimum in winter is very pronounced. The highest absolute maximum air temperature in Afghanistan was measured in Zaranj (+51°C) in the province of Nimroz and the lowest absolute minimum air temperature in

Panjao (−52.2°C) in central Afghanistan (Breckle, 2004). Even allowing for the fact that these data are not based on long-term observations, they do make very clear the huge fluctuations between the thermal maxima of the warmest and the minima of the coldest month. The climate of Afghanistan is highly seasonal and continental, exhibiting a range of more than 100 K.

Afghanistan lies in the sub-tropical dry winter-rain zone of the Old World. Its climate is significantly affected by the high mountain character. In the course of the year, the climate is determined by the elements of the atmospheric circulation which in this region is dominated by two great supraregional systems governed by the shift of the Inner Tropical Convergence Zone (ITCZ) on both sides of the equator during the course of the year, and depends on the virtual course of the sun.

The amount and distribution of precipitation in Afghanistan is also influenced by its mountainous topography as well as by the configuration of the elements of the atmospheric circulation in winter and summer. Figure 2.2 depicts the spatial distribution of precipitation with dry lowlands and relatively moist conditions at higher altitudes.

With the passage of the sun to the south, the ITCZ shifts in *winter* in the same direction. This is accompanied by a southward shift of the entire general circulation system. Consequently, the winter rain zone of the northern

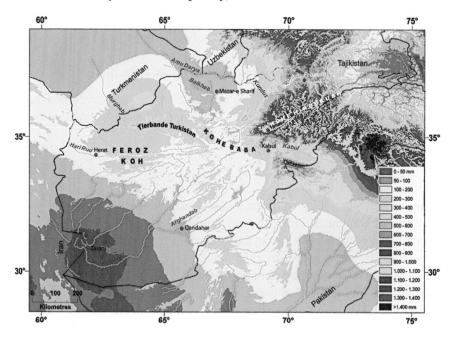

Figure 2.2 Mean annual precipitation in Afghanistan.

hemisphere comes under the influence of the planetary west-wind drift. As a result, the cyclones hike from the Atlantic Ocean eastwards and deploy on their pathway rainfall from the European Mediterranean via Turkey, the east Mediterranean and Iran and reach, in a very reduced form, Afghanistan. Finally, in the Karakorum and the west Himalaya, they completely lose their energy ("cyclone's cemetery").

In *summer*, the eastern parts of Afghanistan come under the influence of the tropical Indian monsoon circulation. The monsoon trough, which bonds strongly with the northward shift of the ITCZ, lies in early May at the southern tip of India in the region of Goa. With a gradual shift to the north, the monsoon trough reaches Pakistan in June. In Afghanistan, June is called "pre-monsoon month" (Sivall, 1977) and is still dry.

In Afghanistan, the influence of the scarce summer monsoon starts in July. The quantity and seasonal distribution of rainfall are of high ecological significance at the monsoon-influenced stations of east Afghanistan, both for the natural vegetation and for agriculture. Summer rainfalls at those stations vary considerably. An analysis of the annual percentages of summer rainfall shows (Figure 2.3) that with increasing distance from the Pakistan border to the west, the rates continuously decline (Rafiqpoor, 1979). While in Khost region, c. 30% of the annual precipitation (>600 mm) falls in summer, that

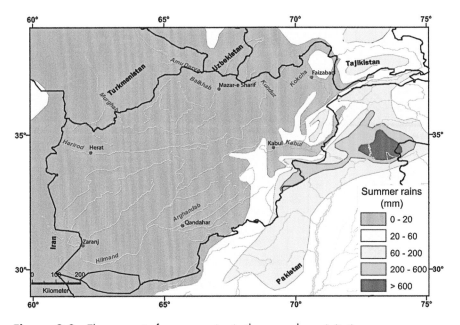

Figure 2.3 The amount of summer rains in the annual precipitation.

decreases at Kabul to nearly <3% (0–20 mm). All climatic stations in the east and southeast between these two receive yearly about 3–30% of their annual rainfall in summer, which also is demonstrated by the corresponding ecological climate diagrams (Figure 2.4). The weather stations of the Afghan

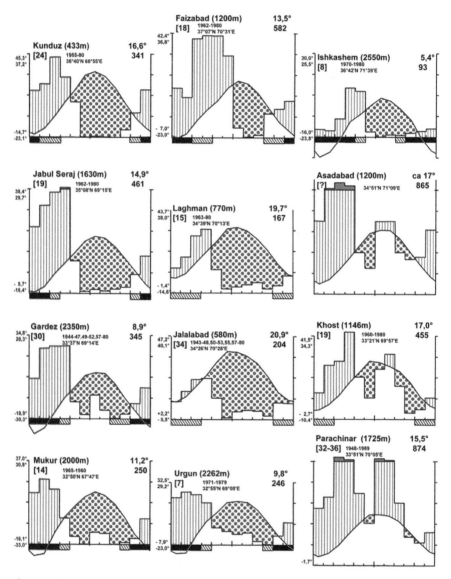

Figure 2.4 Ecological climate diagrams of northeast and east weather stations from Amu Darya to the Afghan/Pakistan border (prepared in OCAD by S.-W. Breckle).

central mountains, as well as those of the desert and steppe regions of the southwest and northern Afghanistan, do not receive summer rains (Figures 2.3 and 2.5).

Regarding the configuration of the atmospheric circulation and the resulting distribution patterns of rainfall, five main landscape units can be defined. They differ from each other depending on their general topography (see Figure 2.1):

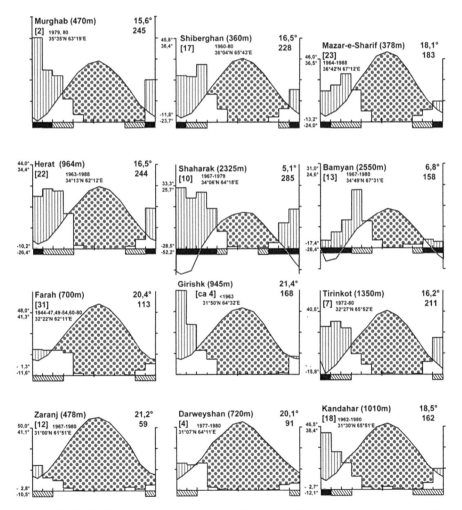

Figure 2.5 Ecological climate diagrams of the semi-desert and desert regions of north, west and southwest Afghanistan (prepared in OCAD by S.-W. Breckle).

1. The high mountains of Hindu Kush (above 4,500 m) with more than 700 mm mean annual rainfall (see Figure 2.2).
2. The mountainous region of central and north Afghanistan including Kohre Baba, Ferozkoh and Tirbandre Turkistan (Paropamesus), etc., between 4,500–1,250 m with 700–200 mm mean annual rainfall.
3. The semi-desert and steppe regions of the southwest and those of north Afghanistan (1,250 m to less than 500 m) with 100–200 mm mean annual rainfall.
4. The arid desert- and semi-desert lowlands of southwest Afghanistan (1000 m to less than 500 m), with less than 100 mm mean annual rainfall.
5. A narrow strip at the east part of the country comes under the influence of the Indian summer monsoon where a secondary maximum of rainfall occurs. This region has 500 mm to more than 1,000 mm mean annual precipitation. In this so-called "summer rain-strip," the basins of Jalalabad and Laghman receive yearly <500 mm precipitation (Figure 2.4).

We can conclude that with an increase of altitude to the interior parts of the country, a west-east change of the hygric conditions is evident with a simultaneous steady increase of rainfall amounts in the same direction (see Figures 2.1 and 2.2). In the high plateaus of the country's central mountain region, we have between 200–300 mm precipitation; in the wind-protected basins (Dashte Nawor, Bamyan, Shahrak: see Figures 2.1 and 2.5) somewhat less. The drought of the foothill areas penetrates from the west and southwest along the major rivers (Helmand, Arghandab, Hari Rud, Farah Rud, and Khash Rud, etc.) into the mountains. Because of their rain-shadow topographic position, these valleys, as well as the closed inter-montane basins, compared with the neighboring mountainous surrounding areas, are much drier. From western lowlands into the Kohe Baba and the central mountainous regions, the precipitation rises to c. 300–500 mm. The same also applies to Ferozkoh and Tirbande Turkistan in the north. In the higher altitudes of Hindu Kush, the amount of precipitation is considerable and may reach >1000 mm (see Figure 2.5).

In winter, large parts of Afghanistan are affected by migratory cyclones, the winter precipitation starting in late autumn (mostly in November – in higher elevations a little earlier) with great variability and low yield, and reach their maximum in March/April (in north Afghanistan partly in May and June).

At the north side of the Hindu Kush, the rainfall slightly increases with altitude (see Figure 2.4: Faizabad and Jabul Seraj). And the one-month later end of the winter rains transforms the fertile loess steppe and the hill country at the northern slopes of the Hindu Kush, the so-called *"Qataghan-Zamin,"* into the "granary" of the country.

2.5 Flora of Afghanistan and Species Diversity of Plants

Afghanistan being a very mountainous country with a high degree of "geo-diversity," it has a wide range of ecological conditions. Correspondingly, it has a wide range of plant species and vegetation types. This diversity is enhanced by its location at the crossroad of several bio-geographical regions. With their widely differing floristic stock, these greatly contribute to the floristic richness and vegetation pattern of the country.

As a part of central Asia, it is a very dry country with sparse precipitation during winter and early spring, followed by a prolonged summer drought, but varying significantly between north and south, west and east, lowlands and mountains. Whereas the mountains get substantial amounts of snow in winter and rain in spring, precipitation is much less and more unreliable in the northern and southern deserts and semi-deserts. Some eastern parts receive episodic or even periodic summer rain from the monsoonal activity. That additional rainfall during the vegetation period provides the pre-condition for the occurrence of various forest types. Today, these forests and woodlands are much degraded or have even completely disappeared.

Although a rather arid country with extensive deserts and semi-deserts (Walter and Breckle, 1994), the number of vascular plant species is distinctly higher than in the more humid central European countries which offer more favorable conditions for plant growth (Heywood and Watson, 1995). This high biodiversity can be partly ascribed to the fact that it did not suffer as much from extinctions by the dramatic changes of climate during the Quaternary, but the main reason is the much greater diversity of habitats. Groombridge (1992) gave an estimate of 3,500 species of vascular plants and 30–35% of endemism, as well as an additional c. 5–10% of species which might be found in future. Our own earlier estimates were about 4,100 species and c. 30% of endemism, but now we can give more precise answers (see below). With these data, Afghanistan comes in the same category of such Mediterranean countries as Italy (Pignatti, 1982) and Greece (Strid and Kit Tan, 1997) with 5,600 and 5,700 plant species, respectively, as well as its biodiversity-index (Barthlott et al., 1996).

As everywhere, the amount and seasonal distribution of precipitation and the altitude determine the distribution patterns of the country's flora and vegetation. Thus, diverse ecological conditions, ranging from hot deserts and humid subtropical regions to high alpine regions, have favored the establishment of a complex and varied flora and vegetation. However, the composition of the flora and the vegetation structure are also greatly influenced by a long history of over-exploitation which has led not only to the almost complete loss of forests but also to widespread degradation of formerly rich woodland and semi-desert ecosystems. Grazing by sheep and especially goats as well as cutting of trees and uprooting of shrubs and even dwarf shrubs have not only greatly reduced the coverage of the vegetation, but also its composition and floristic diversity.

In many plant families, the percentage of Afghan endemic species is high (Hedge and Wendelbo, 1970, Podlech, 2012). Many of these endemics are very isolated. There is strong evidence that the Afghan/central Asiatic area is an ancient and major center of evolution in flowering plants – at all levels – family, genus, and species.

2.5.1 How Many Plant Species?

Answering this question is now easier as a result of the recent definitive checklist (VPA, Breckle et al., 2013). In it, c. 5,000 species of vascular plants are recognized (see also, Gaston, 1998; Heywood, 1999; Jayakumar et al., 2011; Lack and Grotz, 2012; Lozano et al., 2012).

2.5.2 Quantitative Survey on Afghan Flora

The Figures 2.6–2.9 give some statistical data, derived from the 2013 VPA checklist according to species, genera, and families. Figure 2.6 indicates the number of species in the larger plant families. It is obvious that the seven largest plant families comprise more than half of all Afghan species. Figure 2.7 gives the number of genera in the larger plant families. Again, the seven largest families have more than half of all genera; however, the sequence of families differs, Leguminosae (Fabaceae) is only on 5th place now and Gramineae (Poaceae) are in second place. Figure 2.8 indicates the number of species within the larger genera of Afghanistan. There are 12 genera with more than 40 species, by far the largest being *Astragalus* with c. 320 species but *Cousinia* with 147 is also substantial. Whether *Oxytropis* really merits the third rank is questionable; many species are only known from single gatherings and with better material many may turn out to be

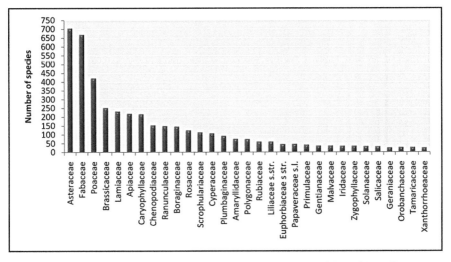

Figure 2.6 Number of species in the larger plant families of the Afghan flora.

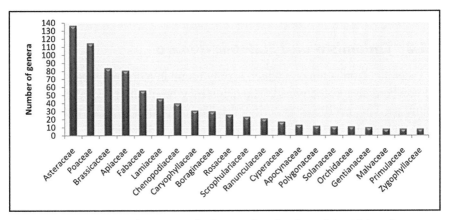

Figure 2.7 Number of genera in the larger plant families of the Afghan flora.

synonyms. Somewhat similar is the situation with *Taraxacum*, but *Acantholimon, Allium,* and *Silene* are other species-rich genera.

Table 2.1 gives a summary of the main statistics of the Afghan flora. The total number of taxa is c. 5,000 and the degree of endemism is 24% (Table 2.1). Only 146 taxa are listed here as being introduced neophytes and/or cultivated. The link between species numbers and altitude is shown on Figure 2.9. It clearly demonstrates the trend of decreasing species numbers with increasing altitude, which is true of many mountainous countries (Körner, 1999; Breckle, 1981; Lauer et al., 2001; Agakhanjanz and Breckle, 2002).

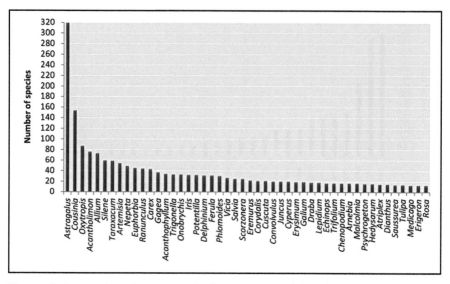

Figure 2.8   Number of species in the larger genera of the Afghan flora.

2.5.3   Endemism and Endemic Genera

An endemic taxon is the one which is only found in a distinct area or location and nowhere else in the world. The taxon may be of any rank, although it is usually used at specific or generic level. The definition requires that the area where the species is endemic is defined: such as a "site endemic" (e.g., just on one hill), a "national endemic" (e.g., found only within the border of the state of Afghanistan), a "geographical range endemic" (e.g., found in the Hindu Kush region, which however covers also parts on Pakistan territory). The opposite of the term *endemism* is *cosmopolitan*, with a worldwide distribution.

Mabberley (2008) recognized worldwide 13,313 vascular plant genera of which 3422 (25.7%) are single-nation endemics. We leave it open as to whether a politically defined area of a state is suitable to define the natural distribution area of an "endemic" species. We tend to use "geographical range endemics," well aware that many Floras are written only for restricted national, mostly political regions.

Endemic species are most likely to develop on biologically isolated areas such as islands or isolated high mountains because of their isolation. The island of Madagascar, separated since ancient times from Africa, provides a striking example of this. There, about 75% of its plants are endemic with even higher totals for its fauna. This, on a much smaller scale, is the case with *Cousinia* in Afghanistan (Table 2.2).

Table 2.1 Number of Families, Genera, Species, Taxa, and Endemics in the Afghan Flora, According to the Recent Checклist.

Taxon group	Number of families	Number of genera	Number of species	Number of taxa	Number of endemics (%)	Number of endemics and sub-endemics (%)	Number of introduced species (%)
Pteridophytes	11	23	50	56	0 (0%)	0	0
Gymnosperms	4	8	24	24	0 (0%)	2 (8%)	3
Monocotyledons	28	195	817	840	57 (6.8%)	75 (8.9%)	40
Dicotyledons	106	860	3,935	4,115	898 (21.9%)	1,138 (27.8%)	148
Total	149	1,086	4,826	5,035	955 (19.0%)	1,215 (24.2%)	191

(Breckle et al., 2013; VPA)

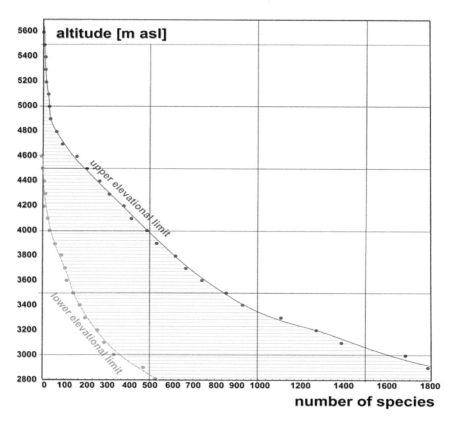

Figure 2.9 The "floristic drainage in mountains." Decrease in species numbers from 3000 m upwards to 5,600 m in Afghanistan. Species with their lower limits are also indicated. The difference between both curves gives the real number of occurring species. Species-numbers depend very much on altitude (Breckle, 2004; Breckle and Rafiqpoor, 2011).

There are two subcategories of endemism – *palaeo-endemism* and *neo-endemism*; though it should be pointed out that the terms are subjective and cannot be given a precise definition. Palaeo-endemism is used to refer to a species that was formerly widespread but is now restricted to a smaller area and is morphologically rather isolated. Neo-endemism refers to a new/ young, species that has recently arisen and become reproductively isolated, or one of hybrid origin and now classified as a separate species. Many of the Afghan endemic species can be designated as neo-endemics, where evolution is still active (*Astragalus, Acantholimon, Oxytropis, Taraxacum*, etc. – cf. Hedge and Wendelbo, 1970). In contrast, four genera that are good candidates for designation as palaeo-endemics are *Halarchon* (Chenopodiaceae),

Table 2.2 Plant Families and Genera in Afghanistan with High Endemism Ratios

Family	Country endemics (Afgh) (%)	Sub-endemics (Afgh + adjac. regions) (%)	Total (%)
Plumbaginaceae	66.7	12.9	79.6
Cousinia	67.5	10.4	77.9
Astragalus	48.8	4.7	53.5
Iridaceae	40.0	5.7	45.7
Lamiaceae	29.3	13.4	42.7
Boraginaceae	26.5	13.6	40.1
Primulaceae	34.9	4.6	39.6
Dionysia	91.7	0.0	91.7
Apiaceae	28.3	9.1	37.4
Amaryllidaceae (incl. *Allium*)	27.6	9.2	36.8
Caryophyllaceae	24.9	8.3	33.2
Ranunculaceae	22.1	3.4	25.5
Brassicaceae	14.6	8.3	22.9

Pseudodraba, and *Veselskya* (Cruciferae = Brassicaceae) *Pyramidoptera* (Umbelliferae = Apiaceae). At least morphologically, they are clearly isolated from other genera, but whether future molecular research will support this designation remains to be seen.

The definition of species and subspecies depends on the opinion of the expert; "splitters" will recognize several dozen more taxa in Afghanistan than "lumpers"; a few examples of genera which when described were considered as Afghan endemics, but which are now subsumed into other genera, are known (Breckle et al., 2013). As Table 2.2 indicates, in some Afghan plant families, the ratio of endemics is rather high. National and political region endemism of large countries or regions has far less significance than endemism of small countries, provinces, and very limited range endemism.

However, if the degree of endemism is used for conservation programmes, then country endemics become important. Conservation can only be carried out under national legislation, and country endemics fully depend on the effort and success of conservation in the country where it grows. The concept of country endemism automatically means that nationally endemic species have a relatively limited range. By IUCN criteria, they must be regarded as being vulnerable.

The concept of endemism very much depends on the knowledge of the geographical range of a species. Quite often, a newly discovered species is found at only one site, and with the then current knowledge, it should be considered a site and country endemic. However, this only lasts until someone discovers it elsewhere or in another country. Then, its original status is lost. So again, our knowledge of the classification and biology of plants is constantly changing with new data. Checklists are a never-ending synthesis!

The list in Table 2.3 cites 19 endemic genera in Afghanistan. This total is certainly not definitive – mainly because taxonomy and classifications are never static and, as new information comes to light, the numbers and names will change. In a recent paper which gives a broader overview of generic endemism throughout all of southwest Asia (Sales and Hedge, 2013), 161 genera were listed as endemic. Southwest Asia was there defined to include, in addition to the usual countries within the term, the Caucasus and parts of central Asia (Pamir-Alai and Tien Shan). Three families, Umbelliferae (Apiaceae), Compositae (Asteraceae) and Cruciferae (Brassicaceae), had by far the greatest number of endemic genera accounting between them for over 120 genera. The grass and legume families had very few. The paper emphasized the major importance of southwest Asia as a global hotspot, especially the Afghanistan/central Asiatic part of it. It also considered a few of the larger non-endemic genera in Afghanistan relative to their ranges in adjacent countries. One surprising fact that emerged from this was the very substantial difference in the distributions of individual species. In, for example, *Acantholimon, Acanthophyllum, Allium,* and *Nepeta* there were remarkably few species that were common to wider areas, such as Afghanistan, Tajikistan, Iran, and Turkey. *Acantholimon* has over 70 species in Afghanistan and none of them are in Turkey. This emphasizes the importance when discussing endemism, of considering it from a broader geographical, not country, viewpoint.

Table 2.3 lists the genera restricted, or almost so, to Afghanistan; it does not include a few genera that are sub-endemic: such as *Kurramiana* Omer & Qaiser [also in Kurram valley] – Gentianaceae; *Calyptrosciadium* Rech. f. & Kuber [northeast Iran] – Umbelliferae (Apiaceae); *Scrithacola* Alava [Pakistan] – Umbelliferae(Apiaceae); *Polychrysum* (Tzvel.) Kovalevsk. [Tajikistan] – Compositae (Asteraceae).

2.5.4 *Areas of Special Interest and Future Collecting*

All regions of the country with a substantial number of endemic organisms deserve conservation and merit further study. At present, there is only a rather vague picture of regions where there are high levels of endemism, but

Table 2.3 Genera Endemic to Afghanistan

Apiaceae (Umbelliferae)

Gongylotaxis Pimenov & Kljuykov: *G. rechingeri* Pimenov & Kljuykov – Ghorat, Bamyan, Kabul; – monotypic; salt marshes

Kandaharia Alava: *K. rechingerorum* Alava – Kandahar; monotypic

Mastigosciadium Rech. f. & Kuber: *M. hysteranthum* Rech. f. – Ghorat; monotypic

Pinacantha Gilli: *P. porandica* Gilli – Kabul/Panjshir; – monotypic; doubtful status

Pyramidoptera Boiss.: *P. cabulica* Boiss., NW, SW, E, C; – monotypic; distinct

Registaniella Rech. f.: *R. hapaxlegomena* Rech. f. – Kandahar; – monotypic

Asteraceae (Compositae)

Chamaepus Wagenitz: *C. afghanicus* Wagenitz - Kataghan; -monotypic

Tiarocarpus Rech.f.: *T. hymenostephanus* Rech.f. – SW; *T. neubaueri* (Rech. f) Rech. f. – C; *T. tragacanthoides* (Rech. f. & Gilli) Rech. f.– C

Brassicaceae (Cruciferae)

Cyphocardamum Hedge: *C. aretioides* Hedge– Ghazni; monotypic; distinct

Pseudodraba Al-Shehbaz, D. German & M.A. Koch: *P. hystrix* Al-Shehbaz, D.A. German & M. Koch – Kabul, Ghazni, Bamyan; also, just, in adjacent Pakistan Baluchistan; monotypic; distinct

Veselskya Opiz: *V. [Pyramidium] griffithiana* (Boiss.) Opiz - Jija, Kandahar, Herat; – monotypic

Caryophyllaceae

Kabulianthe (Rech .f.) Ikonn.: *K. honigbergeri* (Fenzl) Ikonn., – Kabul – monotypic; doubtful status

Ochotonophila Gilli: *O. allochrusoides* Gilli, *O. eglandulosa* Hedge & Wendelbo; *O. flava* Dickoré & Freitag – all mainly in Bamyan

Pentastemonodiscus Rech. f.: *P. monochlamydeus* Rech. f. – Ghorat – monotypic

Scleranthopsis (Rech. f.) Rech.f.: *S. aphanantha* (Rech. f.) Rech. f. – Herat, Kabul, Ghorat; – monotypic

Chenopodiaceae/Amaranthaceae

Halarchon Bunge: *H. vesiculosus* (Moq.) Bunge - Herat, Farah, Kandahar – monotypic, distinct

Papaveraceae

Cryptocapnos Rech. f.: *C. chasmophyticus* Rech. f.– Dilaram, Orozgan – monotypic

Plumbaginaceae

Bamiania Lincz.: *B.pachycormum* (Rech. f.) Lincz. – C; monotypic, doubtful status

Bukiniczia Lincz.: *B. cabulica* (Boiss.) Lincz. – SW–, SE–, Waziristan – monotypic

many are in the middle elevations of various parts of the Hindu Kush. Here, each ridge or valley may have developed neo-endemics as in the thorny cushion belt with *Acantholimon* and *Astragalus,* etc. This is also the case with a few other genera like *Acanthophyllum, Cicer, Nepeta, Oxytropis, Silene, Taraxacum,* etc. But, with our present knowledge, there is no clear pattern of areas with especially higher levels of endemism; endemic species, and genera seems to occur more or less randomly throughout the country.

There are some parts of Afghanistan where collecting plants for herbaria has been very limited, or none, in recent decades: for example, the Ghorat limestone area, the southern slopes of Kohe Baba, some of the Darwaz remote mountains, and some side-valleys in Nuristan and Laghman, where species-rich forests and scrub prevail. The Afghan government will probably concentrate its conservation efforts on the well-known highlight areas of the country: Bande Amir, Ajar valley, Pamire Buzurg, and Dashte Nawar. But, certainly, there are other areas that merit conservation. Some of these regions are remarkable not only for endemic plants, but also for their wildlife or dramatic landscape scenery and geology. They are listed in Breckle and Rafiqpoor (2010, FGA; there on p.135, Table 4.02). These are the prospective nature reserves or national parks in a future peaceful Afghanistan.

2.5.5 Biodiversity of Lower Plants

Concerning lower plant groups as in many countries, our knowledge of diversity, function and interactions of species and number of species in ecosystems is limited and far from complete. While there is a rather good coverage of the mosses and lichens in Afghanistan (see Table 2.4), other groups

Table 2.4 Biodiversity Numbers for Bryophytes

Taxon	Number of species
Bryophyta	310
Mosses	245
Hornworts	1
Liverworts	65
Lichenes	262
Fungi	1,000 s
Algae	>250

Frey, 1969, 1972, 1974, Frey and Kürschner, 1991, 2009, Kürschner and Frey 2011, Kürschner et al., 2017 and for Lichens (Feuerer, 2007) for other groups only estimates (partly from UNEP) are given.

like algae, fungi, etc., are almost not studied. Only random observations and data are available, and only rough estimates on their biodiversity are possible (see below). But with bryophytes (liverworts, mosses, and hornworts) the situation in Afghanistan is much better. They play an important role in various habitats, such as in the understory of forests and woodlands, as pioneers on soil and rock, in steppe and deserts (organisms of "harsh environment") or as epiphytes on tree trunks (Kürschner et al., 2018). Knowledge of them is therefore essential in future studies.

Bryophytes are perfect organisms to study these aspects. As they serve as indicators of ecological disturbances and air pollution (Frey, 1989, Frahm, 1998, Frahm et al., 2008), knowledge of them is of fundamental importance in understanding phytodiversity and ecosystem development (succession stages) and functioning, especially with regard to increasing human impact and global warming effects.

They are among the simplest of the terrestrial plants, in general lacking a complex tissue organization (non-vascular plants) and reproducing by spores. Although they require water for hydration and reproduction, they are widely distributed throughout the world, are common at many different sites and show considerable diversity in form, ecology, and functional adaptations.

They include one of the most ancient lineages of terrestrial plants dating back to the Devonian period some 400 million years ago. Three main evolutionary lines are recognized: the liverworts (Marchantiophyta), the mosses (Bryophyta), and the hornworts (Anthocerotophyta) (Frey and Stech, 2009).

Many bryophytes grow on soil (terrestrial, terricolous), on bare rock surfaces (saxicolous), or on tree trunks (corticolous, epiphytic), roots, rotting wood, fallen logs or are aquatic (immersed, submerged) in springs, by streams, and in mountain flushes. The main requirements for growth appear to be a relatively stable substratum for attachment by rhizoids, a medium that retains moisture for extended periods, adequate sunlight, and a favorable temperature. In mountain regions such as large areas of Afghanistan, bryophytes, especially mosses, form extensive cover, especially in flushes, near watercourses and springs, and in sites where there is snow-melt moisture for a large part of the growing season. There they can dominate the vegetation cover and control the vegetation pattern and dynamics of associated plants (*cf.* Schofield, W.P. www.britannica.com).

Even in semi-humid or semi-arid forests and woodland (e.g., NE, SW Afghanistan) and in arid and semi-arid steppes and deserts (e.g., W, S Afghanistan) bryophytes are widely distributed on marly or clayey soils often forming or being part of sparsely vegetated microphytic crusts. They are typical components in the vegetation, increasing species diversity. Especially Pottiaceae, the "mosses of harsh environments" (Zander, 1993) play an important role in soil and crust stabilization of deserts, hindering soil erosion.

Data about lichens was compiled by T. Feuerer. His checklist lists 208 taxa based on the material of 12 collectors. All 14 relevant publications are evaluated. About 350 species are expected for this country (http://t1p.de/nld3). Mayrhofer counted 262 spec. from Graz herbarium records.

Fungi in Afghanistan are not well known. There are several mushrooms which are sometimes found in bazaars for food. And even on Afghan postal stamps, mushrooms are found (Figure 2.10). But there is no systematic checklist of the fungi of the country nor any Flora. It is also known that several fungi are a threat to wheat and other crops (e.g., the black rust fungi Ug99, or other Ustilaginales and Uredinales parasitic fungi). The very comprehensive book on parasitic fungi on vascular plants of Europe by Brandenburger (1985) is a very useful reference.

A rather detailed study by Geerken (1978) lists observations of 25 fungi (Basidiomycetes) collected during several excursions in various parts of the country (Kabul, Paktia, N Hindu Kush). Additional remarks were published

Figure 2.10 Mushrooms of Afghanistan on stamps by the former Afghan Post.

by Mochtar and Geerken (1979) on the hallucinogenic properties of *Amanita* species and their traditional use.

There is an old paper by Schaarschmidt (1884) dealing with some algae (and cyanobacteria) derived from herbarium collections of paddy field weeds. Hirano (1964) has made valuable contributions to the knowledge of freshwater algae of Afghanistan mainly from Nuristan. His detailed list is still the only basis for a future checklist of algae (Table 2.5). Foged (1959) made observations on the distribution of algae as Sino-Japanese and Eurasian and some cosmopolitans. Several new species have also been recorded; an up-to-date checklist of algae is a desideratum.

2.5.6 Plant Geography

In phytogeographical terms, most of the area is Irano-Turanian; a term which has been much used to describe the overall pattern of plant geography in highland southwest Asia but, with our ever-increasing knowledge of plant distributions and endemism, Irano-Turkestanian is probably a better descriptive term (Sales and Hedge, 2013) for at least the Afghanistan area.

With regard to the distribution patterns of plant species (chorotypes) which are closely related to their history and the general ecological traits of the respective areas, most parts of Afghanistan belong to two very different phytogeographical regions: the *Irano-Turanian floristic region* with c. 92% of the country's surface, and the *Sino-Japanese floristic region* with c. 7%. In both of them, the temperature conditions are rather similar, but they are sharply separated by the amount and seasonal distribution of rainfall. Two smaller parts of the country belong to other floristic regions or show at least strong admixtures of species from adjacent regions: (1) species of

Table 2.5 Number of Taxa on Algae and Cyanobacteriae

Taxon	Number of taxa (+ variants and subspecies)
Cyanobacteriae	47
Euglenophyta	4
Chrysophyta, Heterokontae	2
----------, Chrysophyceae	1
----------, Diatomeae	137 (+c. 35)
Chlorophyta	14
Conjugatae	56 (+c. 5)

Source: Hirano, 1964

the *Saharo-Sindian region* intrude into the lower altitudes of south and east Afghanistan, but only in the hot and dry Jalalabad basin are they so numerous and dominant that the area should be included in that phytogeographical region; (2) in the upper belts of the high mountain areas, the number of *Central Asian elements* increases. The eastern part of the Wakhan is often considered as the southwest extension of the central Asian floristic region. That region extends eastwards along the Karakorum and north of the Himalayas to central and northeast China, and northwards to south Siberia (Freitag et al., 2010).

Within these chorotypes, any particular species has acquired an individual distributional area according to its particular ecological requirements, age, and widely differing dispersal abilities. The topography has greatly contributed to the high number of endemic species that can be found, particularly among mountain plants with less effective modes of dispersal. However, though the overwhelming majority of plant species belong to one or the other chorotype, there are rather many that are distributed in more than one phytogeographical region. For instance, many annuals that are adapted to summer drought occur in both the Mediterranean and the Irano-Turanian region. Other species were able to reach suitable habitats even if they are geographically widely separated from each other (Freitag et al., 2010). This applies in particular to plant species of high mountain areas.

2.5.7 Vegetation Types of Afghanistan

Except for some weeks from spring to early summer, excluding the irrigated areas which cover about 5% of the country's surface and the few areas of forest, the plant cover of Afghanistan has a poor visual appearance and looks rather uniform. For most of the year, when seen from a distance, plant life appears to be almost completely absent, and the monotonous grey or brown colors of the landscapes seem to be caused by the barren soil or rock surfaces. This is caused by the strongly seasonal and predominantly semi-arid climate in combination with the long-lasting destructive influence of man on the plant cover.

Seasonality. Since the Tertiary, the plant species in the Irano-Turanian region have evolved alternative adaptations to the distinct seasonal climate with rainfall or snow during the cold winter, a short favorable spring with moderate temperatures, with some rainfall and moisture stored in the soil, and the very long hot dry summer and autumn. Two groups of life forms are most important (Freitag et al., 2010):

1. The "ephemerals," comprising annuals and geophytes, survive the unfavorable seasons as seeds or in underground perennial organs like bulbs, corms or rhizomes. They start to appear almost simultaneously in spring or (in the lowlands) in late autumn after the first rains and complete their life cycle at the onset of the dry period. Depending on the amount of rainfall, they form just a thin cover or lush, meadow-like stands, even in semi-desert ecosystems, when soil conditions are suitable. The annuals are also perfectly adapted to a repeated failure of the scanty rainfall because in such years the seeds remain dormant in the dry soil. As the ephemerals usually do not suffer from water stress during the rainy season, most of them look like plants from humid regions. The majority of the showy monocots, such as the species of *Tulipa, Iris, Gagea, Fritillaria,* and *Eremurus,* belong to the ephemerals with underground perennial organs (geophytes). Many other herbaceous perennials that continue photosynthesis at least in the first part of the dry season behave in a similar way. Only species with long or deep tap-roots or water-storing tissues may remain active until late summer.

2. The "woody species" continue to be active, though in a much-reduced way, during the dry season when most ephemerals have disappeared. Accordingly, they show a variety of typical adaptations: (a) deep, far and wide reaching root systems; (b) cushion-like habit in dwarf shrubs as protection against strong insolation, dry wind and mechanical damage by sand drift; (c) small leaves covered by grey or white hairs and/or thick layers of epi-cuticular waxes which are likewise effective against strong insolation and drying out; (d) spiny leaves and stems that protect them from grazing; (e) leaves containing essential oils (often in distinct glands) or a variety of poisonous secondary compounds or other elements that render them unpalatable to grazing animals. The two latter features most likely have evolved in parallel with large herbivores like gazelles and wild asses in the lowlands and wild sheep and goats in the mountains. These characters are likewise effective against the flocks of villagers that have replaced the wild herbivores, causing a much higher grazing pressure.

The early start of the growth season in the peripheral lowlands (autumn to early spring) and its much-delayed arrival in the high mountain areas (summer) provide the basis for the economically important transhumance of nomadic or semi-nomadic shepherds in the country (Freitag et al., 2010).

2.5.8   Potential and Actual Vegetation

The survey of the broadly defined vegetation types given here focuses on two aspects: the "potential natural vegetation" (Figure 2.11) as it would exist without the influence of man in natural habitats; and the "actual vegetation" as it is today as a result of man's destructive impact. It is based on the intensive survey by Freitag (1971a, b) which also resulted in two maps of the potential natural vegetation: (1) a country-wide map that was largely adapted by Nedjalkov (1983a,b), slightly modified by Breckle (2007); (2) a detailed map showing the much more diversified eastern-most part of the country in higher resolution both with regard to scale and vegetation types (Freitag, 1971a).

Other authors have mainly dealt with the actual vegetation: Gilli (1969, 1971) studied it mainly around Kabul; Nedjalkov (1983a, b) described some vegetation types from the forest region in Kunar province in eastern Afghanistan. Mountain vegetation and the flora of the central Hindu Kush were studied by Frey and Probst (1978), Pavlov and Gubanov (1983) and some others. The results of the very first surveys were given by Linchevsky and Prozorovski (1949) and Neubauer (1954a, b).

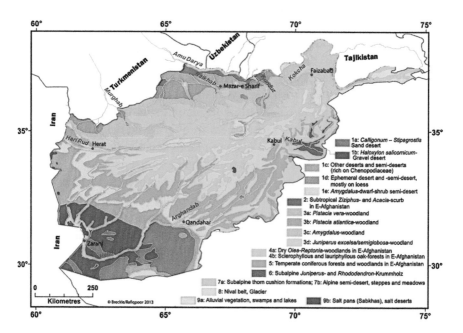

Figure 2.11   Natural vegetation of Afghanistan (modified from Freitag, 1971a, see also Breckle 1973, 1981, 1983, 2007).

The potential natural vegetation is chosen here for classifying the vegetation cover because it indicates the vegetal resources of the various parts of the country as they are determined by climate and soil. Furthermore, it shows to what extent the vegetation can be restored by application of careful and conscious practices of land use, though sometimes, or often, only in the long run. It applies to "normal habitats" which are defined as being flat to moderately sloping, not receiving additional water-supply (plakor sites in Russian literature).

The most important sources for information about the composition, structure and geographical distribution of the natural vegetation were in Freitag (1971a): the few remnants of little-disturbed vegetational over the country but preferably those in remote and inaccessible areas; some intentionally protected small plots (e.g., pistachio and pine woodlands); the surroundings of tombs and shrines ("Ziarat"); and some rare written documents. In contrast to the natural vegetation, the actual vegetation usually reflects predominantly the intensity of anthropogenic influences. Consequently, one type of natural vegetation is often replaced by substitute plant communities.

In fact, the vegetation of the country is much more diversified, because of its mountainous character. The common rocky sites have their peculiar plant communities as well as scree covered slopes or naturally eroding slopes in weak marly sediments. Because of their geographic or topographic isolation, these communities, made up of highly specialized plant species, are particularly rich in narrow-range endemics. They cannot be dealt with in this survey.

The main zonal vegetation categories on normal ecological sites are shown in the map (Figure 2.11) under Nos. 1–8. Those with the predominant influence of one ecological factor, namely additional water supply or high salinity (river valleys, lakes, swamps and saline flats), are summarized in the map under category No. 9. They represent azonal vegetation types, since here the climate is less influential than the outstanding ecological factor (water, salt, gypsum, heavy metals, etc.) causing peculiar habitats. Photographs of characteristic vegetation types are shown in Breckle and Rafiqpoor, 2010 (FGA). The specific vegetation types are listed with legend on Figure 2.11.

2.6 Biodiversity of Animals

The particularly diverse fauna of Afghanistan is supported by the complex geography of the region. The Hindu Kush mountains have been a barrier to a north and westward dispersal of most elements of the Indian fauna, and as a result most of it is typically Palaearctic. The overlap of two major zoo-geographic realms is made more complex by the occurrence of at least five basic major ecological life zones in the country: the central highlands, steppes, southern semi-deserts and deserts, monsoon forests, and interior high and eastern low intra-montane basins, as indicated above. Additionally, some very special sites are important for the fauna of the country: the Amu darya river and riverine forests and scrubs, the Afghan Pamir and Wakhan high mountain area and the often rather isolated deep valleys in Nuristan with their monsoonal forests. Some data can also be found in Encyclopaedia Iranica (http://t1p.de/tibe).

The lists of *vertebrates* is rather complete in contrast to that of the non-vertebrates. But the distribution of many vertebrates is still very incompletely known. Many provinces of Afghanistan have never been visited by zoologists. The zoological literature is substantial, often dealing with special animals and their behavior, but there are no checklists of animal groups. In Table 2.6, we give a preliminary list of species numbers of the main animal taxa in Afghanistan.

From the *Mammalia*, a total of 128 species has been recorded from Afghanistan. The country used to be very rich for mammals, but many species are threatened with extinction or are already extinct mainly through hunting.

The only primate species in Afghanistan is the rhesus macaque (*Macaca mulatta*). It is restricted to the Nuristan and Paktia forests, where it is found in fairly large numbers. Because of their destructive habits, they are not welcomed by local people and are left unharmed in the forests.

The country had harbored a rich assemblage of different carnivores (Gaisler et al., 1968), but as already mentioned, most populations have suffered from drastic reductions in numbers. The cheetah (*Acinonyx jubatus venaticus),* once common in the southern and western steppes, is now apparently extinct due to the decline of its main prey species, *Gazella subgutturosa* and *Equus hemiones onager*. The Caspian tiger (*Panthera tigris virgata*), once found in the wetlands of Āmū Daryā and the Murghab basin, has been exterminated through habitat destruction and intensive hunting. Human predation has depleted the numbers of the snow leopard (*Pantheria [Uncia] uncia),* which is found in alpine valleys of the Pamir plateau and northern Hindu Kush range. Recent studies estimate that there may be c. 200 in Afghanistan and, very recently, a healthy population in the Wakhan corridor has been reported. *Panthera pardus* and *Lynx lynx,* inhabiting the central highlands have also declined in numbers during recent years. Among the smaller cats, *Felis manul* and *Felis lybica,* although not endangered, are suffering from human predation for the expanded fur trade (Niethammer, 1967).

Table 2.6 Species Diversity of Animals

Taxon	Number of species
Mammalia	125; 137
Eulipotyphla (Insectivora)	7
Erinaceidae	3
Chiroptera	34; 32
Lagomorpha	4
Ochotonidae	2
Rodentia	35
Macaca	1
Gazellae	1
Equidae	1
Capra	3
Ovis	2
Carnivore	29
Felidae	10
Canidae	5
Mustelidae	8
Viverridae	4
Ursidae	2
Aves	c. 500; 428; 410
Amphibia	11; 8; 6
Reptilia	<100; 92
Agamae	23
Serpentes	27
Colubridae	15
Gekkonidae	17
Pisces	84; 102; 101
Insecta	Tens of 1000s
Lepidoptera	233
Tenebrionidae	366 (55% end.)
Culicidae	30
Solifugae	34
Scorpiones	8
Mollusca	73
Bivalvia	2
Gastropoda on land	27
Gastropoda in water	10

Mainly acc. to Kullmann, 1972; Hassinger, 1973; Naumann, 1986; Habibi, 2004; UNEP, 2008; Nogge and Arghandewal, 2012

Dogs and foxes (*Canis lupus* and *Vulpes vulpes*) are widely distributed throughout the country. They are found from 300 m in the Seistan basin up to 4,000 m in the alpine valleys of Badakhshan. *Vulpes ruppelli* and *V. cana* most probably are now extinct in Afghanistan. The golden jackal (*Canis aureus*), however, is still present in substantial numbers in the steppe and deserts, in summer months also in the mountains.

The bears of Afghanistan are restricted to mountainous and forested zones. The Asian black bear (*Selenarctos thibetanus*), inhabits the Laghman and Nuristan forests (Povolny, 1966), while the brown bear (*Ursus arctos*), has apparently been exterminated in this region, but is still in the Pamir mountains.

Eight species of Mustelids (otters, badgers, weasels) occur in Afghanistan. They have an extensive range and are found in varying habitats. Trapping by hunters has caused a decline in the numbers of such marten species as *Martes foina, Mustela erminea, M. nivalis,* in the mountain biomes. The otter (*Lutra lutra*), occurs along water-courses of most rivers (Melisch & Ritschel 1996), and its range extends into the forested Kunar region. The hyaena (*Hyaena hyaena*), is distributed in the steppe around Qandahar and in parts of the Kabul river valley.

The pig or wild boar (*Sus scrofa*) has an extensive ecological range and breeds successfully in swamps and reed beds along major river drainages in many parts of the country.

Five ungulate (hoofed) species have their home in the mountains of Afghanistan. Over 2,500 Marco Polo sheep (*Ovis ammon poli*) seasonally occupy the Pamir region (Petocz, 1973). Goats (*Capra ibex sibericus*) not only occur in the Pamir but also in the Darwaz peninsula and the Zebak region of eastern Badakhshan and partially into Nuristan. *Ovis orientalis,* co-inhabit much of the same range as *C. ibix sibericus* in southeastern Badakhshan. *Capra ibex ibex* is found in large numbers in the Hindu Kush, Paghman, and Kohe Baba ranges, while *Capra aegagrus* is mainly found in the southern Hazarajat mountain. *Capra falconeri* is one of the most spectacular and least known species among the country's feral goats; four sub-species occur in Nuristan, Laghman, the Paktia forests, and Kohe Safi region of Kapisa and northern Badakhshan. Local hunting has been a major factor in reducing their numbers in recent years. The red deer (*Cervus elaphus bactrianus*), once common in the wetlands of the Amu Darya, is also endangered because of habitat destruction and hunting. The musk deer (*Moschus moschiferus*), which occurred in Nuristan, has not been reported in recent years and may be extinct.

Of the insectivorous mammals, long-eared hedgehogs (*Hemiechinus auritus* and *H. auritus megalotis*) are sparingly distributed in the steppe and semi-deserts, while *Paraechinus hypomelas* (Brant's hedgehog) is only recorded from the Jalalabad valley in eastern Afghanistan. Besides occurring in the lowlands, *Soricidae* (shrews) are also found in mountainous terrain, in the centrally located Salang and Shibar passes. *Lepus capensis is* the most common Lagomorph species. It has a wide range extending from the western steppes of Herat to the Pamir mountains. The pika (*Ochotona rufescens*) occurs in sub-alpine valleys. It is scattered from the Salang pass to the Orozgan mountains, while the range of *Ochotona macrotis,* is limited to the valleys of Badakhshan.

The long-eared marmot (*Marmota caudata*), is restricted to alpine valleys above 3,000 m. It occurs in the Pamir, Zebak, and Darwaz valleys of Badakhshan and northern Hindu Kush as well as the centrally located mountains around Dashte Nawor.

Two arboreal species of squirrels, *Petaurista petaurista,* and *Hylopetes fimbriatus,* inhabit the Nuristan and Spingar forests. During spring and summer, when not hibernating, the yellow ground squirrel (*Spermophilus fulvus*) is abundant in the Ghazni and Katawaz plains, while *Spermophilopsis leptodactylus* occurs in the clay and loess biotopes of northern Afghanistan.

Serious problems for agriculture in the steppe are posed by the rapidly expanding populations of small rodents – mice and rats – the *Cricetidae* and *Muridae*. An expanding agricultural economy, reduction in predator numbers, especially of wild cats and foxes have fostered the increase.

Thirty-two species of bats (Chiroptera) have been recorded from Afghanistan (Gaisler et al., 1968). Their preferred habitat is in warmer parts of the country, where they may be found in abandoned ruins and caves of the Seistan basin and in the steppe and semi-desert. To the east, common bats (*Myotis* and *Pipistrellus*) have been observed in Laghman and the Kabul river valley.

Birds (Aves) are found everywhere in Afghanistan. There are differing numbers of bird species (Paludan, 1959), depending on the author, but at least 450 species are now recorded, of which nearly half occur in the steppe and semi-deserts.

More than 100 different species of waterfowl and waders pay regular visits to the alkaline lakes of Abe Istada and Dashte Nawor. The rare Siberian crane (*Grus leucogeranus),* visits Abe Istada on its way from India to its breeding grounds on the Ob river in Siberia. Many species also breed at these lakes including shelduck (*Tadorna tadorna),* black-winged stilt (*Himantopus himantopus)*, the elegant avocet (*Recurvirostra avosetta),* and many others. About 20,000 flamingos (*Phoenicopterus ruber roseus)* breed or have been bred at the two lakes. Dashte Nawor at 3,200 m is the world's highest known breeding ground of this species, but the numbers vary greatly from year-to-year according to the water-level (Nogge & Arghandewal, 2012).

The Hamune Saberi and Hamune Puzak lakes extending along the Afghan-Iranian border form an extensive habitat for many water birds that overwinter annually. Again, depending on the very variable water levels the numbers greatly fluctuate. More than half a million waterfowl and waders have been recorded in these lakes. Examples are coot *(Fulica atra)*, various geese *(Anser anser, A. platyrhynchos, A. penelope, A. acuta, A. clypeata, A. crecca)*, pochard *(Aythya farina)*, 3 species of grebe *(Podiceps)*. Besides waterfowl, two species of pelicans, grey heron *(Ardea cinerea)*, egrets *(Egretta alba)*, the unmistakable spoonbill *(Platalea leucorodia)*, cormorant *(Phalacrocorax carbo)*, and hundreds of waders.

About 150 species of birds occur in the central mountains. The Salang Pass is the main flight path during spring and autumn for large numbers of storks *(Ciconia ciconia nigra)*, starlings *(Sturnus vulgaris)*, and numerous species of waterfowl and waders, which migrate from their wintering grounds to northern latitudes (Nogge, 1973) and back. Chukar partridge *(Electoris chukar)*, snowcock *(Tetraogallus himalayensis)*, magpie *(Pica pica)*, hoopoe *(Upupa epops)*, raven *(Corvus corax)*, choughs *(Phyrrhocorax phyrrhocorax and P. graculus)*, and a number of eagles and buzzards, together with lammergeyers *(Gypaetus barbatus)*, are some of the more characteristic birds formerly encountered in the mountains. The kestrel *(Falco tinnunculus)* is by far the most common raptor of this zone. One of the world's rarest and little-known species, the large-billed warbler *(Acrocephalus orinus)*, is a native of the Wakhan corridor.

Many species with Himalayan affinities had been found in the Nuristan and Paktia forests in eastern Afghanistan: monals *(Lophophorus impejanus)*, jays *(Garrulus lanceolatus)*, bulbuls *(Hypsipetes leucocephalus)*, drongos *(Dicrurus macrocerus)* occur in these forests. Parakeets *(Psittacula krameri and P. himalayana)* are summer visitors.

The arid semi-deserts and desert lowlands harbor only a few breeding birds. But during spring and autumn, the avifauna of this region is much enriched by migrants. Sparrows *(Passer montanus, P. domesticus)* and swallows *(Hirundo rustica)* are common all year round in towns and villages; at some times of the year there are swifts *(Apus apus)*.

The herpeto-fauna (Amphibia, Reptilia) of large parts of the country, especially the central highlands, has not been widely studied, and most of the work done was centered on the Kabul river valley and southern Afghanistan

(Kral, 1969, Leviton and Anderson, 1970). The fauna of the northern plains shows strong affinities with that of the deserts and steppe of southern former USSR, while elements of the Indian fauna are included in the herpeto-fauna in the south and east. The most common and abundant amphibian is the green toad (*Bufo viridis*), which is found all over the country. The three species of frogs all belong to the *Rana* genus and are frequent in irrigation streams, although in far smaller numbers than toads. One salamander (*Batrachuperus musteri*) is known from the Paghman range and occurs in mountain streams up to 3,000 m.

Of the two species of turtles and tortoises, the land tortoise (*Testudo horsfieldii*) inhabits arid steppes all over the country up to 2,400 m and *Trionyx gangeticus*, a soft-shelled turtle, is known only from the Indus drainage system in eastern Afghanistan.

The *Agama* family, represented by twenty-three species in four genera, is by far the largest group of lizards. The most characteristic of this group is *Trapelus (Agama) agilis*, which is widespread below 2,500 m throughout the country. *Agama caucasica* and the endemic *A. badakhshana*, inhabit montane biotopes up to 3,200 m. Nine species of *Phrynocephalus* are typical representatives of this family in the southern and north western semi-deserts. The two species of *Uromastyx* are herbivorous and live in long tunnels which they dig in stony soil of scree-covered deserts. There are 15 species of geckos. *Alsophylax pipiens*, a nocturnal animal, is frequently found hovering near lights on the walls of most houses in Kabul. To the east, *Eublepharis macularius* occurs near human settlements.

Among the *Lacertidae,* the most common genus is *Eremias* (ground lizards), representing twelve of the fourteen Lacertid species found in the country. Widely dispersed and abundant are *Eremias guttulata watsonana* and *E. velox persica*. *Ablepharus bivittatus lindbergii*, which has a body length of only 6 cm, is commonly found at higher elevations (2,300 to 3,300 m). Two species of monitor lizards are known from to lower elevations: *Varanus griseus* is found throughout the country; while *V. bengalensis*, an Indian fauna element, is only known from the lower Kabul river valley.

Twenty-seven species of snakes have been recorded from Afghanistan, of which seven are poisonous (five vipers and two cobras). Among the poisonous snakes, the most common is *Echis carinatus*, which occurs at lower elevations north and south of the Hindu Kush. The cobra, *Naja naja oxiana,* occurs in the south and northwest, while *Bungarus caeruleus* is known only east of Jalalabad. Among the non-poisonous snakes, three species of sandboas (*Eryx*) occur all over the country. The *Colubridae* is the largest snake family with 15 species distributed in the southern lowlands, the western and northern steppes. A common Eurasian grass snake, *Natrix tessellata*, reaches its eastern limit in Nuristan and Chitral. In contrast to most other members of this family which inhabit arid areas, *Natrix* is found near watercourses, where it lives on fish and amphibians. Species belonging to the *Psammophis*, *Coluber*, and *Lytorhynchus* genera are other members of this family. The distribution of two worm-snakes, *Typhlops vermicularis* and *Leptotyphlops blanfordi*, is not well known.

The fish fauna (*Pisces*) is astonishingly rich. The Hindu Kush range divides the country's fish into two main assemblages, trout and carp. Brown trout (*Salmo trutta oxiana*), present in northern drainage areas, do not occur in those of the southern slopes. Southern streams are, however, rich in carp species. Already in 1920, extensive ichthyological studies in Seistan were made by Annandale. He separated the fish of this dry basin into two geographical divisions. The *Cyprinidae*, which do not occur in the highlands of central Asia, represent an element derived from the region lying south and southeast of the Helmand basin, while the *Schizothoracinae* and *Cobitidae* are thought to have been carried southward by the Helmand from the Hindu Kush and probably descended from fish of the ancient Oxus system. Tributaries of the Indus river, which drain the eastern portion of Afghanistan, also contain several fish species. Brown trout (*Salmo trutta*) and carp (*Oreinus* sp. and *Schizothorax* sp.) occur in the colder mountain streams of Nuristan and the Konar valley, while two species of *Cyprinidae* occur further downstream. The largest fish in this drainage is the spiny eel (*Mastacembelus armatus*), which can reach 75 cm.

Among the *invertebrates*, only few groups have been studied; some like Platyhelminthes, Nemathelminthes, and Protozoa are almost unknown. The snail fauna is still quite neglected (Naumann, 1986). But some of the butterfly groups like *Zygaena* had been studied intensively. Here only a few examples can be given.

Naumann et al. (1999) gave an extensive survey on the biology, ecology, and behavior of the Zygaenidae (Lepidoptera), a group of diurnal moths that has become one of the most intensively studied models of chemical and evolutionary biology. Iran and Afghanistan have been shown to be very rich countries for these brightly colored, day-flying burnet moths (Naumann et al., 1984).

A few intensive studies on special taxa had been done by Schneider (1975) on woodlice (*Hemilepistes*, Isopoda), which are found in the driest habitats conquered by any species of crustacean. Usually, they are nocturnal and detritivores, feeding mostly on dead plant matter and living underground. Often, they are the prey of Solifugae spiders.

The honey-bee is one of the two old insects who have been domesticated for centuries. The silkworm being the other one. Both are commonly kept in Afghanistan. Schneider (1976) gave a survey on the honey-bees and their culture and found that not only the western honey-bee (*Apis mellifera*) but also the eastern honey-bee (*Apis cerana* = *A. indica*) is found. In all parts of the country, bee-keepers are active and use various types of beehives, which they also transport to suitable places, but sometimes they also collect honey from bee-species in the wild.

The other insect formerly often cultivated is the silkworm, which is the caterpillar of the domestic silk moth (*Bombyx mori*). They feed mainly on mulberry leaves (*Morus alba* and *nigra*). For the production of silk, enormous amounts of mulberry-leaves are needed: from about 100 kg of leaves about 500 g of silk thread can be produced and less than half of this can be reeled (Saberi, 1994). In former times, the production of silk in Afghanistan was more important than today, but new efforts are being made to modernize it. Herat had been an important city for silk trade in the last centuries, or even millennia, along the famous Silk Road.

Anacanthotermes is a harvesting termite restricted to arid and semi-arid habitats. In northeast Afghanistan, termite damage in traditional flat roofs with wood beams was so bad that now houses are built with cupola roofs. Great damage has also been done to railway sleepers (Harris, 1970).

A special topic is animal parasites. Klockenhoff et al. (1973) gave a list on the Mallophagae on more than 40 species of birds in Afghanistan.

A detailed study on the anatomy of larvae, also female and male images, of the goat warble fly (*Crivallia silenus*, Diptera), which infect cattle and goats in Afghanistan, was given by Madel (1970). Apparently, more than 30% of the goats and sheep in Afghanistan had been parasitized by *Crivallia*, causing holes in the skin of animals and thus worthless leather, as well as emaciation, loss of fat and muscle, protruding bones, and overall weakness.

From Fischer (1968) we have a broad survey on all parasitic relevant bacteria and virus-infections, see the following paragraph.

2.7 Species Diversity of Microbes

There are many infectious diseases known from Afghanistan, but they cannot be dealt with here in detail. Studies on microbes, protozoans, etc., are very scarce, but there is a good medicinal survey by Fischer (1968). He deals with the various diseases of the country, e.g., typhus, cholera, amoebiasis, leprosy, tuberculosis, diphtheria, tetanus, leptospirosis, toxoplasmosis, etc., and on anthropozoonoses cases (like brucellosis, Anthrax, rabies, etc.) or diseases transmitted by arthropods (like malaria, leishmaniases, typhus, dengue fever, etc.) as well as with such helminthian diseases as echinococcosis, taeniasis, bilharziosis, filariasis, and the common infections by *Ascaris* or *Oxyuris* and *Trichuris*. During the decades of war, the situation has certainly not improved, but in recent years medical services have been much extended and vaccines more commonly used. However, in such a war-torn country still very frequently resistant strains of bacteria like *Staphylococcus aureus, Klebsiella pneumoniae, Pseudomonas aeruginosa*, and *Acinetobacter calcoaceticus-baumannii* complex are identified in infections of injuries and wounds. Overuse of broad-spectrum antibiotics may be an important factor in building resistant strains (Calhoun et al., 2008).

2.8 Degradation and Desertification

Without the impact of man, in normal habitats, the floristic composition, coverage, and height of vegetation are intimately correlated with the climate throughout the year. However, as in most countries of the area, in Afghanistan, this natural pattern is obscured by a long history of overexploitation. The natural ecosystems have been heavily degraded except for a few inaccessible or inhabitable areas. The forests were exploited for timber and fuel. Woodlands, shrub-lands and even the scarce vegetation of semi-deserts were likewise at first thinned out and then badly devastated or destroyed in the search for firewood and brushwood. Because of the ever-increasing shortage of trees and shrubs, even slightly woody dwarf shrubs (in particular, species of *Artemisia,* many *Cousinia* species and *Alhagi)* are now collected for fuel.

Simultaneously, any area within reach was grazed and often heavily overgrazed by sheep and goats as well as exploited for fodder plants. This also applies to the remote desert and high-mountain areas which are used seasonally as rangelands by nomadic or semi-nomadic shepherds. Consequently, in many rangelands, the structure and floristic composition has been changed fundamentally with an increase or invasion of unpalatable, poisonous or spiny species. This even applies to the monsoon areas.

As overgrazing thins out, the protective ground cover, the fertile upper soil layers become more exposed to wind and water erosion. In periods of low annual rainfall, as happened during the recent exceptionally dry years from 2006 to 2008, wind erosion is much enhanced. The effects of strong episodic rains on over-grazed slopes are even worse as they cause a dramatic loss of soil by erosion and can lead to a complete and partly irreversible change of habitat conditions.

Altogether, these impacts have had disastrous results: (1) the regeneration of trees and shrubs has almost completely ceased; (2) the vegetation cover has decreased, and the composition of the plant communities has changed fundamentally in that palatable species have been widely replaced by spiny, poisonous or unpalatable plants, resulting in a heavy loss of floristic diversity and much reduced productivity; (3) the reduced vegetation cover and trampling effects of animal flocks have intensified erosion in all parts of the country, leading to widespread desertification.

A particular case of degradation in the broader sense is the fallow areas. About 7% of the country's surface, mostly in sloping areas of the montane belts with higher precipitation, during former decades, were cleared for intermittent dry-farming (Lalmi). When the fields are not sown, as in dry years, the fallow fields are quickly invaded by ubiquitous weedy plant species.

Irrigated fields have replaced the natural vegetation almost completely on the terraces of river valleys and adjacent plains. As a result of primitive irrigation techniques, lack of drainage systems and water shortage, the fields suffered from salinization. They were abandoned and subsequently colonized – under heavy grazing pressure – by salt-tolerant plant communities often dominated by *Alhagi maurorum, Aeluropus* and *Prosopis farcta* or by *Peganum harmala*.

In recent years, it has become obvious that in some areas ("protected by mines"), the vegetation has considerably recovered. As the number of nomadic people decreased, pressure on grazing areas has also become less. A better grazing control and management of protected areas may lead to a sustainable use of natural resources in future. The growing demand for energy and water (Beddington, 2010) goes parallel with habitat destruction, alien introductions, over-exploitation and desertification (May, 2010).

The loss of biodiversity and the rate of extinctions are alarmingly steep (Mooney, 2010). There is an urgent need for mankind to grasp the fact that wealth includes not only manufactured capital knowledge and human capital (education and health), but also natural capital in the terms of functioning and sustainable ecosystems. This only can be achieved by sustainable development in terms of inter-generational well-being (Dasgupta, 2010), and by addressing new questions of high relevance for both science and society in order to create ecologically sound economies (Loreau, 2010).

The Year 2010 was designated the "International UN-Year of Biodiversity" and SW Asia has, *inter alia*, a huge biodiversity in its wild progenitors of crop plants. For millennia, this has been known as the many economic plants mentioned in the Holy Quran and the Bible clearly demonstrate (Musselman, 2007; Barthlott et al., 2016). More than 80 plants are mentioned in the Bible or in the Quran. Afghanistan with this natural heritage of fruits, grains, grasses, trees and now has a historic opportunity to integrate environmental issues into all development plans (UNEP Afghanistan Task Force 2003, http://is.gd/8VG9eR).

The UNEP assessment report contains 163 recommendations, covering environmental legislation and enforcement, capacity building, job creation, planning, environmental impact assessment procedures, industry and trade, public participation and education, and participation in international environmental agreements. It also makes recommendations in relation to water supply, waste, hazardous wastes and chemicals, woodlands and forests, energy, air quality, wildlife and protected areas conservation, desertification and food and agriculture resources and identifies actions at specific urban and rural sites for a better environmental management, which has to include nature conservation measures in order to conserve or enhance biodiversity.

2.9 Agriculture, Horticulture, Forestry

Afghanistan has, for centuries, been famous for its various crops. Its melons and grapes have for long been known for their good taste and quality. The huge variety of grapes is also remarkable (Neubauer, 1975). But it is not only grapes and melons, but many other fruits and vegetables that make the agricultural scene so rich (Table 2.7).

Afghanistan is a rather arid country and often it has been said that its agriculture was almost entirely dependent on irrigation. But this ignores the large areas of rain-fed wheat cultivation grown on the steep rolling hills and steppe in the north where rain may persist until May–June and, where there is sufficient soil moisture; agriculture is practiced as "Lalmi," the rain-fed fields. But in most other parts, irrigation is practiced. In all mountain areas, this irrigation is used for the maintenance of the many small fields, along terraces on alluvial fans, by diverting creek water into a small irrigation channel system for fields and orchards. The variety of crops is normally very wide, as with the fruit trees, including the widely cultivated mulberries (*Morus*). The most reliable figures for irrigated areas date back to 1967, giving a total of 2,385,740 ha. Irrigated land can be divided into four classes, according to the origin of the irrigation water. They are rivers, 84.6%; springs, 7.9%; karezes, 7.0%, and the underground Persian wheels, 0.5%.

Irrigation is common in all parts of Afghanistan when any creek or river water is available and can be diverted, sometimes over long distances. Sophisticated terracing, often using the whole valley floor is a part of traditional agriculture. Villages are usually confined to the steeper, often rocky slopes beside the valley bottom. The amount of water for each farmer is (or was) regulated and controlled, whether it is for furrow irrigation or for basin irrigation on larger areas, mainly for maize or even more important for various types of paddy-fields and rice.

The most interesting and best-adapted irrigation technique is certainly that of Karez (Qanats, Foggara) or Turpan water system which, for centuries, has been used throughout SW Asia and N Africa, mainly on the vast peneplains close to mountains. Long, partly underground, channels along the slopes bring the water from creeks to the villages for cattle, sheep and for gardens. The threat of salinization is especially severe in the dry arid lowlands when not enough and less good quality irrigation water is available (Breckle, 2002).

In 1963, some 114,000 ha were reported to be equipped with sprinkler irrigation. From 1967 up to the present, many different figures have been given for irrigated, rain-fed and total cultivated areas. Unfortunately, most of them are unreliable. Some pre-war publications suggest that 2.8 million ha were cultivated, of which 1.4 million ha had sufficient water to support double-cropping.

Figures for rain-fed areas are generally cited as 1.4 million ha, but recent satellite data (1992) indicate that the area might be much larger, and about 3.1 million ha has been suggested (FAO Aquastat). The newest figures from Afghan Statist. Yearbook for 2009/10 is shown in Table 2.7.

Table 2.7 Usable Land Area in Afghanistan 2009/10 (ha)

Total usable land area	65,223,000
Permanent pasture	30,000,000
All other land	25,613,000
Agricultural area	9,610,000 (100%)
Forests and woodland	1,700,000 (18%)
Temporary crop land	4,324,000 (45%)
Irrigated crop areas	1,836,000 (19%)
Cultivated rain-fed area (Lalmi)	1,750,000 (18%)
Permanent crops	117,000

The agricultural crops, vegetables, fruits, as well as common ornamental garden plants, other economic plants in Afghanistan have a great variety of cultivars. Flower pots with ornamental plants can be found in each village, town, and city in Afghanistan. Families normally use flowers for gifts, when visiting relatives, especially in hospitals, but also for other visits. Afghan villagers are often fond of flowers and have small winter-gardens, even under poor conditions between narrow streets and huts. Roses (*Rosa*), carnations (*Dianthus*), marigolds (*Tagetes*), petunias *(Petunia × hybrida)*, *Phlox* and gladiolus (*Gladiolus*) are among the most frequently grown and often are used as cut flowers. Zonal geraniums *(Pelargonium x hortensis)*, are the most common pot flowers. Many varieties are grown and exchanged between neighbors and flower lovers. A small garden was always maintained, with flowers, spices for the kitchen, and always with grapes (*Vitis vinifera*), not only important for their sweet fruits in autumn, but also for leafy shade in summer. With enough space (and water for irrigation), other fruit trees (apricots, apples, pears, pomegranates, etc.) are also cultivated.

Medicinal plants and herbs are part of traditional agriculture in villages. A few of them may even play an increasing role in the future. Formerly, liquorice (*Glycyrrhiza glabra*) was an important crop. It was harvested in some areas, mostly in the wild. The very deep roots and suckers were dug out, cut and dried, and then exported to Pakistan and Iran, and there processed to black liquorice, which is sold worldwide. Other plants and herbal derivatives still play an important role in traditional medicine. At the bazaar, many herbal products are sold. For data on the traditional knowledge of special uses of plants see Aitchison (1890 – a remarkable information source-, Volk (1955, 1961), Pelt et al. (1965), Schapka and Volk (1979); the more recent surveys by Duke (1991), Shawe (2007) and Babury et al. (2009) are also relevant. A pictured book on medicinal plants is used for surveys in villages (Keusgen et al., 2015).

The cultivation of poppy and cannabis has become of considerable importance as *cash crops* to the Afghan economy, but there is a great variety of other plants which are cultivated depending on climatic conditions. The use of many other plants for spices, for traditional medicine, for cooking and vegetables has already been mentioned. Spice or aromatic plants from the umbellifers (Apiaceae), or Labiatae (Lamiaceae) are especially common.

In a dry country like Afghanistan besides agriculture, the economy is also based to a great extent on livestock. Pasture quality in rangelands, however, has declined at an alarming rate by overgrazing, mainly by sheep and goats. *Deforestation* is another old problem. Much of the damage increased during the Soviet Union's occupation of Afghanistan. During this period, which began in 1979, up to 200 trucks a day rumbled over the roads carrying freshly cut timber. Most of the wood was exported to neighboring Pakistan. The Soviet Union began to withdraw its troops in 1989, but timber-smuggling remains a big problem. In some regions ruled by tribal lords, communities have lost control of their resources to the illegal logging operations of foreign traders and "timber mafias."

But before then, there was timber-smuggling to Pakistan, for example, in the 1960s across Paiwar-Kotal, where every day about 60–80 camels crossed, each loaded with 2 heavy logs. The extent of deforestation between 1977 and 2002 in the three eastern provinces of Afghanistan is shown on a map on the internet (http://is.gd/QZeqmX). Deforestation might also have been a contributory factor to the long-lasting droughts in recent years.

Afghans are trying to reverse the effects of environmental abuse. In 2001, the government created the National Environmental Protection Agency of Afghanistan (NEPA). NEPA is working to protect the country's natural resources and rehabilitate the land. *Re-forestation* and tree-planting programmes are underway. Additionally, alternative energy sources, including solar power, are being tested in many villages that never have had access to electricity. Under the present conditions, it is hard to convince Afghans to change destructive practices. Shepherds and goatherds, whose flocks need grazing land, prevent the restoration of forest areas. Loggers and wood sellers continue to cut trees from the shrinking forests. But on the positive side, in the future, a comprehensive, large-scale afforestation programme could create skilled jobs in silviculture and tree nurseries, reduce the level of poverty, reestablish a sustainable forest products industry and provide a range of export commodities.

In the eastern provinces, a large scale afforestation programme is needed, making use of native trees raised in nurseries. A small afforestation project and forestry programme, started decades ago in Paktia (Kotgai) should be re-activated in other provinces to create new jobs and income and preserve the diversity of trees. There are some challenges, however, ranging from insecurity in rural areas and lack of expertise to the unwillingness of international agencies to commit resources and also the problem of a still lucrative drug trade.

There is an old tradition in many regions of central and middle Asia to use poplar (*Populus*) and other trees (*Morus, Platanus*) along irrigation channels, for poles, furniture, and other purposes. Many villages and towns have their own characteristic look with long rows of columnar trees. This tradition needs to be maintained or enlarged. Those trees give shade in summer and can be used for timber; in the case of mulberries not only for fruit production, but also for silk-worm culture and silk production.

As was shown before, Afghanistan has great potential for being autonomous and independent for its food supply, if sustainable agriculture and forestry are re-established. But climatic conditions vary from year-to-year. The rainy season in the post-drought years was very good; wheat production reached a new record in 2009, making the country almost independent from imports. But to be safe, even in drought years, the agriculture and forestry sector needs more varied and new inputs towards sustainability. This is a challenge for the coming decades, not only for Afghanistan but for the whole world facing severe global issues, most notably climate change, population growth and a rapidly growing demand for energy and water (Beddington, 2010).

2.10 Genetic Diversity of Crop and Cultivated Plants

As mentioned above, there is a great variety of crops cultivated in Afghanistan (Table 2.8). They are important not just for what they yield, but as being able, by their biodiversity, to tolerate drought, heat, and frost. There are hundreds of grape cultivars and dozens of melon cultivars.

2.11 Nature Conservation and National Parks

The various natural vegetation formations of Afghanistan present a rich pattern. Land-use (agriculture, grazing, browsing, fuel-collecting, mining, roads, and settlements) have, of course, greatly changed the natural plant cover. Often, overgrazing and agriculture with non-sustainable methods, increasing erosion and salinity, change, to an irreversible extent, the potential productive abilities of valleys and slopes in the close vicinity of villages. The rich vegetation cover of Afghanistan and the rich landscape and ecosystem patterns must be conserved by a system of nature protection areas. This was already proposed by Petocz and Larsson (1977), Petocz et al. (1978), and by various FAO-experts. Here, we give a short description of the proposed protected areas (Table 2.9), being aware that the goal of legal realization may only be reached in the distant future, and probably will be modified according to the remit of a future nature conservation agency.

The system of protected areas should represent most of the typical landscapes and vegetation formations of the country, which are then also the basis for game reserves and wildlife conservation. However, we have to be aware that Afghanistan is a country in which man and his domestic animals have for millennia exercised a strong ecological impact. Man's long-term

Table 2.8 List of Major Crops in Afghanistan*

Crops	Area harvested (ha)	Yield (t/ha)
Almonds	11,029	7,000
Apples	8,550	8,500
Apricots	8,170	nf
Barley	267,000	1,798
Cotton	33,000	1,300
Citrus fruits	200	nf
Figs	1,924	nf
Grapes	60,832	15,000
Linseed	1,780	nf
Maize	140,000	2,143
Millet	10,000	nf
Nuts	nf	nf
Olives	2,000	4,500
Oranges	1,000	nf
Other Melons	35,400	nf
Peaches, Nectarines	1,675	10,001
Pears	225	nf
Pistachios	nf	nf
Plums and sloes	3,500	nf
Pomegranate	8,413	nf
Potatoes	21,600	14,000
Pulses	0	nf
Rice, Paddy	200,000	2,345
Seeds Cotton	33,000	nf
Sesame seeds	9,660	nf
Stone fruit	2,757	nf
Sugar beet	1,100	15,000
Sugar cane	3,080	40,000
Sunflower seed	170	nf
Vegetables, fresh	72,200	nf
Walnut	2,382	4,200
Water melons	25,100	nf
Wheat	2,575,000	1,967

*Economic plants in Afghanistan, area and yield, data from FAO for 2009/2010 (data from Afghan Statist. Yearbook 2009/10).

nf: no figure given.

presence is everywhere to be seen in overgrazed landscapes, abandoned historical cities and irrigation networks, impoverished fauna communities, and impoverished weedy pastures. According to the semi-desert climate with sufficient rainfall, it may well have been that most parts of the country were once covered by trees (Freitag, 1971a,b) prior to human settlements.

Two areas in Afghanistan are especially critical in biodiversity preservation: the "Wakhan Corridor" with some of the last relatively pristine wildlife habitats and wildlife populations left in Afghanistan, while the "Hazarajat Plateau" has some of the most important existing and potential protected areas in the country.

The "Great Pamir" extends over about 5,500 km^2 of Wakhan. A considerable part of the western Big Pamir was once included in the so-called "Great Pamir Wildlife Reserve" encompassing about 679 km^2. Although designated as a reserve, it has never been legally established, and between 1968 and 1977 functioned as a hunting reserve for foreigners, managed by the Afghan Tourist Organization. Before then, part of the area was a royal hunting reserve of the former King Mohammad Zahir Shah. In a 2004 survey of wildlife in the Wakhan, it was recommended that the eastern tip of the Little Pamir should be designated a strictly protected area (about 250 km^2). This area is at present not used by herders, and thus the habitat is in excellent condition and does not conflict with human use patterns. There is also no barrier between it and the proposed Shaymak Reserve in Tajikistan, enabling the Marco Polo sheep to move freely back and forth.

The eastern tip of the "Waghjir Valley" (about 300 km^2), is at present uninhabited and used only for yak grazing in winter. There, Marco Polo sheep cross the Yuli Pass between China and Afghanistan in winter. The presence of snow leopards and species such as wolf, brown bear, and Asian ibex was assessed. It was also recommended that this area should be designated as a reserve with yak grazing allowed to continue but other activities prohibited.

Hunting pressure – mainly for meat and casual trade in wildlife furs – was much reduced during the period of Soviet occupation, but has increased since then.

UNEP noted that the Wakhi have responded positively to recent calls by the Afghan Transitional Authority to hand in their arms and stop hunting, and the area escaped much of the recent conflicts and is free of landmines.

In the Hazarajat Plateau region, "Bande Amir" is often described, and rightly so, as one of the great wonders of the world. Consisting of six crystal blue lakes separated by a series of natural white travertine dams in a unique step-like lock system, and surrounded by spectacular reddish cliffs – the best example of this landscape type in the world. Bande Amir deserves protection not only as a major source of future revenue from international eco-tourism for the country of Afghanistan, but also as a hotspot of biodiversity, as well as a geo-morphologically unique landscape in Afghanistan. Even today, Bande Amir is regularly visited by Afghanis on holidays. Alongside the natural beauty of the lakes, there is also a shrine dedicated to the Caliph Ali, son-in-law of the prophet Mohammad. Bande Amir has been identified as having appropriate components to qualify as a UNESCO World Heritage Site, yet while it was identified as a National Park in 1973 (Afghanistan's first national park), it only obtained in June 2009 formal legal status.

The assessment team of WCS (Wildlife Conservation Society) found that Bande Amir has enough water to support healthy populations of urial sheep and ibex (*Ovis orientalis*). During 2001, it was one of the front lines of fighting between Taliban and resistance forces and some areas remain heavily mined.

Finally, the "Ajar Valley" is a long gorge created by the Ajar River and the sheer-sided Jawzari canyon. The area has long been known for its good populations of ibex, urial sheep, Bactrian deer and other wildlife, and for many years it was a royal hunting reserve. Although Ajar was gazetted as a wildlife reserve in 1977, there is only a preliminary management plan that has never been implemented, and no recent accurate border assessment, wildlife surveys, socio-economic surveys, or enforcement has taken place. Geologically, in the Ajar valley the mesozoic red-formation of north Afghanistan outcrops in an excellent manner with the typical Jurassic fossil flora and fauna in the so-called Doab-Saighan-Ajar-Series which are majestic in appearance and with Jurassic coal bands of economic importance interest.

Some future areas which deserve a protection status encompass the metamorphic massif of the "Kabul crystalline," the granite massif of Salang, as well as the Kohe Baba summit (peak of Shah Fuladi 5143 m). The two last massifs show in their north-facing slopes signs of the last glacial maximum in the form of well-developed side and end moraines. Recently in these regions, the periglacial processes form the so-called "mesoforms of periglacial": smooth slopes, rock glaciers, etc. In Afghanistan, both areas are geologically and geo-morphologically the *locus typicus* of the "sub-recent" and recent active forming processes, and merit protection.

One of the important "geo-heritages" is undoubtedly the *Tectonic Lineament* of *Chaman-Moqor* as an example of the fragmentation of the pre-Cambrain Gondwana continent. This lineament plays an important role in the geological history of the whole region (Iran, Afghanistan, and Indian). This lineament today separates the Iran-Central Afghanistan micro-continent from the East-Afghan-Baluchestan-Plate; in the landscape today, it can be seen as a marked tectonic step.

But even close to the capital of Kabul, there are important wildlife areas. Heavy snowfalls in past years replenished water levels at the "Kole Hashmat Khan" wetland on the outskirts of Kabul, which had been dry for much of the past five years. The wetland was declared a waterfowl reserve by King Zahir Shah in the 1930s and in the 1960s supported tens of thousands of ducks, as well as wintering and migrant birds, though later water diversions damaged these wetlands.

Some of the necessary next steps, which were identified by WCS are:

- Carry out wildlife surveys, socio-economic surveys, and range-land and biodiversity assessments at each existing or potential protected area site.
- Develop the Wakhan protected area.
- Initiatives, including updating the "Great Pamir" Wildlife Reserve.
- Management plan and officially designating the "Little Pamir" Protected Area and Waghjir Protected Area.
- Develop and enact Hazarajat protected area initiatives, including updating the Bande Amir National Park management plan and officially designating the Ajar Valley Wildlife Sanctuary.
- Develop a training programme for local park rangers.
- Further develop a Transboundary Peace Park Initiative between Afghanistan, Pakistan, Tajikistan, and China.
- Review policies and legislation affecting wildlife, wild lands, and protected areas.
- Develop an environmental services valuation project.
- Develop a plan for forest and wildlife assessments for Nuristan, Kunar, Paktika, Khost, and Paktia.
- Help raise capacity within Afghanistan's environmental sector through a series of training courses and a Conservation Study Exchange Programme.
- And finally, develop and enact a geo-heritage programme to identify and preserve the geologically and geo-morphologically unique areas in Afghanistan (see the proposal list in Breckle and Rafiqpoor, 2010).

Conservation of endangered plants and animals and protected areas in Afghanistan:

It is desirable that the protected areas cover regions with a high plant and animal biodiversity and attractive landscapes. The realization of the below list of proposed nature reserves (Table 2.9) would guarantee such a preservation of the great biodiversity and impressive landscapes of the country.

Table 2.9 List of Proposed Protected Areas of Afghanistan

Reserves	Afghan Province, Area	Altitude (m)	Approx. Size (km²)	Main characteristics (IUCN Category, Year of establishment)	Vegetation formation (see Chapter 2)	References
Ajar Valley	Bay	2000–3800	400	Wildlife Reserve (IUCN: Ib, 1978) unique nature monument, Geo-heritage	1e, 3c, 7	Shank, 2006
Great Pamir	Bak, Wakh	4000–6000	679.38	Wildlife Reserve (IUCN: IV, 1978)	7, 8	Petocz et al., 1978; Podlech and Anders, 1977; Gratzl, 1972; Dor and Naumann, 1978; de Grancy an Kostka, 1978
Dashte Nawor	Gha	3200–4800	75 (lake) 700 NP	Wildlife Reserve, Wetland, waterfowl sanctuary, (IUCN: IV, 1977)	7, 9a, 9b	Shank and Rodenburg, 1977; Petocz and Habibi, 1975
Abe Istada	Gha	1950–2100	270	Wildlife Reserve, Wetland, waterfowl sanctuary, (IUCN: IV, 1977)	9b, 3b	Shank and Rodenburg, 1977; Petocz and Habibi, 1975; Niethammer, 1971
Bande Amir	Bay	2900–3832	410 (NP)	Wildlife Reserve, Wetland, NP, (IUCN: Ib, 1973); Unique natural Monument: Geo-Heritage	7, 9a	Dieterle, 1973; Shank and Larsson, 1977; Evans, 1994; Jux and Kempf, 1971
Western Nuristan	Kun, Lag, Bak	1100–6300	5200	Monsoonal influenced forests types, up to high mountain belts unique flora, wildlife	4b, 5, 2, 4a, 6, 7, 8	Petocz and Larsson, 1977; Rodenburg, 1977; Habibi, 1977
Hamune Puzak	Nim	475	>350	Wildlife Reserve, Wetland	1b, 9a, 9b	Petocz et al., 1976

Table 2.9 (Continued)

Reserves	Afghan Province, Area	Altitude (m)	Approx. Size (km²)	Main characteristics (IUCN Category, Year of establishment)	Vegetation formation (see Chapter 2)	References
Imam Sahib and Darqad	KDZ, Tak	350–470	400	Wildlife Reserve, Wetland, Tugai Forests of AmuDarya	1a, 1d, 9a,	Evans, 1994
Northwestern Badakhshan, Darwaz	Bak	2500–4400	800	Semi-humid woodlands and mountains, unique flora	3a, 3a, 7	Reports by Naumann
Little Pamir including Lake Zorkol, Lake Chaqmatin	Bak, Wakh [Taj, Pak, China]	4000–5900	2000	Transboundary Reserve, Wildlife Reserve, Wetland	8, 7	Petocz et al., 1978; Nogge, 1973; Evans, 1994; Dor and Naumann, 1978; de Grancy and Kostka, 1978
Waghjir Valley Pamir	Bak, [Pak, China]	3800–5500	300	See above	8, 7	See above
Registan Desert	KDH	900–1100	18000	Arid sand-desert area, desert wildlife	1a, 1c, 1e	Freitag, 1971a, b
North Salang	Bal	1600–3600	350	Open *Juniperus* woodlands	3d	Freitag, 1971a, b
Lake Hashmat Khan	KBL	1793	1.91	Wildlife Reserve, Wetland, Waterfowl sanctuary, (IUCN: IV, 1973)	9a, 9b	Niethammer, 1967; Rahim and Larsson, 1978
Taiwara, Kohe Malmond	Gho, Far	2000–4200	800	Unique flora and vegetation, Limestone massifs, Geo-Heritage	3b, 3c, 7	Rechingerk, 1963

Reserves	Afghan Province, Area	Altitude (m)	Approx. Size (km²)	Main characteristics (IUCN Category, Year of establishment)	Vegetation formation (see Chapter 2)	References
Paghman	KBL, Paw	1800–4000	200	Open *Cercis* woodlands and high mountains	3c, 7	Freitag, 1971
Mir Samir	Kap	3500–6800	450	Arid high mountain, glaciation	8, 7	Breckle, 1971, 1973; Maeder, 1972
Safed Koh	Pay, [Pak]	2600–4700	250	Humid high mountain, unique forests and Krummholz	4, 5, 6, 7, 8	Breckle, 1972, 1974, 1975
Gulran Reserve	Her, Bag	250–1000	10000	Wildlife Reserve, *Pistacia vera* woodlands	1d, 3a	Freitag, 1971a, b
Moqor-Chaman-Lineament, from Moqor to Spinboldak	Zab, Pak	1300–2500	400	Geo-Heritage Marked Tectonic Lineament	3b, 3c	Krumsiek, 1980
Salang Granite at the summit of Kotale Salang	Paw/Bal	1600–3600	200	Geo-Heritage, Granit, recent glacial and periglacial activity	3c, 3d, 7	
Shah Fuladi The summit of Kohe Baba	Bay	2000–5500	200	Geo-Heritage, old glacial and recent periglacial activity	7, 8	

*Existing designated Reserves are underlined. The total area of all proposed reserves would cover about 42,360 km² equivalent to about 6.5% of the total land area of Afghanistan.

(Based on Various Sources and Our Own Experience)

This is only realistic by the involvement of the rural population, but it is known that there are many attempts on the communal village level to save traditional uses by modern means. The former ziarats are an excellent grid system of rather untouched small ecosystems. All this should be enhanced by provincial and governmental aids and conservation programmes in the sense of the UN CBD (Chandra and Idrisova, 2011).

2.12 Conclusions

Afghanistan's biodiversity is high and important but declining. Centuries of grazing by settlers as well as by nomads, hunting and increase of agricultural fields by more and newer irrigation systems has caused a steady decline and some extinctions. As in many other developing countries focusing on ecological aspects, projects should be parallel with substantial economic measures; only then can it provide a long-term security of biodiversity conservation in the country, even more so because people depend on those resources for their existence. Local community people should be actively involved in natural resources management to secure the future of the country's remaining high biodiversity despite year-long wars. A long-term programme on monitoring and management of conservation issues is essential and can be in concordance with the international CBD (UN Convention on Biological Diversity).

Keywords

- climate
- degradation
- desertification
- ecosystem diversity
- endemism
- nature conservation

References

Agakhanjanz, O. E., & Breckle, S. W., (2002). Plant diversity and endemism in high mountains of Central Asia, the Caucasus and Siberia. In: *Mountain Biodiversity-A Global Assessment*, Körner, C., & Spehn, E., (eds.). Parthenon Publ. Group, Boca Raton, New York etc., 117–127.

Aitchison, J. E. T., (1890). Notes on the products of western Afghanistan and northeastern Persia. *Trans. Bot. Soc. Edinb., 18*, pp. 228.

Annandale, N., (1920). Aquatic fauna of Seistan, *Rec. Indian Museum, 18*, 150–253.

Babury, M. O., & Seddiqui, M. N., (2009). Diagnosis of another *Glycyrrhiza* sp. in Paktia province, Afghanistan. Darmal, *J. Faculty Pharmacy, Kabul Univ., 2*, 15–21.

Barthlott, W., Erdelen, W., & Rafipoor, M. D., (2014). Biodiversity and technical innovations: Biomimicry from the macro- to the nanoscale. In: *Concept and Value in Biodiver-*

sity, Lanzerath, D., & Friele, M., (eds.). Routledge Studies in Biodiversity Politics and Management, pp. 300–315.

Barthlott, W., Lauer, W., & Placke, A., (1996). Global distribution of species diversity in vascular plants: Towards a world map of phytodiversity. *Erdkunde, 50,* 317 327.

Beddington, J., (2010). Food security: Contributions from science to a new and greener revolution. *Phil. Trans. R. Soc. B., 365*(1537), 61–71.

Brandenburger, W., (1985). *Parasitic Fungi of Vascular Plants in Europe (Germ.).* Fischer/Stuttgart, pp. 1248.

Breckle, S.-W., & Rafipoor, M. D., (2010). *Field Guide Afghanistan – Flora and Vegetation.* Scientia Bonnensis/Bonn, New York, pp. 864.

Breckle, S.-W., & Rafipoor, M. D., (2011). Diversity of Flora and Vegetation of Afghanistan (Germ.). *Geogr Rundschau, 63*(11), 40–50.

Breckle, S.-W., (1972). Alpenroses in the Hindu Kush? (Germ.) *Jahrb. Ver. Schutz Alpenpflanzen und Tiere, 37,* 140–146.

Breckle, S.-W., (1973). Microclimatic Measurements and Ecological Observations at the Alpine Belt of the Afghan Hindu Kush (Germ.). *Bot. Jahrb. Syst., 93,* 25–55.

Breckle, S.-W., (1974). Notes on alpine and nival flora of the Hindu Kush. East Afghanistan. *Bot. Not. (Lund), 127,* 278–284.

Breckle, S.-W., (1975). Ecological observations above the treeline of Safed Koh (E-Afghanistan) (Germ.). *Vegetatio, 30,* 89–97.

Breckle, S.-W., (1981).Current State of Research on Flora and Vegetation of Afghanistan (Germ.). In: Rathjens, C. (eds.). *Neue Forschungen in Afghanistan* (pp. 87–104), Leske, Opladen.

Breckle, S.-W., (1983). Temperate deserts and semi-deserts of Afghanistan and Iran. *Ecosystems of the World, 5,* 271–319.

Breckle, S.-W., (2002). Salt deserts in Iran and Afghanistan. In: *Sabkha Ecosystems,* Böer, B., & Barth, H. J., (eds.). Kluwer, 109–122.

Breckle, S.-W., (2004). Flora, vegetation and ecology of the alpine–nival belt of the Hindu Kush (Afghanistan) (Germ.). In: Breckle, S.-W., Schweizer, B., & Fangmeier, A. (eds.). *Proceedings of 2nd Symposium AFW Schimper-Foundation,* Results of World-Wide Ecological Studies. Stuttgart-Hohenheim, 97–117.

Breckle, S.-W., (2007). Flora and vegetation of Afghanistan. *Basic and Applied Dryland Research (BADR online), 1*(2), 155–194.

Breckle, S.-W., Hedge, I. C., & Rafipoor, M. D., (2013). *Vascular plants of Afghanistan – An augmented checklist,* Scientia Bonnensis/Bonn, New York, pp. 598.

Calhoun, J. H., Murray, C. K., & Manring, M. M., (2008). Multi-drug resistant organisms in military wounds from Iraq and Afghanistan. *Clin. Orthop. Relat. Res., 466,* 1356–1362.

Chandra, A., & Idrisova, A., (2011). Convention on biological diversity: A review of national challenges and opportunities for implementation. *Biodivers. Conserv., 20,* 3295–3316.

Dasgupta, P., (2010). Nature's role in sustaining economic development. *Phil. Trans. R. Soc. B., 365,* 5–11.

De Grancy, S., & Kostka, R., (1978). Great Pamir (Germ.). Österr. Forschungsunternehmen 1975 in den Wakhan-Pamir/Afghanistan. Akad Druck Verlagsanstalt Graz, 400 pp.

Dieterle, A., (1973). Studies on Vegetation in the Bande-Amir Region (C-Afghanistan) (Germ.). *Diss. Bot., 24,* pp. 87.

Dor, R., & Naumann, C. M., (1978). *The Kirgises of the Afghan Pamir (Germ.).* Akad. Druck Verlagsanstalt Graz, pp. 124 + 81 Abb.

Duke, J. A., (1991). Medicinal plants of Pakistan. In: *Plant life of South Asia,* Ali, S. I., & Ghaffar, A. (eds.). Karachi, 195–225.

Evans, M. I., (1904). *Important bird areas in the Middle East.* Birdlife Conservation Series No. 2, Cambridge UK, pp. 410.

Feuerer, T., (2007). https://web.archive.org/web/20071118080034/, http://www.biologie.uni-hamburg.de/checklists/lichens/asia/afghanistan_1.htm.

Fischer, L., (1968). *Afghanistan. Medicinal Area Studies (Germ.).* Springer/Berlin, pp. 168 + maps.

Foged, N., (1959). Diatoms of Afghanistan. *Kong. Dans. Vidensk. Biology Skrift, Vol 2/1.*

Frahm, J. P., (1998). *Mosses as Bio-indicators (Germ.).* Quelle & Meyer Verlag, Wiesbaden.

Frahm, J. P., Stapper, N. J, & Franzen-Reuter. I., (2008). *Epiphytic Mosses as Indicators for Environmental Quality (Germ.).* Kommission Reinhaltung der Luft im VDI und DIN, KRdL Schriftenreihe 40, Düsseldorf. 152 pp.

Freitag, H., (1971a). The Natural Vegetation of Afghanistan (Germ.). Beiträge zur Flora und Vegetation Afghanistans. *Vegetatio, 22*(4–5), 255–344.

Freitag, H., (1971b). Studies on the natural vegetation of Afghanistan. In: *Plant Life of South West Asia,* Davis, P. H., Harper, P. C., & Hedge, I. C. (eds.). Bot. Soc. Edinburgh, 89–106.

Freitag, H., Hedge, I. C., Rafipoor, M. D., & Breckle, S. W., (2010). Flora and vegetation geography of Afghanistan. In: *Field Guide Afghanistan – Flora and Vegetation*, Breckle, S. W., & Rafipoor, M. D., (eds.). Scientia Bonnensis, Bonn, Manama, New York, Floríanópolis, 79–115.

Frey, W., & Kürschner, H., (1991). Conspectus bryophytorum et arabicorum. An annotated catalogue of the bryophytes of SW Asia. *Bryoph. Bibl., 39,* 1–181.

Frey, W., & Kürschner, H., (2009). New records of bryophytes from Afghanistan – with a note on the bryological exploration of the country. *Nova Hedwigia, 88,* 503–511.

Frey, W., & Probst, W., (1978). Vegetation and Flora of the Central Hindu Kush (Afghanistan) (Germ.). Beihefte Tübinger Atlas des Vorderen Orients, *Reihe A (Naturwissenschaften)* 3, Reichert, 126 pp, Wiesbaden.

Frey, W., & Stech, M., (2009). Marchantiophyta, Bryophyta, Anthocerotophyta. In: *Syllabus of Plant Families*, Frey, W., (ed.). *Adolf Engler's Syllabus der Pflanzenfamilien,* 13th ed. Gebr. Borntraeger Verlagsbuchhandlung, Stuttgart, pp. 1–257.

Frey, W., (1969). Contributions to the Mossflora of Afghanistan I. (Germ.). *Nova Hedwigia, 17,* 351–357.

Frey, W., (1972). Contributions to the Mossflora of Afghanistan II. (Germ.). *Bryologist, 75,* 125–135.

Frey, W., (1974). The Liverworts of Iran and Afghanistan (Germ.). *Bryologist, 77,* 48–56.

Frey, W., (1989). The bryophytes of Saudi Arabia and the importance of their conservation. Studies in Arabian bryophytes 9. – In: *Wildlife Conservation and Development In Saudi Arabia,* Abu-Zinada, A. H., Goriup, P. D., & Nader, I. A., (eds.). NCWCD Publ., Riyadh, 3, 209–219.

Gaisler, J. D., Povolny, D., Sebek, Z., & Tenora, F., (1968). Faunal and ecological review of mammals occurring in the environs of Jalalabad: Chiroptera. Carnivora, Lagomorpha. *Zool. Listy, 17*(1), 41–48, 185–189.

Gaston, K. J., (1998). Biodiversity – The road to an atlas. *Progr. Physic. Geogr., 22*, 269–281.

Geerken, H., (1978). Mycology of Afghanistan (Germ.). *Afgh. J., 5*(6–8), 114–115.

Gilli, A., (1969). Afghan Plant Communities I. Xerophilic Plant Communities (Germ.). *Vegetatio, 16,* 307–375.

Gilli, A., (1971). Afghan Plant Communities II. The Mesophytic and Hygrophylic Plant Communities in the Summer dry Region (Germ.). *Vegetatio, 23,* 199–234.

Gratzl, C., (1972). Hindu Kush (Germ.). Österr Forschungsexpedition in den Wakhan. *Akad Druck Verlagsanstalt Graz,* 140 pp.

Groombridge, B., (1992). *Global Biodiversity.* Status of the earth's living resources. Chapman & Hall, London, pp. 585.

Habibi, K., (1977). *The Mammals of Afghanistan. Their distribution and status.* UNDP/FAO Report, Field Work Doc. *1,* pp. 91.

Habibi, K., (2004). *Mammals of Afghanistan,* pp. 168.

Harris, W. V., (1970). Termites of the palearctic region. In: *Biology of Termites,* Krishna, K., Weesner, F. M., (eds.). Acad. Press, New York, pp. 295–313.

Hassinger, J. A., (1973). Survey of the mammals of Afghanistan. *Fieldiana: Zoology, 60.*

Hedge, I. C., & Wendelbo, P., (1970). Some remarks on endemism in Afghanistan. *Israel, J. Bot., 19,* 401–417.

Heywood, V. H., & Watson, R. T., (1995). *Global Biodiversity Assessment.* UNEP, Cambridge Univ. Press, pp. 1140.

Heywood, V. H., (1999). The importance of inventories in biodiversity studies. In: *Biodiversity, Taxonomy, and Ecology,* Tandon, R. K., & Prithipal, S., (eds.). Sci. Publ. (India), 65–76.

Hirano, M., (1964). Freshwater algae of Afghanistan. In: *Plants of West Pakistan and Afghanistan,* Kitamura, S., (ed.). Res Kyoto Univ. Sci. Expedition to the Karakoram and Hindukush, 1955, *vol. 3,* 167–245.

Jayakumar, S., Kim, S. S., & Heo, J., (2011). Floristic inventory and diversity assessment – A critical review. *Proc. Intern. Acad. Ecology and Environm. Sciences, 1,* 151–168.

Jux, U., & Kempf, E. K., (1971). Water Reservoirs by Travertine Sedimentation in the Central Afghan high Mountains (Germ.). *Z. Geomorph. Suppl., 12,* 107–137.

Keusgen, M., Breckle, S. W., & Rafiqpoor, M. D., (2015). *Medicinal Plants of Afghanistan.* Univ. Marburg, pp. 247.

Klockenhoff, H., Rheinwald, G., & Wink, M., (1973). Mallophagenbefall bei Vögeln, Massenbefall als Folge von Schäden an den Wirten. *Bonn. Zoolog. Beiträge, 24,* 122–133.

Körner, C., (1999). *Alpine Plant Life.* Functional plant ecology of high mountain ecosystems. Springer, Berlin, pp. 344.

Kral, B., (1969). Notes on the herpetofauna of certain provinces of Afghanistan, *Zool. Listy, 18*(1), 55–66.

Krumsiek, K., (1980). On the Plate Tectonic Development of the Indo-Iranian Region (Results of Paleomagnetic Measurements in Afghanistan) (Germ.). *Geotektonische Forschungen, 60*(2), 223.

Kullmann, E., (1972). *The Animal World. Afghanistan. (Germ.).* Ländermonographie Erdmann/Tübingen, 68–78.

Kürschner, H., & Frey, W., (2011). Liverworts Mosses and Hornworts of SW-Asia (Marchantiophyta, Bryophyta, Anthocerotophyta). A systematic treatise with keys to genera and species occurring in Afghanistan, Bahrain, Iraq, Iran, Israel, Jordan, Kuwait, Lebanon,

Oman, Qatar, Saudi Arabia, Sinai Peninsula, Syria, Turkey, United Arab Emirates and Yemen (incl. Socotra island). *Nova Hedwigia, Beiheft, 139*, pp. 240.

Kürschner, H., Lüth, M., & Frey, W., (2018). Field Guide Afghanistan – Mosses and Liverworts. *Flora and Augmented Checklist* (in press), UNEP, Geneva, c.320 pp.

Lack, H. W., & Grotz, K., (2012). *Flora's Treasures–Capturing the Green World (Germ.).* Botanisches Museum Berlin-Dahlem, 132 pp.

Lauer, W., Rafiqpoor, M. D., & Theisen, I., (2001). *Physiogeography, Vegetation and Syntaxonomy of the Flora of the Páramo de Papallacta (Eastern Cordillera Ecuador) (Germ.).* Erdwissenschaftliche Forschung, 31,142 pp., Franz Steiner Verlag, Stuttgart.

Leviton, A. E., & Anderson, S. C., (1970). The amphibians and reptiles of Afghanistan. *Proc. California Acad. Sc., 38*, 163–206.

Linchevsky, I. A., & Prozorovski, A. V., (1949). The basic principles of the distribution of the vegetation of Afghanistan. *Kew Bull.*, 179–214.

Loreau, M., (2010). Linking biodiversity and ecosystems: Towards a unifying ecological theory. *Phil. Trans. R. Soc. B, 365*, 49–60.

Lozano, F. D., Rebelo, A. G., & Bittman, R., (2012). How plant inventories improve future monitoring. *Biodivers. Conserv., 21*, 1937–1951.

Mabberley, D. J., (2008). *Mabberley's Plant-Book.* A portable dictionary of plants, their classification and uses, 3rd edn. Cambridge University Press, (2009 – reprint with corrections).

Madel, G., (1970). Anatomie der Larven und Imagines der Ziegendasselfliege *Crivellia silenus* Brauer (Diptera, Oestridae). *Z. Parasitenk., 34*, 158–170.

Maeder, H., (1972). *Mountains, Horses and Bazaars. Afghanistan, the Country in the Hindu Kush (Germ.).* Walter/Olten, p. 180.

May, R. M., (2010). Ecological science and tomorrow's world. *Phil. Trans. Roy. Soc. B, 365*, 41–47.

Melisch, R., & Rietschel, G., (1996). The Eurasian otter *Lutra lutra* in Afghanistan. *Bonn. Zool. Beitr., 46*, 367–375.

Mochtar, S. G., & Geerken, H., (1979). The hallucinogenics muscarine and ibotenic acid in the Middle Hindu Kush (Germ.). *Afgh. J., 6*, 62–65.

Mooney, H. A., (2010). The ecosystem-service chain and the biological diversity crisis. *Phil. Trans. Roy. Soc. B, 365*, 31–39.

Musselman, L. J., (2007). *Figs, Dates, Laurel, and Myrrh: Plants of the Bible and the Quran.* Timber Press, Portland, USA, pp. 336.

Naumann, C. M., & Tremewan, W. G., (1999). *The Western Palaearctic Zygaenidae.* 1st edn. Apollo Books, Stenstrup, pp. 304.

Naumann, C. M., (1986). To the Wildlife of Afghanistan (Germ.). *Stiftung Bibliotheca Afghanica*, Band 4: Afghanistan Ländermonographie, 89–101.

Naumann, C. M., Feist, R., Richter, G., & Weber, U., (1984). Verbreitungsatlas der Gattung *Zygaena* Fabricius, (1775). (Lepidoptera, Zygaenidae). *Theses Zoologicae, 5*, 1–97.

Nedjalkov, S. T., (1983a). Ecological classification of vegetation cover of Afghanistan. In: *Ecology and Biogeography in Afghanistan*, Sokolov, V. E., (ed.). Moscow [in Russian], pp. 30–53.

Nedjalkov, S. T., (1983b). Woody shrub vegetation of region Konarha. In: *Ecology and Biogeography in Afghanistan*, Sokolov, V. E., (ed.). Moscow [in Russian], pp. 76–100.

Neubauer, H. F., (1954a). The Forests of Afghanistan (Germ.). *Angewandte Pflanzensoziol.* Festschr Aichinger, 494–503.

Neubauer, H. F., (1954b). Attempt to characterize the vegetation conditions of Afghanistan (Germ.). *Ann. Naturhist. Mus. Wien, 60*, 77–113.

Neubauer, H. F., (1975). The Nuristan Vines and their Cultural and Historical Significance (Germ.). *Angew. Botanik, 49*, 123–130.

Niethammer, G., (1971). Bird life at the Abe-Istada (Afghanistan) (Germ.). *Die Vogelwarte, 26*, 221–227.

Niethammer, J., (1967). Fur animals skins in the bazaar of Kabul/Afghanistan (Germ.). *Pelztiergewerbe, 18*, 7–19.

Nogge, G., & Arghandewal, E., (2012). Afghanistan zoologically considered (Germ.). *Scientia Bonnensis*, pp. 128.

Nogge, G., & Niethammer, J., (1976). Die Vögel auf den Basaren von Kabul und Charikar. *Afghan. J., 3*, 150–157.

Nogge, G., (1973). Bird hunting at Hindu Kush (Germ.). *Natur und Museum, 103*, 276–79.

Paludan, K., (1959). On the birds of Afghanistan. Vidensk. Medd. Dansk Naturhist. *Foren Copenhagen, 122*, pp. 332.

Pavlov, V. N., & Gubanov, I. A., (1983). Geobotanic features of the mountain Afghanistan. In: *Ecology and Biogeography in Afghanistan*, Sokolov, V. E., (ed.). Moscow [in Russian], 54–75.

Pelt, J. M., Hayon, J. C., & Younus, M. C., (1965). Plantes medicinales et drogues de l'Afghanistan. *Bull. Soc. Pharm. Nancy, 66*, 16–61.

Petocz, R. G. et al. (1978). *Report on the Afghan Pamir (3 parts)*. Unpubl. FAO Report, Doc, 5–7, pp. 33, 42, 35.

Petocz, R. G., & Habibi, K., (1975). The flamingoes (*Phoenicopteris ruber roseus*) of Abe-Istada and Dashte-Nawor, Ghazni Prov., Afghanistan. *Direct. Wildlife and Nat. Parcs. Min. Agric. Kabul.* pp. 115.

Petocz, R. G., & Larsson, J. Y., (1977). *National Parks and Utilization of Wildlife Resources*, Afghanistan. Ecological reconnaissance of western Nuristan with recommendations for management. UNDP, FAO, Field Doc. 9, 63 pp. Kabul.

Petocz, R. G., (1973). Marco Polo Sheep (*Ovis ammon poli*) of the Afghan Pamir, mimeo. report, United Nations Development Program, Kabul.

Petocz, R. G., Rodenburg, W. F., & Habibi, K., (1976). *The Birds of Hamune-Puzak*. Unpubl. FAO Report.

Pignatti, S., (1982). *Flora d'Italia I, Edagricola*, Bologna.

Podlech, D., & Anders, O., (1977). Florula of the Wakhan (Northeast-Afghanistan) (Germ.). *Mitt. Botan. Staatss. München, 13*, 361–502.

Podlech, D., (2012). *Checklist of the Flowering Plants of Afghanistan*. Published at www.sysbot.biologie.uni-muenchen.de/de/personen/podlech/flowering_plants_afghanistan, pdf.

Povolny, D., (1966). The discovery of the bear *Selenarctos thibetanus* in Afghanistan. *Zoologicke Listy, 15*, 305–316.

Rafiqpoor, M. D., & Breckle, S. W., (2010). The physical geography of Afghanistan. In: *Field Guide Afghanistan –Flora and Vegetation*, Breckle, S. W., & Rafiqpoor, M. D. Scientia Bonnensis. Bonn, Manama, New York, Florianópolis. pp. 23–78.

Rafiqpoor, M. D., (1979). *Precipitation analyzes in Afghanistan - Attempt to differentiate the country in terms of climate (Germ.).* Unveröff. Diplomarbeit, Geogr. Inst. Univ. Bonn.

Rahim, A., & Larsson, J. Y., (1978). *A Preliminary Study of Lake Hashmat Khan with Recommendations for Management.* UNDP, FAO, Field Doc. Min. Agric., 74/016, pp. 23.

Rechinger, K. H., & Rechinger, W., (1963–2016). *Flora Iranica*, Bd. 1–181. Akad. Druck-und Verlagsanstalt Graz, Naturhist. Museum Wien.

Rodenburg, W., (1977). *The Trade in Wild Animal Furs in Afghanistan*. Min. Agriculture, Field Doc., pp. 51.

Saberi, H., (1994). Silk kebab and pink tea. In: *Look and Feel, Studies in Texture, Appearance and Incidental Characteristics of Food,* Walker, H., (ed.). Oxford Symposium 1993 on food and cookery, Prospect books, Devon, 187–202.

Sales, F., & Hedge, I. C., (2013). Generic endemism in southwest Asia, an overview. *Rostaniha, 14*(1), 22–35.

Schaarschmidt, J., (1884). Notes on Afghanistan algae. *J. Linn. Soc. London, Botany, 21,* 241–250.

Schapka, U., & Volk, O. H., (1979). A directory of common plant names in Afghanistan (Germ.). *Afgh. J., 6,* 3–14. Graz.

Schneider, P., (1975). Sozial Lebende Asseln in Afghanistan. *Afghan. J., 2,* 148–150.

Schneider, P., (1976). Honigbienen und ihre Zucht in Afghanistan. *Afghan. J., 3,* 101–104.

Shank, C. C., (2006). *A Biodiversity Profile of Afghanistan in 2006*, A component of the National Capacity Self-Assessment (NCSA) in National Adaption Programme of Action (NAPA) for Afghanistan, pp. 169.

Shank, C., & Larsson, J., (1977). A strategy for the establishment and development of Band-e Amir National Park. *Min. Agriculture, Field Doc. 8*, pp. 37.

Shank, C., & Rodenburg, W., (1977). Management plan for Abe-Istada and Dashte-Nawor flamingo and waterfowl sanctuaries. *Min. Agriculture, Field Doc.*, pp. 43.

Shawe, K., (2007). *The Medicinal Plant Sector in Afghanistan*. FAO Report, Shawe medic plants mission, Kabul, pp. 46.

Sivall, T. R., (1977). Synoptic-climatological study of the Asian summer monsoon in Afghanistan. *Geografiska Annaler, 59*, 67–87.

Strid, A., & Kit Tan, (1997). *Flora of Greece,* Königstein, vol. 1.

UNEP, (2003). *Afghanistan. Post Conflict Environmental Assessment*, Switzerland, pp. 178.

UNEP, (2008). *Biodiversity Profile of Afghanistan*. An output of the national capacity need self-assessment for global environmental management (NCSA) for Afghanistan, pp. 149.

Volk, O. H., (1955). Afghan medicinal plants (Germ.). *Planta Medica, 3,* 171–178.

Volk, O. H., (1961). A survey of Afghan medicinal plants. *Pak. J. Sci. Industr. Research, 4,* 232–238.

Walter, H., Breckle, S. W., & Agakhanjanz, O. E., (1994). Ecology of the earth. Vol. 3 – Special Ecology of the Moderate and Arctic Zones of Eurasia (Germ.). *UTB Große Reihe, 2 Aufl., Fischer Stuttgart,* pp. 726.

Zander, R. H., (1993). Genera of the Pottiaceae: Mosses of harsh environments. *Bull. Buffalo Soc. Nat. Sci., 32,* 1–378.

Plate 2.1   Some typical vegetation types and landscapes in Afghanistan: A: Open dry *Pinus gerardiana* forest in Bashgal-Valley/Nuristan (Photo: SWBr); B: Oasis with *Phoenix dactylifera* in Seistan/S-Afghanistan (Photo: Kulke); C: Steep mountain slopes of N Hindu Kush with *Eremurus korshinskyi* (Photo: ICHedge); D: The famous Bande Amir lakes in C-Afghanistan, first Afghan National Park (Photo: I. C. Hedge); E: Gravel desert with mobile barkhan sand dunes, Dashte Margo/S-Afghanistan (Photo: SWBr); F: Subalpine thorny cushion slope with *Acantholimon* and *Acanthophyllum*, Dashte Nawor (Photo: H. Freitag); G: Mountain meadow with *Primula auriculata* along a creek, Maidan-Valley/West of Kabul (Photo: Azizi).

Plate 2.2 Some examples of rare or typical Afghan plant species: A: Hamada
desert near Qandahar with the endemic chenopod *Halarchon
vesiculosum* (Photo: H. Huss); B: *Iris korolkovii*, a rare mountain species
from NE-Afghanistan and Pamir (Photo: F. Joisten); C: *Rhododendron
afghanicum*, very rare treeline species, possibly now extinct , Safed
Koh (Photo: SWBr); D: *Fritillaria imperialis*, a famous garden plant in
Europe, grows wild in shady rocky places of Hindu Kush mountains
(Photo: Azizi); E: *Cousinia chionophila*, one of the alpine members
of this large genus, Hindu Kush, Mir Samir (Photo: SWBr); F: *Primula
macrophylla*, with a Himalayan distribution, the highest Afghan plant
species, up to 5600m (Photo: SWBr); G: *Anemone brecklei*, hybrid of
A. tschernjaewii x A. *bucharica*, Tangi Tashkhurgan, N-Afghanistan
(Photo: SWBr).

Plate 2.3 Some examples of Afghan animals: A: *Varanus bengalensis* is still common in the hot basin of Jalalabad, E-Afghanistan (photo: I. C. Hedge); B: *Testudo horsfieldii*, a common tortoise of steppe and semi-deserts in Afghanistan (photo: I. C. Hedge); C: *Mantis* searching for prey in the loessic semi-desert (photo: SWBr); D: *Acanthoderes* sp (Cerambycidae), an example of a desert beetle, Jalalabad, E-Afghanistan (photo: SWBr); E: postal stamps from Afghanistan, exhibiting the great variety of Afghan wild animals (photo: SWBr), with 1: *Eublepharis macularius*; 2: *Garrulus lanceolatus*; 3: *Chaimarrornus leucocephalus*; 4: *Lophophorus impejanus*; 5: *Testudo horsfieldii*; 6: *Agama caucasica*; 7: *Podiceps cristatus*; 8: *Parnassius autocrator*; 9: *Callimorpha principalis*; 10: *Aquila chrysaetos*; 11: *Hyaena*; 12: *Sus scrofa*; 13: *Cervus elephas bactrianus*; 14: *Hystrix indica*; 15: *Gazella subgutturosa*; 16: *Macaca mulatta*; 17: *Aegyptius monachus*; 18: *Bubo bubo*; 19: *Phoenicopterus ruber*. F: *Hemilepistus aphghanicus*, a desert woodlouse living underground, SE-Afghanistan (photo: SWBr); G: Nest of wasps, *Vespa*, N-Afghanistan (photo: F. Joisten).

Plate 2.4 Examples of some common and typical crops of Afghanistan: A:
Cannabis indica, the female plants yield hashish and marihuana (photo:
SWBr); B: *Papaver somniferum*, the milky juice of unripe capsules is
tapped, dried and yields opium (photo: H. Kreuzmann); C: a great
variety of fruits is found in the bazaars, especially many types of melons
(photo: FJoisten); D: Afghan grapes are famed as sweet and aromatic,
hundreds of breeds are known (photo: Baba ola); E: Pomegranate
(*Punica granatum*), famous from all parts of the country, as a postal
stamp (photo: SWBr); F: *Armeniaca vulgaris*, a great variety of apricot
breeds are found in many parts of Afghanistan (photo: C. Naumann).

Plate 2.5 Some examples of environmental problems in Afghanistan: A: logging and deforestation of the remnant woods and forests, smuggling of *Cedrus deodara* poles by camels to Pakistan (photo: SWBr); B: *Nannorhops ritchieana*, the palm leaves are used for chairs, beds and many woven items ; palms have become rare in SE-Afghanistan (photo: SWBr); C: Overgrazing is the most serious problem; all parts of the

country show signs of overgrazing by sheep and goats, as can be seen here with their many tracks, C-Afghanistan (photo: W. Obermayer); D: Overgrazing has much to do with the large number of nomads, moving in spring to the mountains, in fall to the lowlands, with all their herds of camels, cows, horses, donkeys, sheep and goats (photo: SWBr); E: Hunting is very traditional in Afghanistan, wildlife has suffered greatly and many species are now extinct (photo: SWBr); F: forest fires are not common, but in the remnant forests of SE- and E-Afghanistan, they can damage whole mountain slopes (photo: SWBr); G: Collecting fuel and fodder for winter-time including all woody material, even small thorny cushions and thistles are brought to the villages, N-Afghanistan (photo: SWBr); H: Large piles of collected fodder and fuel for winter-time are stored on the roofs of village houses in the mountains, Panshir-Valley (photo: SWBr).

CHAPTER 3

Biodiversity in Bangladesh

SHARIF A. MUKUL,[1-3] SHEKHAR R. BISWAS,[4] and A. Z. M. MANZOOR RASHID[5]

[1]Department of Environmental Management, School of Environmental Science and Management, Independent University, Bangladesh, Dhaka 1229, Bangladesh
[2]Tropical Forests and People Research Centre, University of the Sunshine Coast, Maroochydore DC, Qld 4558, Australia
[3]School of Geography, Planning and Environmental Management, The University of Queensland, Brisbane, Qld 4072, Australia, E-mail: sharif_a_mukul@yahoo.com; smukul@iub.edu.bd
[4]Department of Biology, Lakehead University, Thunder Bay, ON P7B 5E1, Canada
[5]Department of Forestry and Environmental Science, School of Agriculture and Mineral Sciences, Shahjalal University of Science and Technology, Sylhet 3114, Bangladesh

3.1 Introduction

Bangladesh is situated in the world largest deltaic plain – the Ganges–Brahmaputra delta, in the northeastern part of South Asia between 20°34′ and 26°38′ North latitude and 88°01′ and 92°41′ East longitude. The country mostly consists of floodplains with some hilly areas, with a sub-tropical monsoon climate (Islam, 2003). In the country, about 80% of the land is low-lying and/or flooded at least during the monsoon, makes the country the single largest flood-basin in South Asia. The majority of country's land is formed by river alluvium from the Ganges and the Brahmaputra and their tributaries (Sohel et al., 2015). Geographically, the country falls near the Indo-Burma region – one of the global biodiversity hot-spot and believed to have more than 7,000 endemic plant species (Mittermeier et al., 1998). Bangladesh, due to its unique geophysical location and a suitable climatic condition is exceptionally endowed with a rich variety of biodiversity (Nishat et al., 2002). Nevertheless, in last decades, like most other regions of the world, Bangladesh also went through a critical period unsuitable for country's biodiversity and ecosystem. The government so far along with various international conservation agencies has also been trying to improve and manage this overwhelming situation. This chapter aims to provide an insight of the biodiversity of Bangladesh, from ecosystem to species level, genetic diversity, and major threats to the biodiversity in the country with key initiatives so far taken for biodiversity conservation.

3.2   Major Ecosystems of Bangladesh

Ecologically, Bangladesh supports a diverse set of ecosystems. The country, has the world's largest continuous mangrove forest – The Sundarbans on its southwestern part – habitat of the world's largest surviving population of the Royal Bengal tiger (*Panthera tigris*); in its eastern part it has a large tract of evergreen to semi-evergreen hill forests, once very rich in biodiversity but mostly degraded now; besides in the northeastern part there are many wetlands, locally called haors that harbor a huge number of aquatic plants, migratory birds and freshwater fish species (Khan et al., 2007).

The forests of Bangladesh cover three major vegetation type occurring in three distinctly different ecosystems, that is, hill forests (evergreen to semi-evergreen); plain land Sal (*Shorea robusta*) forests and mangrove forests. Although, once very rich in biodiversity during the last few decades all forest and ecosystems of the country have been heavily degraded (Mukul et al., 2008). There have some contradictions on the actual forest coverage of the country. Although the official forest coverage is 2.53 million ha (see Table 3.1) representing nearly 17.5% of the country's total land area, only 1.52 million ha of them are under the jurisdiction of the Forest Department (Khan et al., 2007; FAO, 2006). In addition to that, most of the forests of the country are geographically located only in few districts and are poorly stocked (Figure 3.1).

3.3   Plant and Wildlife Diversity in Bangladesh

In Bangladesh, some 2,260 species of plant reported alone from the Chittagong Hill Tracts, which falls between two major floristic regions of Asia (MoEF, 1993). Until now, an estimated 5,700 species of angiosperms alone, including 68 woody legumes, 130 fibre yielding plants, 500 medicinal plants, 29 orchids, 3 gymnosperms and 1,700 pteridophytes have been recorded from the country (Firoz et al., 2004).

The country also possesses a rich faunal diversity. Bangladesh is home of about 138 mammal species, more than 566 species of birds (passerine and non-passerine), 167 species of reptiles, 49 species of amphibians (IUCN, 2015). In addition to that, at least 253 species of fish (inland freshwater), 305 species of butterflies, 305 species of shrimp/prawn, 2,493 species of insects, 362 species of mollusks, 66 species of corals, 15 species of crabs, 19 species of mites, 164 species of algae, 4 species of echinoderms are believed to exist in the country (IUCN 2015 & 2000; Islam et al., 2003).

Table 3.1 Major Forest Types and Areas in Bangladesh

Forest type	Location	Area (million ha)	Remarks
Hill forests			
Managed reserved forest (evergreen to semi-evergreen)	Eastern part of the country (Chittagong, Chittagong Hill Tracts and Sylhet)	0.67	Highly degraded and managed by the Forest Department.
Unclassed State Forest (USF)	Chittagong Hill Tracts	0.73	Under the control of district administration and denuded mainly due to faulty management and shifting cultivation. Mainly scrub forest.
Plain land Sal forests			
Tropical moist deciduous forest	Central and northwestern region (Dhaka, Mymensingh, Tangail, etc.)	0.12	Mainly *Sal* forest but now converting to exotic short rotation plantations. Managed by the Forest Department.
Mangrove forests			
Sundarbans	Southwest (Khulna, Satkhira)	0.57	World's largest continuous mangrove forest and including 0.17 million ha of water.
Coastal plantations	Along the shoreline of twelve districts	0.10	Mangrove plantations along the shoreline of 12 districts. Managed by Forest Department
Village forests	Homestead forests all over the country	0.27	Diversified productive system. Fulfill majority of country's domestic timber, fuelwood and bamboo requirements.
Tea gardens and rubber plantations	Chittagong Hill Tracts and Sylhet	0.07	Plantations of tree and rubber with various short rotation species as shade tree.
Total		2.53	17.49 % of country's landmass

Source: Mukul et al. (2014a); Khan et al. (2007).

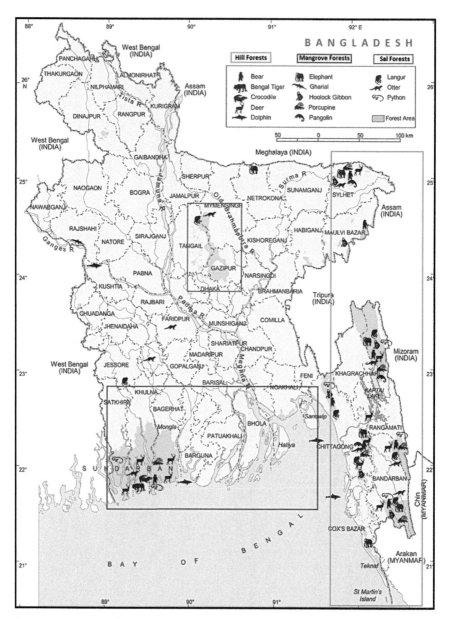

Figure 3.1 Major forest types and distribution of wildlife in Bangladesh.

3.4 Genetic and Crop Diversity in Bangladesh

Bangladesh has a very rich in agro-biodiversity and supposed to have more than 8,000 rice varieties along with 3,000 crop varieties. Other than these more than 3000 varieties of pulses, 781 varieties of oilseed, 3516 vegetables (both species and cultivar), 156 spices, 89 fruits, are recorded alone by the Bangladesh Agricultural Research Council (Chowdhury, 2012). There are also more than 5000 varieties (both species and cultivar) of jute, and 475 varieties of tea recorded from the country. Many of these species varieties are however currently under threat due to the massive cultivation of certain high yielding rice and crop varieties and excessive use of agrochemicals in crop fields.

3.5 Threatened Biodiversity in Bangladesh

A great number of plant and wildlife species have already gone extinct from Bangladesh over the last decades (Rahman, 2004). Table 3.2 outlined the extinct mammal species in the country. A substantial number of country's remaining plants, mammals, birds, and reptiles are currently under tremendous pressure. IUCN (2015) has listed a total of 156 species of mammals, birds, reptiles, and amphibians under various degrees of risks in the country (Table 3.2). A reliable statistic on country's plant diversity is still unavailable, nevertheless, it is anticipated that already 10% of country's plant species gone extinct. A recent inventory by Bangladesh National Herbarium identified 106 vascular plants with risks of various degrees of threats (Khan et al., 2001). Table 3.3 gives the present status of inland and resident vertebrates in Bangladesh.

3.6 Biodiversity Conservation Initiatives in Bangladesh

Despite a rapid loss and degradation of wild habitats, biodiversity conservation has received a wider attention in Bangladesh in the present years (Mukul et al., 2017; Mukul, 2007). The government of Bangladesh, as a signatory party to various regional and international conservation related agreement and conventions, are now increasingly committed to conserving country's remaining biodiversity. Already the government has ratified five major biodiversity-related conventions (i.e., Convention on Biological Diversity, Convention on International Trade in Endangered Species, Convention on the Conservation of Migratory Species, Ramsar Convention, and World Heritage Convention).

The country has also adopted various *in situ* and *ex situ* conservation measures to maintain its rich biological heritage. Declarations of protected areas, ecologically critical areas, World Heritage Site, Ramsar Sites are among the widely used ways for *in situ* conservation (Mukul et al., 2008). At present,

Table 3.2 Mammalian Species Extinct from Bangladesh in the Past Decades

Common name	Scientific name
Great one-horned rhinoceros	*Rhinoceros unicornis*
Javan rhinoceros	*Rhinoceros sondiacus*
Asiatic two-horned rhinoceros	*Didermocerus sumatrensis*
Blue bull/nilgai	*Boselaphus tragocamelus*
Wild buffalo	*Bubalus bubalis*
Gaur	*Bos gaurus*
Banteng	*Bos banteng*
Swamp deer	*Cervus duvauceli*
Marbled cat	*Canis lupus*
Sloth beer	*Melursus ursinus*

Sources: IUCN (2000, 2015).

Table 3.3 Present Status of Inland and Resident Vertebrates in Bangladesh

Group	Total no. of species	Extinct	Threatened			
			Critically endangered	Endangered	Vulnerable	Total
Amphibians	49	0	2	3	5	10
Reptiles	167	1	17	10	11	39
Birds	566	19	10	12	17	58
Mammals	138	11	17	12	9	49
Total	920	31	46	37	42	156

Source: IUCN (2015).

the country has 38 protected areas including 17 national parks and 21 wild-life sanctuaries distributed across the country. Together, the protected areas of Bangladesh cover nearly 17.5% of the forest area and 1.8% of country's total land area (Mukul et al., 2017; Mukul et al., 2008). In addition to that, the country has 7 eco-parks, 2 safari parks and botanical gardens which also contribute significantly to the conservation of country's dwindling biodiversity.

3.7 Challenges and Major Threats to Biodiversity in Bangladesh

Biodiversity loss in Bangladesh is attributed to several socio-economic, biophysical and organizational factors (Mukul et al., 2012a, 2014b).

Following are some key reasons behind the rapid biodiversity loss in the country.

3.7.1 High Population Density, Extreme Poverty, and Unemployment

Bangladesh is one of the world's densely populated countries with an extreme poverty and high unemployment rate. More than 85% population of the country are living in rural areas and somehow depends on various natural resources which lead to exploitation of plant and animal products for people's livelihood and income (Mukul et al., 2012a). Rural fuel consumption pattern, which is strongly concerned with degradation of natural forest area is another important issue related to biodiversity depletion in the country (Mukul et al., 2014c).

3.7.2 Climate Change and Sea Level Rise

Bangladesh is one of the largest victims of climate change and associated sea level rise. The majority of the country will go under water if the water level rises by 50 cm. The country has already experienced severe change in precipitation pattern, temperature, etc. The climate change in the country will largely impact the persistence of large living animals and the ecosystems of which they are part (Alamgir et al., 2015).

3.7.3 Habitat Loss, Degradation, and Fragmentation

Biodiversity conservation is strongly associated with the intact ecosystems and natural landscape, however, transformation of land use patterns, expansion of agricultural lands, changes in cropping pattern, introduction of high yielding varieties, urbanization, expansion of road networks, embankments, and other man-made factors have caused immense damage to wild habitats in all ecosystem types in the country. Following are some common reason of habitat loss, degradation, and fragmentation:

- Land use change and agricultural expansions;
- Encroachment;
- Shifting cultivation;
- Urbanization; and
- Commercial shrimp cultivation in coastal areas.

3.7.4 Illegal Poaching, Logging, and Fuel Wood Collection

There is a big international market (largely illegal) of unregulated wild animals and their parts (e.g., teeth, bones, fur, ivory) mainly for their aesthetic and medicinal value (see Mukul et al., 2012b, 2014b). Besides, illegal logging, fuelwood collection, unsustainable harvest of non-timber forest products including medicinal plants are also responsible for the depletion of biodiversity in the country (Mukul et al., 2010; Khan et al., 2009).

3.7.5 Environmental Pollution and Degradation

One of the major threats to aquatic biodiversity in Bangladesh is pollution of soil and water. The aquatic ecosystem is the greatest victim and is polluted by toxic agrochemicals (i.e., chemical fertilizers, insecticides) and industrial effluents that cause depletion of aquatic and/or marine biodiversity.

3.7.6 Invasive Alien Species

A large number of exotic and non-native plant species have been introduced to the country since British colonial period for agriculture, horticulture, forestry, and fisheries (Mukul et al., 2006). Some of the species have become escapes accidentally and have adapted to local conditions proliferated profusely. Some species although have naturalized but many have become invasive over local flora and fauna. Besides, replacing natural plantation with the monoculture of short rotation and fast growing species have threatened the existence of local fauna as they have not adapted to those species (Uddin et al., 2013).

3.7.7 Limitations in Legal and Policy Framework

Lack of adequate institutional or administrative frameworks and suitable policies, weak implementation of existing policies, lack of integration of sectoral activities are other additional challenges to the biodiversity conservation in Bangladesh (Chowdhury et al., 2014; Rashid et al., 2013). Besides, poor coordination and cross-sectoral integration, weak national information system and inadequate knowledge on ecosystem structure and function are the vital reason for biodiversity loss in the country.

3.7.8 Lack of Political Commitments and Willingness

Unfortunately, there are no dealings of biodiversity, forestry or other relevant issues in the political campaign of the parties. In some cases, politically influential and elite persons are found involved in environmental degradation and illegal forest activities (e.g., encroachment).

3.7.9 Lack of Public Awareness

Lack of biodiversity-related information and knowledge inevitably leads to poor awareness and capacity for biodiversity conservation. Gaps in awareness have identified at various levels from policymakers to grass root people that sometimes even leads to misappropriation of existing law. Besides, the lack of appropriate implementation of existing biodiversity laws and regulation are common in the country.

3.8 Concluding Remarks

Bangladesh is one of the most vulnerable countries due to global climate change and consequential sea level rise. The government should immediately seek for proper adaptation measures to cope with this inevitable event. Besides, focusing only on ecological aspects will not provide a long-term security of biodiversity conservation in the country since people still substantially depend on these resources for their existence. The government should actively involve local community people in natural resource management to secure the future of country's biological diversity. A well-coordinated program on monitoring and management of country's biodiversity is an urgent task. Inadequate institutional capacities and lack of trained manpower are the attributes of biodiversity and conservation issues malfunction in dealing with. Finally, government laws concerning biodiversity issue requires urgent reform considering country's changing socio-political context and the environment.

Keywords

- conservation initiatives
- crop diversity
- plant and wildlife diversity
- threatened biodiversity

References

Alamgir, M., Mukul, S. A., & Turton, S., (2015). Modelling spatial distribution of critically endangered Asian elephant and Hoolock gibbon in Bangladesh forest ecosystems under a changing climate. *Applied Geography, 60*, 10–19.

Chowdhury, M. K. A., (2012). *Conservation and Sustainable Use of Plant Genetic Resources in Bangladesh*. Bangladesh Agricultural Research Council (BARC), Dhaka.

Chowdhury, M. S. H., Gudmundsson, C., Izumiyama, S., Koike, M., Nazia, N., Rana, M. P., Mukul, S. A., Muhammed, N., & Redowan, M., (2014). Community attitudes toward forest conservation programs through collaborative protected area management in Bangladesh. *Environment, Development and Sustainability, 16*, 1235–1252.

FAO, (2006). Global Forest Resource Assessment 2005, Progress towards sustainable forest management, FAO forestry paper 147. Rome (Italy): Food and Agriculture Organization of the United Nations (FAO).

Firoz, R., Mobasher, S. M., Waliuzzaman, M., & Alam, M. K., (2004). *Proceedings of the Regional Workshops on National Biodiversity Strategy and Action Plan*. IUCN Bangladesh Country Office, Dhaka.

Islam, M. M., Amin, A. S. M. R., & Sarker, S. K., (2003). National report on alien invasive species of Bangladesh. In: *Invasive Alien Species in South-Southeast Asia: National Reports & Directory of Resources,* Pallewatta, N., Reaser, J. K., & Gutierrez, A. T., (eds.). Global Invasive Species Programme (GISP). Cape Town (South Africa), pp. 7–24.

Islam, S. S., (2003). State of forest genetic resources conservation and management in Bangladesh. *Forest Genetic Resources Working Papers*, Working Paper FGR/68E. Rome (Italy): Forest Resources Division, FAO.

IUCN, (2000). *Red List of Threatened Animals of Bangladesh*. The World Conservation Union, Dhaka.

IUCN, (2015). *Red List of Bangladesh – A Brief on Assessment Result 2015.* The World Conservation Union, Dhaka.

Khan, M. A. S. A., Mukul, S. A., Uddin, M. S., Kibria, M. G., & Sultana, F., (2009). The use of medicinal plants in health care practices by Rohingya refugees in a degraded forest and conservation area of Bangladesh. *Intern. J. Biodiversity Sci. Management, 5*, 76–82.

Khan, M. A. S. A., Uddin, M. B., Uddin, M. S., Chowdhury, M. S. H., & Mukul, S. A., (2007). Distribution and status of forests in the tropic: Bangladesh perspective. *Proc. Pakistan Acad. Sci., 44*, 145–153.

Khan, M. S., Rahman, M. M., & Ali, M, A., (2001). *Red Data Book of Vascular Plants of Bangladesh*. Bangladesh National Herbarium, Dhaka.

Mittermeier, R. A., Myers, N., Thomsen, J. B., Da Fonseca, G. A., & Olivieri, S., (1998). Biodiversity hotspots and major tropical wilderness areas: Approaches to setting conservation priorities. *Conservation Biol., 12*, 516–20.

MoEF, (1993). *Forestry Master Plan-Main Report,* ADB (TA No. 1355-BAN), UNDP/FAO BGD 88/025. Ministry of Environment and Forest (MoEF), Dhaka.

Mukul, S. A., Biswas, S. R., Rashid, A. Z. M. M., Miah, M. D., Kabir, M. E., Uddin, M. B., Alamgir, M., Khan, N. A., Sohel, M. S. I., Chowdhury, M. S. H., Rana, M. P., Rahman, S. A., Khan, M. A. S. A., & Hoque, M. A., (2014a). A new estimate of carbon for Bangladesh forest ecosystems with their spatial distribution and REDD+ implications. *Intern. J. Res. Land-Use Sustainability, 1*, 33–41.

Mukul, S. A., Herbohn, J., Rashid, A. Z. M. M., & Uddin, M. B., (2014b). Comparing the effectiveness of forest law enforcement and economic incentive to prevent illegal logging in Bangladesh. *Intern. Forestry Review, 16,* 363–375.

Mukul, S. A., Rashid, A. Z. M. M., & Khan, N. A., (2017). Forest protected area systems and biodiversity conservation in Bangladesh. In: *Protected Areas: Policies, Management and Future Directions,* Mukul, S. A., & Rashid, A. Z. M. M., (eds.). Nova Science Publishers, USA, pp. 157–177.

Mukul, S. A., Rashid, A. Z. M. M., & Uddin, M. B., (2012b). The role of spiritual beliefs in conserving wildlife species in religious shrines of Bangladesh. *Biodiversity, 13,* 108–114.

Mukul, S. A., Rashid, A. Z. M. M., Quazi, S. A., Uddin, M. B., & Fox, J., (2012a). Local peoples' response to co-management in protected areas: A case study from Satchari National Park, Bangladesh. *Forests, Trees and Livelihoods, 21,* 16–29.

Mukul, S. A., Tito, M. R., & Munim, S. A., (2014c). Can homegardens help save forests in Bangladesh? Domestic biomass fuel consumption patterns and implications for forest conservation in south-central Bangladesh. *International Journal of Research on Land-Use Sustainability, 1,* 18–25.

Mukul, S. A., Uddin, M. B., & Tito, M. R., (2006). Study on the status and various uses of invasive alien plant species in and around Satchari National Park. Sylhet, Bangladesh. *Tigerpaper, 33,* 28–32.

Mukul, S. A., Uddin, M. B., Rashid, A. Z. M. M., & Fox, J., (2010). Integrating livelihoods and conservation in protected areas: Understanding role and stakeholders' views on the prospects of non-timber forest products, A Bangladesh case study. *International Journal of Sustainable Development and World Ecology, 17,* 180–188.

Mukul, S. A., Uddin, M. B., Uddin, M. S., Khan, M. A. S. A., & Marzan, B., (2008). Protected areas of Bangladesh: current status and efficacy for biodiversity conservation. *Proc. Pakistan Acad. Sci., 45,* 59–68.

Mukulnn, S. A., (2007). Biodiversity conservation strategies in Bangladesh: The state of protected areas. *Tigerpaper, 34,* 28–32.

Nishat, A., Huq, S. M. I., Barua, S. P., Reza, A. H. M. A., & Khan, A. S. M., (2002). *Bio-Ecological Zones of Bangladesh.* Dhaka, The World Conservation Union.

Rahman, M. M., (2004). Forest resources of Bangladesh with reference to conservation of biodiversity and wildlife in particular for poverty alleviation. In: *Forests for Poverty Reduction: Opportunities with Clean Development Mechanism, Environmental Services and Biodiversity,* Sim, H. C., Appanah, S., & Youn, Y. C., (eds). FAO-Regional office for Asia and the Pacific (FAO-RAP), Bangkok. (Thailand), pp. 139–48.

Rashid, A. Z. M. M., Craig, D., Mukul, S. A., & Khan, N. A., (2013). A journey towards shared governance: Status and prospects of collaborative management in the protected areas of Bangladesh. *J. Fores. Res., 24,* 599–605.

Sohel, M. S. I., Mukul, S. A., & Chicharo, L., (2015). A new ecohydrological approach for ecosystem service provisions and sustainable management of aquatic ecosystems in Bangladesh. *Ecohydrology & Hydrobiology, 15,* 1–12.

Uddin, M. B., Steinbauer, M. J., Jentsch, A., Mukul, S. A., & Beierkuhnlein, C., (2013). Do environmental attributes, disturbances, and protection regimes determine the distribution of exotic plant species in Bangladesh forest ecosystem? *Forest Ecology and Management, 303,* 72–80.

Plate 3.1   Ecosystem diversity in Bangladesh: from top left – (a) hill forests
dominated by dipterocarp species, (b) mangrove forests of Sundarbans,
(c) Sal (*Shorea robusta*) forests in central Bangladesh, (d) homestead
forests managed by the rural landowners, (e) Lotus (*Nelumbo nucifera*)
flower in wetlands, and (f) Tea (*Camellia sinensis*) gardens in northeast
Bangladesh. (Photo credits: Sharif A. Mukul).

Plate 3.2 Wildlife diversity in Bangladesh; from top left – (a) Capped langur (*Presbytis pileatus*), (b) Rhesus macaque (*Macaca mulatta*), (c) Bengal monitor (*Varanus bengalensis*), (d) Twin-spotted tree frog (*Rhacophorus bipunctatus*), (e) Red-whiskered bulbul (*Pycnonotus jocosus*), and (f) Knight Butterfly (*Labadea martha*). (Photo credits: Sharif A. Mukul).

Plate 3.3   Plant diversity in Bangladesh; from top left – (a) Kadam (*Anthocephalus chinensis*), (b) Akanda (*Calotropis gigantea*), (c) Ban chalta (*Dillenia pentagyna*), (d) Bhat (*Clerodendrum fragrans*), (e) Lantana (*Lantana camara*), and (f) Bon jam (*Ardisia colorata*). (Photo credits: Sharif A. Mukul).

Plate 3.4 Major causes of forest loss and disturbances in Bangladesh; (a) illegal logging in forests in Sylhet, (b) unsustainable non-timber forest products collection, (c) transportation through natural forests, (d) fuelwood collection, (e) encroachment of forest lands, and (f) agricultural expansion in forest lands. (Photo credits: Sharif A. Mukul).

CHAPTER 4

Biodiversity in India

T. PULLAIAH¹ and RAMAKRISHNA²

¹Department of Botany, Sri Krishnadevaraya University, Anantapur 515003, A.P., India,
E-mail: pullaiah.thammineni@gmail.com
²Former Director of Zoological Survey of India, Kolkata, India,
E-mail: ramakrishna.zsi@gmail.com

4.1 Introduction

According to Udvardy (1975), the world of plants and animals occurs within the earth biosphere in the form of an intertwined network of individuals, populations, and interacting systems. This world is classified based on (i) taxonomy, (ii) ecology, (iii) phylogeny, and (iv) biogeography. The tropical ecosystems of the Eurasian continent, and the neighboring southeast Asian (Australasian) Archipelago form a subkingdom of the Palaeotropical kingdom of Engler; Sclater named this area as an "Indian region," but Wallace's term, viz., Oriental region stuck. This was named as Indomalayan region by Udvardy (1975). It consists of the mainland of southeast Asia, south of the temperate Palaeoarctic Himalayas chain and the continuing Szechwan mountains. The Indo-Malayan realm of the biogeographic province includes: (i) Rainforests of Malabar, Ceylon, Bangladesh, Burma, Indo-China, South China and Malaya, (ii) Monsoon forests of Indo-Gangetic region, Burma, Thailand, Mahanadian, Coromondal, Ceylonese; (iii) Thorn forests of Deccan, (iv) Thar desert, and (v) Islands of Seychelles Islands, Amarantes Islands, Laccadives islands, Maldives and Chago Islands' Cocos keeling and Christmas Islands, Andaman and Nicobar Islands, Sumatra, Java, Lesser Sunda islands, Celebes, Borneo, Phillipines, and Thaiwan.

India is located in the northern hemisphere between 8° 4' to 37° 6' N latitude and 68° 7' to 97° 25' E longitude; stretched about 3,214 km from North to South and about 2,933 km from East to West; covers an area of 32,87,263 km². India has a rich and varied heritage of biodiversity, encompassing a wide spectrum of habitats from tropical rainforests in the Andaman and Nicobar Islands to dry alpine scrub high in the Himalaya. Between these two extremes, the country has semi-evergreen rainforests, deciduous monsoon forests, thorn forests, and subtropical pine forests in the lower montane zone and temperate montane forests. India has four biodiversity hotspots namely: the Western Ghats, the Himalaya, Indo Burmese including the

Andaman Islands and the Sundaland including the Nicobar Islands. In addition, India is one of the 19 mega biodiversity countries of the world and one of the five in Asia. The whole of the Indian sub-continent is rich in biological and ecological diversity. It lies at the confluence of African, European and Indo-Malayan realms of the biota, therefore, includes African, European, and Eurasian and Mediterranean elements. On the basis of the ecosystems, India has been divided into 10 biogeographic zones and 26 biotic provinces (Rodgers et al., 2002). Physically, the country is divided into four relatively well-defined regions:

- the Himalayan Mountains;
- the Gangetic Plains;
- the southern (Deccan) Plateau; and
- the islands of Lakshadweep, Andaman and Nicobar.

The 10 biogeographic zones according to Rodgers et al. (2002) are:

1. Trans-Himalaya (2 provinces).
2. The Himalaya (4 provinces).
3. The Indian Desert (2 provinces).
4. The Semi-Arid Zone (2 provinces).
5. The Western Ghats (2 provinces).
6. The Deccan Peninsula (5 provinces).
7. The Gangetic Plain (2 provinces).
8. The Coasts (3 provinces).
9. North East India (2 provinces).
10. The Islands (2 provinces).

4.2 Biodiversity in India

India occupies an area of 32,87,263 km^2 including nearly 8200 kms of coastline showing a notable habitat diversity of habitats due to significant variations in rainfall, altitude, latitude, and topography, that induce seasonal vegetation changes. India includes a wide range of ecozones like desert, mountains, highlands, tropical and temperate forests, swamplands, plains, grasslands, riverine areas as well as island archipelago. It hosts four biodiversity hotspots: the Himalayas, the Western Ghats, the hilly ranges that straddle the India-Myanmar border and the Nicobar group of Islands. These hotspots have numerous threatened and endemic species. The great diversity of the Indian ecoregions, ranging from permanent ice covered the Himalayas to tropical rainforests. The Himalaya, which runs across India's northern tier, is the boundary between two of the earth's great ecozones—the Palaearctic, which covers most of temperate-to-arctic Eurasia, and Indo-Malaya, which covers most of the Indian subcontinent and extends into Indochina, Sundaland (Malaysia and western Indonesia), and the Philippines.

India is one of 19 megabiodiversity countries in the world for their exceptional number of unique species, endowed with an immense variety of natural resources by way of its rich animal (96,614 species) and plant (47,147 species). According to Nayar (1996), about 5725 species of flowering plants are broadly considered as endemics and represent 33.5% of the flora, of which, 3471 species are found in the Himalayas, 2051 species in Peninsular India and 239 species in Andaman & Nicobar Islands. Besides, 147 genera are endemic to India, out of these 71 genera are endemic to the Himalaya, 60 genera to Peninsular India and one genus (*Pubistylus* Thoth.) is endemic to Andaman & Nicobar Islands, whereas 15 genera are widely distributed throughout the country. There are three megacenters of Endemism (Eastern Himalaya, Western Himalaya, and Western Ghats) and 26 microcenters of endemism. The geographical location of the country in the tropical latitude with greater solar energy, heat, and humidity promote more plant growth to support more organisms. The structurally complex habitats provide more niches and diverse ways of exploiting environmental resources and thus increase species diversity. It is known that habitats, plant communities determine the physical structure of the environment and therefore have a considerable influence on the distribution and interaction of animal species. According to world biogeographic classification, India represents two of the major realms (the Palaearctic and Indo-Malayan) and three biomes (Tropical Humid Forests, Tropical Dry/Deciduous Forests, and Warm Deserts/Semi-Deserts). These include 10 biogeographic regions and 20 biogeographic zones. The rich vegetation wealth and diversity in the country is enormous due to the massive diversity in ecosystems. Champion and Seth (1968) recognized 16 major forest types comprising 221 minor types. Of this tropical moist deciduous forms the major percentage (37%), followed by tropical dry deciduous (28.6%), tropical wet evergreen (8%), tropical thorn forest (2.6%) and others of minor values. Recent Forest Cover Assessment by Forest Survey of India (2011) indicates that a dense forest covers 416,809 km² (12.68%), open forest of 258,729 km² (7.87%), thus totaling 20.55%. While the non-forest cover constitutes a major portion (79.45%) with Scrub forest depicting only 1.44% of the total area. Among the states, Madhya Pradesh has the maximum area of forest with 77,265 km² followed by Arunachal Pradesh (68,045 km²),

Chhattisgarh (56,448 km²). In terms of the forest cover percentage, Lakshadweep has the maximum (85.91%) because it includes Coconut plantations, followed by Andaman & Nicobar Islands (84.01%), Mizoram (82.98%), Arunachal Pradesh (81.25%), and Nagaland (80.49%) (Table 4.1).

4.2.1 Ecosystems of India

India is the 7th largest country in the world and Asia's 2nd largest nation with an area of 3,287,263 km² (329 million hectare) encompassing a varied

Table 4.1 Major Forest Types and Their Distribution

No.	Forest types	Area (%)	Distribution
	Tropical Forests		
1	Tropical wet evergreen	5.8	N-E & South India and Andaman & Nicobar Islands
2	Tropical semi evergreen	2.5	South and East India
3	Tropical moist deciduous	30.3	Central and East India
4	Tropical littoral & swamp	0.9	Along the East and West Coast
5	Tropical dry deciduous	38.2	Western & Central India
6	Tropical thorn	6.7	Western and Central India
7	Tropical dry evergreen	0.1	Central and South India
	Subtropical Forests		
8.	Subtropical broad leaved	0.4	South Indian Hill forests
9	Subtropical pine forests	5.0	Sub Himalayan tract
10.	Subtropical dry evergreen forests	0.2	North-East and South India
	Temperate Forests		
11	Montane wet temperate	2.0	Eastern Himalaya located between 1800 m and 3000 m elevation in high rainfall areas
12	Himalayan moist temperate	3.4	Between 1500 m and 3000 m elevations in the Himalaya
13	Himalayan dry temperate	0.2	In low rainfall areas of Himalaya
	Sub-alpine and Alpine Forests		
14	Sub alpine		Throughout the Himalaya above 3000 m elevation up to the tree limit
15	Moist alpine scrub	4.3	Tree line up to about 5500 m elevations in Himalaya
16	Dry alpine scrub		

landscapes rich in natural resources. The country has a great diversity of natural ecosystems from the cold and high Himalayan ranges (28,000 ft. above sea level) to the sea coasts (7,500 km), from the wet northeastern green forests to the dry northwestern arid deserts. Of the only 2.4% of the total land mass of the world sharing about 11% of the total floristic diversity and is one of the 19 (3rd in Asia and 11th in World) Megadiversity Centers having four hotspots. 3.9% of the total geographical area occupied by the grassland ecosystem having semi-arid pastures of Deccan peninsula, Humid semi-water-logged areas of Terai belt, Rolling Shola grasslands in Western Ghats and the High altitude alpine pastures in the Himalayas. The dominant grass group includes *Sehima – Dicanthium* Type, *Dichanthium – Cenchrus – Lasiurus* Type, *Phragmites – Saccharum – Imperata* type, *Themeda – Arundinella* Type and Temperate alpine Type. Equally important is the wetland ecosystem having nearly 4.1 m ha (excluding paddy fields and mangroves) of which 1.5 m ha natural fresh and salt water with 2.6 million ha manmade reservoirs, canals, fish ponds, 2175 natural wetland, ca 65,000 manmade wetlands and 25 Ramsar sites. The Desert Ecosystem consists of Hot Desert which occupies 2% of India's landmass: located in the states of Rajasthan, Gujarat, Punjab and Haryana and the Cold Desert located in the states of Jammu & Kashmir, Himachal Pradesh at altitudes ranging between 4000 and 6000 m. Climatic variations are the prime factors with – 45°C temperature and 10–70 mm rainfall. The Coastal and Marine Ecosystem is of Mangroves having 14% of the Geographical Area of India, i.e., 7% of World's Mangroves and the Seagrasses and Seaweeds.

4.2.2 Floral Diversity in India

In terms of plant diversity, India ranks tenth in the world and fourth in Asia. With over 50,200 plant species, India represents over 12% of the world's known floral diversity. As elsewhere in the world, many organisms, especially in lower groups such as bacteria, fungi, algae, lichens, and bryophytes, are yet to be described and remote geographical areas are to be comprehensively explored. The richness of Indian plant species as compared to the world is shown in Table 4.2.

Flowering plants are the dominant taxon on the earth today and in terms of evolutionary significance, flowering plants are relatively recent, with fossil evidence indicating their first emergence at around 140 m years ago in the early Cretaceous followed by rapid diversification and radiation in the mid-Cretaceous (Willis and Mc Elwain, 2002). Amongst the various families of flowering plants in India the dominant are Orchidaceae, Fabaceae

Table 4.2 Number of Species in Different Plant Groups in Comparison with World

Group	No. of Species in India	Endemic species	%	No. of Species in World	% with reference to Indian species
Angiosperms (249/390 Families)	18,043 (2984 genera)	4,036 (147 Genera)	22.57	352,000 (14,044 genera)	07.17
Gymnosperms	74	08	10.81	1088	11.38
Pteridophytes	1,267	47	15.47	9,294	10.55
Bryophytes	2,523	629	25.64	15,344	17.27
Lichens	2,390	568	23.76	13,500	17.70
Fungi	14,883	4,100	24.57	75,000	10.50
Algae	7,284	1,924	26.91	40,000	17.25
Virus & Bacteria	11,813	Not known	09.1	8,050	12.25
Total plant species diversity = 50,204.					

Source: Singh and Dash, 2014.

3471 species endemic to Himalayas

240 species are endemic to Andaman & Nicobar Islands

3 Megacenters of endemism (E. Himalaya, W. Himalaya & W. Ghats)

24 Microcenters of endemism. No flowering plant Family is endemic to India:

Nicobar group of Islands, Agasthyamalai hills, Anamalai and High ranges (Cardamom hills), Palni hills, Nilgiris – Silent Valley, Wyanad, Kodagu, Shimoga – Kanara, Mahabaleshwar – Khandala ranges, Konkan – Raigad, Marathwada – Satpura ranges, Tirupati – Kadapa – Nallamalai hills, Visakhapatnam – Ganjam – Jeypore hills, Southern Deccan (Leeward side), Chotanagpur plateau, Kathiawar Kutch, Rajasthan – Aravalli hills, Khasia – Jaintia hills, Patkoi – Manipur – Lushai hills, Assam, Arunachal Pradesh Himalaya, Sikkim Himalaya, Garhwal – Kumaon Himalaya, Lahaul – Himachal Pradesh Himalaya, Kashmir – Ladak Himalaya.

(Leguminosae), Poaceae (Gramineae), Rubiaceae, Euphorbiaceae, Acanthaceae, Asteraceae (Compositae), Cyperaceae, Lamiaceae (Labiatae), and Urticaceae. Amongst the families, Lamiaceae and Asteraceae are more abundant in the temperate regions while the rest are largely tropical in distribution. One interesting feature of the Indian flora is the presence of Asteraceae, in relatively low position, however, it is the richest family of the flowering plants in the world. The diversity of Dicot and Monocot are given in Tables 4.3–4.6.

4.2.2.1 Endemicity

The endemic species are those taxa whose distribution is confined to a restricted area due to their specific ecological niches and edaphic gradients. Therefore, the habitats of endemic species are far more vulnerable than other

Table 4.3 Diversity of Dicotyledons (203/2282/12575) from India

Families	No. of Genera & Species	Remarks
Asteraceae	166 & 892	Out of 203 dicot families, 10 families include 40% genera and c. 42% species.
Fabaceae	133 & 975	
Rubiaceae	113 & 616	
Acanthaceae	49 & 472	The remaining genera and species belong to 193 families of which 84 families are monogeneric/monotypic.
Euphorbiaceae	84 & 527	
Lamiaceae	72 & 435	
Apiaceae	72 & 288	
Brassicaceae	64 & 432	
Scrophulariaceae	62 & 368	
Asclepiadaceae	45 & 255	

Table 4.4 Diversity of Monocotyledons (46/702/4448) from India

Family	No. of genera	No. of species	Remarks
Poaceae	263 (32 monotypic and 17 endemic)	1291	Out of 46 monocot families these 7 families include 87% of the monocot genera and species. The remaining 39 families include 23% of the monocot genera and species.
Orchidaceae	184	1229	
Liliaceae	45	214	
Cyperaceae	38	545	
Araceae	29	126	
Zingiberaceae	24	191	
Arecaceae	24	124	

Table 4.5 Prolific Plant Genera in India

Dicotyledons		Monocotyledons	
Genus	*No. of species*	*Genus*	*No. of species*
Impatiens	205	*Carex*	117
Primula	135	*Habenaria*	100
Crotalaria	104	*Dendrobium*	100
Pedicularis	98	There are 189 monotypic genera of which 133 genera under Dicot and 56 genera under Monocot.	
Rhododendron	97		
Astragalus	90		
Syzygium	91		

Table 4.6 Distribution of Wild Relatives of Crop Plants in Different Phytogeographical Regions of India (Modified from Arora, 1991)

Phytogeographic region	No. of species	Dominant genera
Western Himalaya	125	*Pyrus, Prunus, Allium, Carum,Cicer, Cucumis, Hordeum*
Eastern Himalaya	82	*Pyrus, Prunus, Sorbus, Rubus, Ribes, Hordeum.*
Northeastern region	132	*Citrus, Musa, Mangifera, Myrica, Morus, Solanum, Brassica.*
Gangetic plains	6	*Emblica, Syzygium, Artocarpus, Ziziphus*
Indus/N-W plain	45	*Ziziphus, Carissa, Capparis, Grewia, Sesamum, Cordia.*
Malabar/Western Peninsular region/ Western Ghats	146	*Artocarpus, Syzygium, Euphorbia, Mangifera, Mimusops, Oryza, Vigna, Dolichos, Solanum, Luffa, Cucumis, Zingiber.*
Deccan/Eastern Peninsular region/ Eastern Ghats/Islands	91	*Artocarpus, Garcinia, Syzygium, Vitis, Mangifera, Vigna, Oryza, Musa, Piper. Mangifera*

species. Endemic species once lost, it is a loss of biodiversity of these species forever. In India, there are about 5725 endemic taxa of angiosperms (33.5% of Indian flora) which are located in four hotspots. The major hotspots in India which contain the largest number of endemic plant species are the Southern Western Ghats and Eastern Himalayas with 1286 and 1808 endemic species, respectively. There are about 1272 species of endemic angiosperms out of 3800 species occurring in Kerala (33.5% of Kerala flora), representing 22.6% of Indian endemics. Seventy% of the 1272 species of endemics have the major areas of distribution in Kerala with spillovers in adjacent regions. On the basis of the study of the distributional range, about 102 endemic species occur exclusively in Kerala (http://www.jntbgri.in/endemicplants/). The families with high percentage of endemic species include Berberidaceae (3 genera/68 species) constituting 98% of the plants, members of Saxifragaceae (17 genera/140 species) forming 92%, species belonging to Ranunculaceae (28 genera/191 species) constituting 72% of plants and Rosaceae (40 genera/4320 species) with 70% of the plants of the family located in the Himalayan belt. In respect of Peninsular India, the endemicity is Melastomaceae (having 21 genera/150 species) constituting 56% of plants, Balsaminaceae (with 2 genera/215 species) having 44%, Acanthaceae (49 genera/472 species) with 38% and members of Asclepiadaceae (with 46/255 species) constituting 32% of the Indian floral families. It is also interesting to note that the country has 40 species of insectivorous plants, 130 species of parasites, 70 species saprophytes and 130 species of primitive plants.

India is one of the 12 centers/regions (Vavilovian center) of the diversity of crop plants in the world (Zeven and de Wet, 1982). Besides, India exhibits the diverse agro/eco-climate of the 'Hindustani Center,' apart from possessing about 166 species of agro-horticultural crop plants, also has a rich diversity in wild relatives of crop plants numbering about 320 species. In addition, the Indian agriculture has been enriched by a continuous stream of introductions of new crops and their cultivars by man since the ancient times. The Mughals, Spaniards and the Portuguese brought several new crops to this country. Prominent amongst these were pear, grape, nut fruits, date palm, garlic, opium, maize, potato, sweet potato, tomato, chillies, French bean, peas, etc. The Arabs possibly brought clove, coriander, cumin, ammi or lavage (*ajwain*) and fennel. This process continued during the British period which saw the introduction of coffee, cocoa, cashew nut, strawberry, litchi, *Cinchona*, etc., besides collection of tea in Assam and adjoining region. The current matrix of diversity consists of the gene pool of indigenous crop plants, their wild and/or weedy relatives and the well-adapted introductions from practically all over the globe. The introduced types, depending on the time of introduction, extent of material introduced and the areas of introduction, exhibit enormous build-up of diversity within the Indian subcontinent. Some good examples are: *cereals* – wheat, barley, oats, maize; *pulses* – gram, French bean (*rajmah*) and peas; *vegetables* -potato, onion, cauliflower, cabbage, carrot and tomato; *fruits* – apple, pear, grapes, cherry, peach and apricot; *oilseeds* – soybean, sunflower and groundnut; *fiber crop* – cotton; to a very limited extent also in *medicinal plants*-mint, liquorice, *Digitalis* (foxglove), *Cinchona* (quinine), *Hyoscyamus* (henbane) and others such as *Humulus lupulus* (hops) (Arora, 1990). Thus India is endowed with 18,043 species of flowering plants having very high intraspecific variability. Mango with 5,000 recorded varieties, *Oryza sativa* (Rice) with over 60,000 recorded landraces are unique. About 33% of the higher plant species are endemic, 356 major and minor crops plant species and 326 wild related species of crop plants as detailed in Table 4.6.

Indian sub-continent is the home of 166 species of crops and 334 species of wild relatives of cultivated plants. Of this, India's major contribution to world's crops and among the wild relatives are rice (*Oryza*), legume *(Vigna, Cajans, Crotalaria)*, Mango *(Mangifera)*, Banana *(Musa)*, Pepper *(Piper)*, Turmeric *(Curcuma)*, Ginger *(Zinger)*, Garlic *(Allium.)*, Cardamom *(Elettaria, Amomum)*, *Citrus* fruits, Sugar *(Saccharum)*, Melon *(Cucumis)*,

Ridged gourd *(Luffa)*, Jack fruit *(Artocarpus)*, Yams *(Dioscorea)*, Vines *(Trichosanthes)*, Sesame *(Sesamum)*, Hibiscus *(Abelmoschus)*, Cinnamon *(Cinnamomum)*, Nutmeg *(Myristica)*, Amaranthus, Brassica, Corchorus, and several others (Table 4.7).

4.2.2.2 Gymnosperms

Gymnosperms are currently placed in five distinct and widely divergent orders namely Cycadales, Ginkgoales, Taxales, Coniferales, and Gnetales. According to Maarten et al. (2011). There are 1026 species in 84 genera in the world in almost all continents, except Antartica. Singh and Srivastava (2013), reported the presence of 146 species and 8 varieties (indigenous/ introduced) of gymnosperms in 46 genera belonging to 12 families from India.

The conifers flora of India is dominated by the genera of Northern hemisphere, viz. *Pinus, Abies, Cedrus,* and *Picea.* All Indian conifers except *Podocarpus wallichianus* (Peninsular India and Andamans) and *Podocarpus*

Table 4.7 Cultivated Plants and Their Wild Relatives in India

Sl. No.	Category	No of Cultivated plants	No of Wild relatives
1.	Cereals	15	37
2.	Millets	13	33
3.	Grain Legumes	18	36
4.	Vegetables	105	168
5.	Fruits & Nuts	117	176
6.	Oil Seeds	19	13
7.	Sugar yielding crops	03	18
8.	Fibre crops	12	23
9.	Forage & Fodder Crops	96	33
10.	Spices and Condiments	46	123
11.	Plantation crops	20	21
12.	Medicinal and aromatic plants	89	58
13.	Ornamental plants	182	90
14.	Agroforestry species	35	31
15.	Other crops	41	42
	Total	**811**	**902**

neriifolius (Eastern Himalaya and Andamans) are restricted to Himalaya. The species of *Cycas* are distributed widely in Eastern and Western Ghats, Northeast India and Andaman & Nicobar Islands. Majority of the species of *Ephedra* are distributed in higher elevations of Himalaya preferably with alkaline soils. Similarly, the species of *Gnetum* inhabit evergreen tropical rainforests of Eastern and Western Ghats, Northeast India and Andaman & Nicobar Islands. *Ephedra gerardiana* and *E. nebrodensis* (good source of alkaloid ephedrine) and *Taxus wallichiana* (which yield anticancerous taxol) are highly valued and exploited medicinal plants (http://www.bsienvis.nic. in/Database/Status_of_Plant_Diversity_in_India).

Families of Gymnosperms (Genera/Species)	Dominant genera (species)
Pinaceae (7/39)	*Ephedra* (8)
Cupressaceae (29/28)	*Cycas* (7)
Ephedraceae (1/8)	*Pinus* (6)
Gnetaceae (1/5)	*Juniperus* (8)
Cycadaceae (10/27)	*Gnetum* (5)
Cephalotaxaceae (1/4)	133 (1088) species (11 endemic) belonging
Podocarpaceae (2/4)	to 42 (88) genera and 11 (14) families (72
Taxaceae (2/4)	introduced species)
Araucariaceae (2/6)	
Taxodiaceae (6/7)	
Ginkgoaceae (1/1)	

4.2.2.3 Pteridophytes

On a very conservative estimate 500 species of ferns and 100 species of fern-allies are on record from India. According to a census, the Pteridophytic flora of India comprises of 67 families, 191 genera and more than 1,000 species (Dixit, 1984) including 47 endemic Indian ferns, less than 10% of those reported previously and 414 species of Pteridophytes (219 at risk, of which 160 critically endangered, 82 Near-threatened and 113 Rare), constituting 41–43% of the total number of 950–1000 Pteridophytes of India. Chandra (2000) recorded 34 families, 144 genera and more than 1100 species of ferns with about 235 endemic species from Indian region. However, the estimate of recent Pteridophytes (Sanjappa Per. Com. 2016) indicated the presence of 1236 (9294) species (endemic 193) belonging to 204 (568) genera and of which 33 (35) with Families 67% (179/845) are represented in Indo-Burmese hotspots of Northeast Indian states.

Dominant families (genera)	Dominant genera (species)
Polypodiaceae (137)	*Selaginella* (62)
Dryopteridaceae (125)	*Pteris* (62)
Athyriaceae (97)	*Dryopteris* (53)
Thelypteridaceae (83)	*Asplenium* (45)
Selaginellaceae (62)	*Polysticum* (45)

4.2.2.4 Bryophytes

The less known group of plants, comprising about 2800 species, is the second largest group of green plants in India, next only to the angiosperms. The details are as given below:

Mosses	Liverworts & Hornworts
1576 species (endemic 483); 338 genera, 57 families	875 species (endemic 146); 144 genera; 50 families
Eastern Himalaya (1031)	Eastern Himalaya (548)
Western Himalaya (751)	Western Himalaya (235)
Western Ghats (540)	Western Ghats (280)
Dominant Families:	Dominant Families:
Pottiaceae (29/129)	Lejeuneaceae (36/155)
Dicranaceae (17/119)	Jungermanniaceae (16/76)
Bryaceae (11/98)	Plagiochilaceae (4/69)
Dominant Genera:	Dominant Genera:
Fissidens (67)	*Plagiochilla* (64)
Bryum (37)	*Frullania* (63)
Brachythecium (36)	*Jungermannia* (42)

4.2.2.5 Algae

Represented by over 6500 species in ca. 666 genera, they are found growing in a variety of habitats ranging from freshwater, marine, terrestrial, and to soil, of which 1924 species are endemic to the country. According to Rao and Gupta (1997), freshwater algae are dominated by Chlorophyceae (green algae), Bacillariophyceae (diatoms), and Cyanophyceae (blue green algae) representing the major portion of Indian algal flora accounting for ca. 390 genera and 4500 species followed by terrestrial algae (125 genera and 615 spp.); soil algae (80 genera and 1500 spp.); marine algae (169 genera and

680 spp.). The economically important algae in the freshwater ecosystem are *Chlorella, Spirulina, Syncchocccus* (Single Cell Protein), *Euglena, Chlamydomonas, Scenedesmus, Ankistrodesmus, Nitzschia* (diatom rich in vitamin A), diatomaceous earth; production of Diotomite (Fire proof material). Species of *Sargassum* yield Sodium alginate – a raw material for textile and paper industries, *Gelidiella, Gracilaria, Hypnea* – yield agar agar and carrageenans used in pharmaceutical industries, which are distributed in marine ecosystem.

4.2.2.6 Medicinal Plants

India abounds in an unmatched variety as well as a bulk availability of medicinal herbs. As the largest producer of medicinal plants, India is called the "'Botanical Garden of the World." Singh et al. (2001, Flora of India) has estimated nearly 7,000 species of Angiosperms are medicinal, besides India is second largest (11%) medicinal plant exporter at the global level. A total of 960 species of medicinal plants are in trade having source of 1289 botanical drugs and with an annual demand of raw drugs is to the tune of 3,19,500 MT. The annual production of raw drugs is to the tune of 1,81,483 MT, from the 960 medicinal plant species in trade, 178 species are consumed in high volumes exceeding 100 MT per year. The Indian medicinal plants and their products also account of exports in the range of Rs. 80–90 billion. Major herbal exports from India are *Psyllium* seeds and husk, castor oil and opium extracts, which together account for 60% of the exports. 80% of the exports are to the developed countries viz. USA, Germany, France, Switzerland, UK, and Japan. A large percentage of the exports is as crude formulation and not as pure phytochemicals. India is the largest grower of *Psyllium* and *Senna* plants and one of the largest growers of castor plant. Twenty other plants are commonly exported as crude drugs worth US$ 8 million. Five of these, namely *Glycyrrhiza glabra, Commiphora mukul, Plantago ovata, Aloe barbadensis,* and *Azadirachta indica* are even used in modern medicine.

India has one of the richest and oldest medicinal plant cultures of the world. The so far estimated number of 6560 species of medicinal plants of India are a great bio-cultural resource (Foundation for Revitalization of Local Health and Traditions, FRLHT; http://envis.frlht.org/traded-medicinal-plants-database.php#). The uniqueness of the Indian medical heritage draws from two streams of knowledge, folk, and codified stream, which are coexisting living traditions that have historically enjoyed a symbiotic relationship. India is a global leader in *in-situ* conservation of medicinal plants having established the largest *in-situ* conservation network for medicinal plants in the tropical world. So far, 110 MPCAs, each of an average size of 200 ha, have been set up across 13 States of India. There is a felt need for expansion of the MPCA network at least three-fold in order to capture breeding populations of all the currently known 315 species of threatened medicinal plants and its further strengthening through long-term conservation research activities and sustainable community involvement programs. Medicinal plants play a vital role in the life of ethnic tribes in India and elsewhere, as many as 900 species are used in Ayurveda, 700 species used in Unani, 600 species in Amchi and a large number in modern system formulations.

4.2.3 Forests

Forests are an important reservoir of biodiversity. Ancient and frontier forests, because of their long standing and relatively lower levels of human disturbance, are typically richer in biodiversity than other natural or semi-natural forests. An illustration of the conservation importance of forests relative to biodiversity is found in the recent analysis of biodiversity hotspots. Loss of forest will inevitably result in reduction of biodiversity as a direct result of loss of habitat. There are many anthropocentric reasons to preserve biodiversity (direct use value, option value, etc.) but one such principal reason considered here is the indirect use value in the form of ecosystem services. A loss in biodiversity affects the stability of an ecosystem resulting in a reduction of its resistance to disruption of the food web (by loss of the weak interaction effect), resistance to species invasion and resilience to global environmental change. Therefore, the argument for conserving biodiversity is, twofold:

1. biodiversity is essential for maintaining ecosystem services, and
2. biodiversity increases the stability and resilience of an ecosystem to a disturbance.

Forest based programmes by conservation of forests through the Forest Act guarantee economically, ecologically, socially, and culturally sustainable utilization of the forests. It aims at promoting a rational use of forest resources. A key element of the Forest Act, with regards to safeguarding biodiversity, is defining certain habitats of special importance and giving guidelines as to how these habitats may be managed. Joint Forest (Protected Area) Management an Act under Participatory Management under Wildlife Act, silvicultural practices, co-operation between forest dwellers and forest authorities are added to the forest protection and conservation. This has been further substantiated by Raven et al. (2004) that locally driven indigenous efforts incorporating participatory approach, conflict resolution, and mutual learning are more likely to garner the relatively broad support and participation of local communities and strengthening of the local institutions for conservation. Tenth meeting of the Conference of the Parties (COP 10) was held in Nagoya, Japan, from 18 to 29 October 2010. The conference included a high-level ministerial segment organized by the host country to discuss the issues in respect of Inland waters biodiversity, marine, and coastal biodiversity, mountain biodiversity, protected areas, sustainable use of biodiversity, and biodiversity and climate change. These events included discussion of the Group of 77 on biodiversity for Development and Poverty Alleviation, high level panel discussion on "Local Biodiversity Strategies and Action Plan."

4.2.4 Biosphere Reserves

Biosphere reserve is a specific type of a conservation area which accommodates and benefits both the natural environments and communities living around it. This is possible because the biosphere reserve consists of three different but associated zones: core, buffer, and the transitional zone. Core zone is ecologically sensitive but a pristine area, where nature conservation is a priority with allowed low impact activities. Buffer zone is less ecologically sensitive but a natural area where recreational and sustainable utilization of a natural product can be accommodated. Finally, the transition zone is less ecologically sensitive where a great variety of land user occur. All zones are interdependent and managed according to the definitions above. It is here, by linking conservation, sustainable utilization, and utilization of natural resources occur. The details of Indian Biosphere reserves are given below:

S. No.	Name of the Biosphere Reserve & Area (km²)	Date of Designation	Location in the State(s)/Union Territory
1	Nilgiri (5520)	01.08.1986	Part of Wyanad, Nagarhole, Bandipur and Madumalai, Nilambur, Silent Valley and Siruvani hills in Tamil Nadu, Kerala and Karnataka.
2	Nanda Devi (5860.69)	18.01.1988	Part of Chamoli, Pithoragarh and Almora districts in Uttarakhand.
3	Nokrek (820)	01.09.1988	Part of East, West and South Garo Hill districts in Meghalaya.
4	Manas (2837)	14.03.1989	Part of Kokrajhar, Bongaigaon, Barpeta, Nalbari, Kamprup and Darang districts in Assam.

S. No.	Name of the Biosphere Reserve & Area (km²)	Date of Designation	Location in the State(s)/Union Territory
5	Sunderban (9630)	29.03.1989	Part of delta of Ganges & Brahamaputra river system in West Bengal.
6	Gulf of Mannar (10,500)	18.02.1989	India part of Gulf of Mannar extending from Rameswaram island in the North to Kanyakumari in the South of Tamil Nadu.
7	Great Nicobar (885)	06.01.1989	Southern most island of Andaman and Nicobar Islands.
8	Similipal (4374)	21.06.1994	Part of Mayurbhanj district in Orissa.
9	Dibru-Saikhova (765)	28.07.1997	Part of Dibrugarh and Tinsukia districts in Assam.
10	Dehang-Dibang (5111.5).	02.09.1998	Part of Upper Siang, West Siang and Dibang Valley districts in Arunachal Pradesh.
11	Pachmarhi (4981.72)	03.03.1999	Part of Betul, Hoshangabad and Chhindwara districts in Madhya Pradesh.
12	Khangchendzonga (2931.12)	07.02.2000	Part of North and West districts in Sikkim.
13	Agasthyamalai (3500.36)	12.11.2001	Part of Thirunelveli and Kanyakumari districts in Tamil Nadu and Thiruvanthapuram, Kollam and Pathanmthitta districts in Kerala.
14	Achanakmar-Amarkantak (3,835.51)	30.03.2005	Part of Anuppur and Dindori districts of Madhya Pradesh and Bilaspur district of Chhattisgarh
15	Kachchh (12,454)	29.01.2008	Part of Kachchh, Rajkot, Surendranagar and Patan districts in Gujarat.
16	Cold Desert (7,770)	28.08.2009	Pin Valley National Park and surroundings; Chandratal & Sarchu; and Kibber Wildlife sanctuary in Himachal Pradesh.
17	Seshachalam (4755.997)	20.09.2010	Seshachalam hill ranges in Eastern Ghats encompassing part of Chittoor and Kadapa districts in Andhra Pradesh
18	Panna (2998.98)	25.08.2011	Part of Panna and Chhattarpur districts in Madhya Pradesh.

Among the above 18 biosphere reserves in India, Nilgiri, Nandadevi, Nokrek, Sunderban, Great Nicobar, Gulf of Mannar, Simlipal, Pachmarhi, Achanakmar-Amarkantak's have been included in the World Network of Biosphere Reserves of UNESCO.

4.2.5 Protected Area Network in India

Since the Vedic times, the forest management in India is founded on the strong concept of conserving nature and its pristine wealth for posterity. This practice has been inherited through centuries and has evolved into scientific management, thereby continuously addressing the needs of the time. Conservation and sustainable use of biological resources based on local knowledge systems and practices are ingrained in Indian ethos. The country has a number of alternative medicines, like Ayurveda, Unani, Siddha, and Homeopathic systems which are predominantly based on plant- and animal-based raw materials in most of their preparations and formulations. Herbal preparations for various purposes including pharmaceutical and cosmetic purposes form part of the traditional biodiversity uses in India.

The strategies for conservation and sustainable utilization of biodiversity have comprised providing special status and protection to biodiversity-rich areas by declaring them as national parks, wildlife sanctuaries, biosphere reserves, ecologically fragile and sensitive areas. India's first national park was established in 1936 as Hailey National Park, now known as Jim Corbett National Park, Uttarakhand. By 1970, India only had five national parks. In 1972, India enacted the Wildlife Protection Act and Project Tiger to safeguard the habitats of conservation dependent species. In 1988 there were 54 national parks covering 21,003 km^2 and 372 sanctuaries covering 88,649 km^2 giving a combined 109,652 km^2 or 3.34% of the country's geographical area. This network has grown steadily and in 2009 India has a total of 99 national parks covering 39,155 km^2 and 513 sanctuaries covering 118,417 km^2 giving a combined coverage of 612 (NPs + WLS) covering 157,572 km^2 country or 4.79% of the geographical area. In addition to this, there are 43 Conservation Reserves (total area 1155 km^2) and three Community Reserves (total area 17.8 km^2). Thus on 1st January 2009 there are 658 PAs (National Parks, Wildlife Sanctuaries, Conservation Reserves and Community Reserves) covering 158,745 km^2 and 4.83% of country's geographical area. Recent analysis indicates, In India, a network of 733 PAs has been established, extending over 1,60,901.74 km^2 (4.89% of total geographic area), comprising 103 National Parks, 537 Wildlife Sanctuaries, 67 Conservation Reserves, and 26 Community Reserves (Table 4.8).

The creation of two additional categories of PAs viz. Conservation Reserve and Community Reserve; the establishment of the National Tiger Conservation Authority (NTCA) and the Wildlife Crime Control Bureau (WCCB) and the launching of the XI Plan Scheme on 'Integrated Development of Wildlife Habitats' are some of the important recent initiatives of

Table 4.8   Protected Areas of India (as on July, 2016 (Source: www.wii.nic.in, Open Source)

Protected Areas	No.	Total Area (km²)	Coverage % of Country
National Parks (NPs)	103	40500.13	1.23
Wildlife Sanctuaries (WLSs)	537	118005.30	3.59
Conservation Reserves (CRs)	67	2349.38	0.07
Community Reserves	26	46.93	0.001
Protected Areas (PAs)	733	160901.74	4.89
Geographical Area of India (http://knowindia.gov.in/)	=	32,87,263 km²	
Forest cover of India (FSI, 2015)	=	7,01,673 km²	
Percentage Area under Forest cover	=	21.34 % of Geographical Area of India	

the Ministry of Environment & Forests, Government of India. There are 49 tiger reserves in India which are governed by Project Tiger which is administered by the National Tiger Conservation Authority. India is home to 70% of tigers in the world. In 2006, there were 1,411 tigers which increased to 1,706 in 2011 and 2,226 in 2014. Recognizing the importance of the natural and cultural importance, UNESCO has designated five Protected Areas as World Heritage Sites.

It is a well-known fact that the ecosystems and species do not recognize political boundaries; the concept of Transboundary Protected Areas has been initiated for coordinated conservation of ecological units and corridors with bilateral and/or multilateral cooperation of the neighboring nations. There are four categories of the Protected Areas viz, National Parks, Sanctuaries, Conservation Reserves, and Community Reserves. Sanctuary is an area which is of adequate ecological, faunal, floral, geomorphological, natural or zoological significance. The Sanctuary is declared for the purpose of protecting, propagating or developing wildlife or its environment. Certain rights of people living inside the Sanctuary could be permitted. Further, during the settlement of claims, before finally notifying the Sanctuary, the Collector may, in consultation with the Chief Wildlife Warden, allow the continuation of any right of any person in or over any land within the limits of the Sanctuary. National Park is an area having adequate ecological, faunal, floral, geomorphological, natural or zoological significance. The National Park is declared for the purpose of protecting, propagating or developing wildlife or its environment, like that of a Sanctuary. The difference between a Sanctuary and a National Park mainly lies in the vesting of rights of people living inside. Unlike a Sanctuary, where certain rights can be allowed, in a National Park, no rights are allowed. No grazing of any livestock shall also be permitted inside a National Park while in a Sanctuary; the Chief Wildlife Warden may regulate, control or prohibit it. In addition, while any removal or exploitation of wildlife or forest produce from a Sanctuary requires the recommendation of the State Board for Wildlife, removal etc., from a National Park requires recommendation of the National Board for Wildlife (However, as per orders of Hon'ble Supreme Court dated 9th May 2002 in Writ Petition (Civil) No. 337 of 1995, such removal/ exploitation from a Sanctuary also requires recommendation of the Standing Committee of National Board for Wildlife). Conservation Reserves can be declared by the State Governments in any area owned by the Government, particularly the areas adjacent to National Parks and Sanctuaries and those areas which link one Protected Area with another. Such declaration should be made after having consultations with the local communities. Conservation Reserves are declared for the purpose of protecting landscapes, seascapes, flora and fauna and their habitat. The rights of people living inside a Conservation Reserve are not affected. Community Reserves can be declared by the State Government in any private or community land not comprised within a National Park, Sanctuary or a Conservation Reserve, where an individual or a community has volunteered to conserve wildlife and its habitat. Community Reserves are declared for the purpose of protecting fauna, flora and traditional or cultural conservation values and practices. As in the case of a Conservation Reserve, the rights of people living inside a Community Reserve are not affected.

The National Board for Wildlife (NBWL), chaired by the Prime Minister of India provides for the policy framework for wildlife conservation in the country. The National Wildlife Action Plan (2002–2016) was adopted in 2002, emphasizing the people's participation and their support for wildlife conservation. India's conservation planning is based on the philosophy of identifying and protecting representative wild habitats across all the ecosystems. The Indian Constitution entails the subject of forests and wildlife in the Concurrent list. The Federal Ministry acts as a guiding torch dealing with the policies and planning on wildlife conservation, while the provincial Forest Departments are vested with the responsibility of implementation of national policies and plans. The Protected Area Network includes National Parks Wildlife Sanctuaries, Conservation Reserves, Community Reserves, Tiger Reserves and Elephant reserves. Besides above, Biosphere reserves are recognized that do the areas comprise terrestrial, marine and coastal ecosystems. Each reserve promotes solutions reconciling the conservation of biodiversity with its sustainable use. Biosphere reserves are 'science for sustainability support sites' – special places for testing interdisciplinary approaches to understanding and managing changes and interactions between social and ecological systems, including conflict prevention and management of biodiversity (www.moef.nic.in).

4.2.6 Indian Hotspots

Conservation priorities are assigned based on the exceptional distribution and concentration of endemic animals and plants and experiencing the exceptional loss of habitat. To qualify as a biodiversity hotspot (Myers 2000 edition of the hotspot-map), a region must meet two strict criteria: Hot-spots are regions with a large number of endemics and rich in biological diversity. It must contain at least 0.5% or 1,500 species of vascular plants as endemics, and it has to have lost at least 70% of its primary vegetation and biodiversity hotspot is a biogeographic region with a significant reservoir of biodiversity that is threatened with destruction. Biodiversity hotspots are a method to identify those regions of the world where attention is needed to address biodiversity loss and to guide investments in conservation. The idea was first developed by Norman Myers in 1988 to identify tropical forest 'hotspots' characterized both by exceptional levels of plant endemism and serious habitat loss, which he then expanded to a more global scope. Conservation International adopted Myers' hotspots as its institutional blueprint in 1989, and in 1999, the organization undertook an extensive global review which introduced quantitative thresholds for the designation of biodiversity hotspots. A reworking of the hotspots analysis in 2004 resulted in the system in place today.

As on date, 35 hotspots have been identified throughout the world, many of them are distributed in tropics. In India, four hotspots have been identified viz. (i) Western Ghats hotspot, wherein as many as 7402 species of plants have been identified with over 1273 species of plants exclusively confined to the Western Ghats. The importance of this hotspot is that it exhibits climatic diversity and geodiversity with reference to minerals, Controls climate as they are adjacent to Arabian Sea (South West Monsoon), covers 6 states and 51 districts of the country, home for more than 40 tribal groups, more than 40% of the Indian rivers originate from this area and having highest Biological diversity. Based on the geologic history, the rainforests of Western Ghats is unique because of its richness in biodiversity ('Hotspot'), having evolutionary continuity for more than 50 million years and is the only area in India having an admixture of the following biotic elements, resultant of the biogeographic evolution of the country *viz,* Gondwana derivatives, Peninsular endemics, Madagascan elements, Indo-Chinese derivatives, Malayan elements as well as Palaearctic elements. (ii) Nicobar Islands is a part of Sundaland Hotspot. This hotspot covers the western half of the Indo-Malayan archipelago, an arc of some 17,000 equatorial islands and is dominated by two of the largest islands in the world: Borneo and Sumatra. The occurrence/distribution of Orchidacean members *Bentinckia condopana* an endemic species of Western Ghats and *Bentinckia nicobarica* in Katchal island of Nicobar indicates the Gondwana affinity and later by Continental drift. Secondly, the distribution of *Dipterocarpus, Pterocarpus, Coelogyne, Habenaria,* in Andaman group of Islands while the distribution of *Cyathia, Cyrtandra, Rhopaloblaste, Bentinckia, Kibara, Stemorus* are confined to Nicobar group and do not extend to Andaman Islands. The Nicobar Islands were isolated during all times by the Great Channel from Sumatra in the southeast and by the Ten Degree Channel in the north. The uniqueness of Nicobar Islands is the presence of Crab Eating Macaque, Salt Water Crocodile, Giant Leather Back Turtle, Malayan Box Turtle, Nicobar Tree Shrew, Nicobar Megapode, Reticulated Python, and the Giant Robber Crab. The Great Nicobar Biosphere Reserve of this hotspot is included in UNESCO-MAB Network during May 2013. The third in the series is The Indo-Burma Hotspot is ranked in the top 10 hotspots for irreplaceability and in the top five for threat, with only 5% of its natural habitat remaining and with more people than any other hotspot (Con. Int. 2011). The Indo-Burma Hotspot begins at the evergreen forests in the foothills of Chittagong in Bangladesh and extends through the Garo and Khasi Hills of Meghalaya, then eastwards through Manipur, Mizoram, and Nagaland to the Andaman Islands of India. The topography of the hotspot is complex and characterized by a series of north-south mountain ranges, which descend from the Himalayan chain and its southeastern extensions. This diversity is enriched by the development of areas of endemism as a result of the hotspot's geological and evolutionary history. Fourthly the Himalayan Hotspot includes the highest mountains in the world. The mountain rise abruptly, resulting in the diversity of ecosystems that range from alluvial grassland and subtropical broadleaf forests to alpine meadows above the tree line.

4.2.7 Faunal Resource

India is rich in terms of biological diversity due to the unique geolocation, diversified climatic conditions, and enormous ecodiversity. In fact, within slightly more than 2% of the total geographic area of the world, has more than 97,000 species of animals distributed in India, of which insects alone constitute more than 63,000 species (Ramakrishna and Alfred, 2007). The details of the groups of animals distributed in India are given below:

Taxonomic group	No. of species		% in India
	World	India	
PROTISTA (Protozoa)	31259	3509	11.23
ANIMALIA			
Mesozoa	71	10	14.08
Porifera	5000	500	10.00
Cnidaria	10107	1052	10.41
Ctenophora	100	12	12.00
Platyhelminthes	17513	1653	9.43
Rotifera	2044	370	18.10
Gastrotricha	3000	100	3.33
Kinorhyncha	100	10	10.00
Nematoda	30034	2911	9.69
Acanthocephala	800	229	28.62
Sipuncula	145	35	24.14
Mollusca	66535	5178	7.78
Echiura	127	43	33.86
Annelida	17002	1002	5.89
Onychophora	100	1	1.00
Crustacea	60003	3565	5.94
Insecta	1020169	63706	6.24
Arachnida	73470	5882	8.01
Pycnogonida	600	17	2.83
Chilopoda	8000	101	1.26
Diplopoda	7500	162	2.16
Symphyla	120	4	3.33
Merostomata	4	2	50.00
Phoronida	11	3	27.27
Bryozoa (Ectoprocta)	4000	200	5.00
Entoprocta	60	10	16.66
Brachiopoda	300	4	1.33
Chaetognatha	111	30	27.02

Taxonomic group	No. of species		% in India
	World	India	
Tardigrada	514	30	5.83
Echinodermata	6600	779	11.80
Hemichordata	120	12	10.00
Chordata	64738	5894	9.10
Protochordata	2106	119	5.65
Pisces	32156	3092	9.62
Amphibia	6802	428	6.29
Reptilia	9232	530	5.74
Aves	9026	1302	14.42
Mammalia	5416	423	7.81
Total (Animalia)	**1,398,972**	**93,505**	**6.68**
Grand Total **(Protista + Animalia)**	**1430231**	**97014**	**6.78**

The mammalian fauna of the world is represented by 5416 species belonging to 1229 genera, 153 families and 29 orders (Wilson and Reeder, 1993). Of these, 423 species belonging to 191 genera, 48 families, and 14 orders are found in India. Another 15 orders do not occur in our country. The details are as given below:

Orders	Families		Genera		Species	
	India	World	India	World	India	World
Insectivora	3	7	11	66	28	428
Scandentia	1	1	2	5	3	19
Chiroptera	7	17	34	177	123	1117
Primates	3	13	6	60	16	233
Pholidota	1	1	1	1	2	7
Carnivora	7	11	35	129	60	271
Proboscidia	1	1	1	2	1	2
Sirenia	1		1	3	1	5
Perissodactyla	2	3	2	6	3	18
Artiodactyla	5	10	21	81	31	220
Lagomorpha	2	2	3	13	11	80
Rodentia	4	33	46	481	118	2277
Cetacea	7	10	19	41	26	78
Total	**44**	**111**	**182**	**1065**	**423**	**4755 (8.53%)**

4.2.7.1 Primates

Primates include 16 species belonging to 6 genera the family Loridae with a lone species of the genus *Loris* and *Nycticebus,* the family Cercopithecidae, subfamily Cercopithicinae, includes 7 species of the genus *Macaca.* The sufamily Colobinae includes the genus: *Semnopithecus* and 4 species of *Trachypithecus.* The ape family Hylobatidae includes a lone species viz., *Bunopithecus hoolock* (Harlan, 1834), *(Hyalobates hoolock).* 60 species of the order Carnivora are reported from Indian region under seven families viz., Canidae (one species of *Cuon,* 2 species of *Vulpes),* Felidae (includes species of *Acionyx jubatus* and *Caracal caracal, Felis chaus, Felis silvestris, Catopuma temminckii, Lynx lynx, Otocolobus manul, Prionailurus bengalensis, Prionailurus rubiginosus, Prionailurus viverrinus, Neofelis nebulosa, Panthera leo, Panthera pardus, Panthera tigris, Pardofelis marmorata,* and *Uncia uncia),* family Herpestridae (includes 7 species of *Herpestes),* family Hyaenidae *(Hyaena hyaena),* family Mustelidae *(Lutra lutra, Lutrogale perspicillata, Amblonyx cinereus, Arctonyx collaris, Melogale moschata, Melogale personata, Mellivora capensis,* 3 species of *Martes,* 5 species of *Mustella, Ailurus fulgens, Helaractos malayanus, Melursus ursinus, Ursus arctos, Ursinus thibetanus,* and family Viverridae includes *Arctictis binturong, Arctogalidea trivirgata, Paguma larvata, Paradoxurus hermophroditus, Paradoxurus jerdoni, Prionodon pardicolor, Viverra civettina, Viverra zibetha,* and *Vivericula indica).*

4.2.7.2 Rodentia

Rodentia is a largest order of mammals in the world comprising 2277 species in 481 genera under 33 families (Wilson and Reeder, 2005), includes squirrels, rats, mice, voles, gerbils, hamsters, dormice, porcupines etc. checklist from Pradhan and Talmale (2009) revealed, valid Indian rodent taxa up to subspecies level includes 118 species and 89 subspecies under 46 genera belonging to 7 families. According to Padmanabhan and Dinesh (2011), Marine mammals are a diverse group of animals, globally accounting to 119 species (Jefferson et al., 1993). They include cetaceans (whales, dolphins, and porpoises), sirenians (manatees and dugong), pinnipeds (true seals, eared seals, and walrus), a few otters (the sea otter and marine otter) and the polar bear (usually grouped with the marine mammals). In India, 31 species of marine mammals (30 species of Cetacea and one species of Sirenia) are documented accounting to one-fourth of the world's marine mammalian fauna and almost 8% of the total Indian mammalian fauna (George et al., 2011).

4.2.7.3 Aves

Birds evolved about 150 million years ago, occupied diverse ecological niches and are distributed in all available habitats. The Indian avifauna represents Palearctic, Oriental, Ethiopian, and Australasian zoogeographic elements. It includes about 176 endemic forms. 350 migrating species and subspecies winter in the Indian Territory, while a few migrate from India. The Indian subcontinent has 2123 species and subspecies under 78 families and 20 orders thus representing 13.66 % of the world avian diversity (ZSI, 2010). The recent estimate of aves, based on the discovery as well as the new records to India, is 1302 species (Grewal et al., 2016). About 176 species under 106 genera, 39 families, and 11 orders are endemic to the Indian subcontinent, including Pakistan, Nepal, Bhutan, Myanmar, Sri Lanka and Bangladesh. One genus *Ophrysia* and 50 species under 11 genera are exclusively endemic to India. There are 25 monotypic genera.

4.2.7.4 Reptilia

An updated checklist of 530 species of reptiles includes 3 species of crocodiles, 34 species of turtles and tortoises, 202 species of lizards and 279 species of snakes belonging to 28 families recorded till date from India. The marine reptiles include 70 species belonging to 5 sub-families inhabiting the world oceans and estuaries. Of these, 22 species belonging to 3 families and 3 sub-families have been documented from Indian waters (Das, 2003). A current listing based on distributional records and review of published checklists revealed that 199 species of lizards (Reptilia: Sauria) are currently validly reported on the basis of distributional records within the boundaries of India (Venugopal, 2010). The world fauna of reptiles includes 9230 known species.

4.2.7.5 Amphibia

One of the main problems of establishing the evolutionary lineage of this group is the sparse fossil record. However, the earliest frog fossil *Vieraella herbstii* is about 160 million years old and it came from Argentina. The modern-day frogs, toads, and caecilians are anatomically highly modified; however, the frogs and toads have adopted for jumping while the caecilians have lost their limbs and adopted for fossorial life. The present-day frogs evolved around 250–280 million years ago. *Triadobatrachus masenotti* found in Madagascar and believed to be the ancestor of the modern-day frog is 200 million years. Amphibian estimates is found to be around 6,771 species of which 5,966 species of frogs and toads, 186 species of Caecelians and 19 species of Salamanders. The systematic list of Indian Amphibia includes 438 species comprising of the order Anura includes 356 species, order caudata with a lone species of *Tylototriton verrucosus* Anderson, 1871 and the order Gymnophia with 35 species.

4.2.7.6   Fishes

Nelson (2006) estimated 27,977 valid species of fishes under 62 orders, 515 families, and 4,494 genera, and the eventual number of extant fish species was projected to be close to 32,500. Estimates from ZSI (2012) accounts for 32,120 valid species, 4495 genera under 62 orders and 515 families. About 11,952 species or 42.72 %, normally live in freshwater lakes and rivers that cover only 1 % of the earth's surface and account for a little less than 0.01% of its water. The secondary freshwater species number 12,457 and the remaining 3568 species are exclusively marine. The Indian subcontinent, occupying a position at the confluence of three biogeographic realms, viz., the Palaearctic, Afro-Tropical, and Indo-Malayan, exhibits a great variety of ecological habitats, harboring rich ichthyofaunal diversity, comprising about 3200 species of which 930 species are freshwater inhabitants and the rest are marine. The Indian species represent about 8.9% of the known fish species of the world. The class chondrichthyes comprises 131 species under 67 genera, 28 families and 10 orders and the class osteichthyes is represented by 2419 species belonging to 902 genera, 226 families, and 30 orders. Recent estimates from ZSI (2012) includes 2314 species, 151 cartilaginous under 74 genera, 32 families and 11 orders and the remaining 2165 species belonging to 808 genera and 193 families under 29 orders are bony.

4.2.7.7   Protochordata

Protochordata includes two subphyla viz., Cephalochordata and Tunicata exclusively distributed in the marine environment. The total number of species from the Indian marine waters is 119, which forms 5.65% of the estimated taxa of the world. Cephalochordates are represented by two families, Belostomatidae represented by a single genus (*Branchiostoma* with three species) and Epigonythidae represented by the genus *Epigonycyhys* (with two species) which includes the larval form of the genus viz., *Amphioxys* (previously thought to be a different genus). The Urochordates are represented in three classes (Ascidacea, Thaliacea, and Larvacea) are usually the burrowing forms and many of them act as biofoulers.

Among the Invertebrates, echinoderms includes 779 species from India, the class Holothuroidea (Sea cucumber) is the largest with 163 species followed by Asteroidea (158 species), Ophiroidea (152 species), Echinoidea (113 species), and Crinoidea (65 species). Global diversity is estimated to have 6600 extant species and 13,000 extinct ones. Molluscs are very ancient organisms believed to have evolved from a flatworm-like ancestor during the Precambrium about 650 million years ago. Because many species secrete a shell of some sort, the fossil record is good. Different classes of molluscs have been predominant in the past and the Ammonites represent a group of Cephalopods which were extremely abundant for millions of years before they became extinct. Their close relatives the Nautiloid cephalopods were also once very successful but are now only represented in the world by one species, *Nautilus*. The molluscs are of great diversity with nearly 5000 species that account for about 8% of the world fauna. At the family level, 62.8% of the families known from the world are represented in India. The richness of the species is due to the diverse ecosystems and habitats and the table shows the species diversity at different levels. Freshwater molluscs includes 199 species and subspecies under 59 genera, 29 subgenera, and 26 families. Among them, 19 families belong to gastropodas and 7 families belonging to Bivalvia. Fauna of land molluscs includes 1129 valid described species under 138 genera and 34 families. Among the marine forms, Cephalopoda includes 56 species falling under 24 genera, 14 families, 4 orders and 2 subclasses. Order Nautiloidea includes one family, single genus *Nautilus* represented by a *Nautilus pompoides* (Monotypic). Members of marine bivalvia include 606 species under 171 genera, 69 families, 11 orders, and 4 subclasses. The class Scaphopoda includes 2 orders, 2 families, and 2 genera and 17 species. The class gastopoda includes rest of the species.

A highly diversified group of roundworms occurring in all types of habitat. Globally, 26,646 species are recognized with 8359 parasitic in vertebrates and 10,681 free living, 4105 plant-parasitic worms and 3501 species parasites on invertebrates. The Indian representative includes 2991 known species. Parasitic nematodes: total 72 species under 2 subgenera, 27 genera, 7 subfamilies, 10 families, 6 superfamilies one suborder and 3 orders have been recorded. Out of three orders, Tylenchida includes 2 superfamilies, 3 families, 6 genera, and 12 species. Order Oxyurida possesses 3 superfamilies, 4 families, 7 subfamilies, 18 genera, and 44 species. Order Rhabditida includes one suborder, one superfamily, 3 families, 3 genera, 2 subgenera, and 16 species.

Commonly called thorny-headed worm or spiny-headed worm, is a phylum highly adapted to a parasitic mode of life. Typically, the acanthocephalans have complex life cycles, involving a number of hosts which include invertebrates, fishes, amphibians, birds, and mammals. Adult acanthocephalans are found in the intestines of vertebrates and the larvae are found in the body cavity of arthropods which function as intermediate hosts. The phylum includes 229 species.

4.2.7.8   Rotifers

Rotifers are microscopic, multicellular pseudocoelomate organisms occurring in three subclasses viz., Monogononta with 1500 species, Bdelloidea with 350 species and Seisonidae with 2 known species. 31,250 species are known from the world, however, ZSI estimate is 1817 species under 126 genera and 34 families. The Indian rotiferan fauna includes 370 species in 63 genera and 25 families from India. Annelids include earthworms, leaches, pot worms and rag worms included under the class Oligochaeta and Clitellata. The world annelid fauna is estimated to be 17,000 species while the Indian fauna is estimated to be over 1000 species with 400 species of Polychaeta, 537 species of Oligochaeta and 64 species of leaches.

4.2.7.9   Cnidarians

Cnidarians include diploblastic multicellular, usually radially symmetrical diverse group of organisms. 10,105 species of Cnidarians are recognized world over of which the class Anthozoa includes 6,142 species (Crowther et al., 2011), Cubozoa includes 42 species, Hydrozoa includes 3,643 species and Schyphozoa with 228 species. The recently created class Staurozoa includes 50 species. The Scleractenia, commonly known as reef building corals, includes 828 hermatypic or reef-building coral species belonging to 112 genera and 18 families around the world. A total of 464 species of corals included under 72 genera and 17 families are distributed in Andaman & Nicobar Islands, Lakshadweep, Gulf of Mannar and Gulf of Katchchh area.

4.2.7.10   Porifera

Porifera commonly known as sponges distributed in fresh, marine and brackish water. Nearly 25,000 species are recognized the world over, however, includes 350 species of marine and 150 species of freshwater habitat in India. 257 species belonging to 113 genera are endemic to the country.

4.2.7.11 Protozoa

Protozoa live under all natural conditions and have been reported from freshwater, brackish and marine, soil and moss as well as free-living forms. The world protozoan fauna are represented by 31,250 species of which 10,000 are parasitic. Parasitic protozoans occur in epizoic, luminicolous, coelozoic, histozoic and coprozoic environs. Protista (Protozoa) are represented by 2519 parasitic species belonging to 284 genera, 118 families under 32 orders. 226 symbiotic species represented by 226 species belonging to 2 phyla under 3 classes, 5 orders, 15 families and 40 genera and 1247 free-living totaling 3509 species in India.

4.2.7.12 Arachnida

Arachnida includes spiders, scorpions, ticks and mites as well as a few smaller groups of Pseudoscorpions, and phalangids. Globally 42,751 spider species from 110 families have been reported under 3,859 genera. Besides about 30,000 species of Acarina, 2,300 species of Pseudoscorpions, 1,752 species of Scorpions, 1600 species of Opilions 1,000 species of Solpugidas, and 4 species of King Crabs have been recorded. 1447 species from 59 spider families have so far been reported under 362 genera from India, of which 1006 spider species of 288 genera under 55 families are endemic to India. Scorpions: were known to occur during the prehistoric Silurian period about 450 million years before. The scorpion fauna of India was first explored by Pocock (1900). Globally, 1988 species of scorpions are known to occur of which, 113 valid species of 25 genera under 6 families exist in India.

4.2.7.13 Crustacea

Globally, the number of crustacean species has been estimated to be around 60,000, comprising of 17 faunal groups which is 5.5% of the total extant arthropod species occurring in the world. Amongst the groups, Decapoda represents the highest diversity of 14,756 species, followed by Copepoda (14,000 species) and Isopoda (11,000 species), while two crustacean groups *viz.,* Notostraca and Euphausiacea are represented by only 16 and 90 species respectively all over the world. The number of crustaceans estimated earlier was 2934 species. However, as per the recent estimate, there are about

3,549 Crustacean species so far reported from India which comprise 5.9% of the global stock. Group-wise, Decapoda contributes the highest diversity of 1,550 (43.67%) species in India, followed by Copepoda 767 species (21.6%), while Notostraca represents only 2 species in the country. Out of the total of 3,549 species occurring in India, 318 species (8.96%) were described by scientists of the Zoological Survey of India.

4.2.7.14 Insecta

The phylum Arthropoda alone includes 12,42,040 species, constituting about 80% of the total number of insect species. The most successful group, Insecta, accounts for about 66% (10,20,007 species in 39 orders) of all animals. The most successful insect order, Coleoptera, represents about 38% (3,87,100 species) of the insect species of the world (Zhang, 2011). Compilations on the insect fauna of India have been produced from time to time. Maxwell and Howlett (1909) published the book Indian Insect Life, wherein 25,700 species of insect were reported from the Indian region, including adjacent countries. Beeson (1941) and Menon (1965) estimated the number of species from India to be 40,000 and 50,000, respectively. In recent times, Varshney (1997) reported 51,450 species under 589 families. Subsequently, Alfred et al. (1998) reported the occurrence of 59,353 species of insect belonging to 619 families in India.

4.2.8 Endemism

Insect diversity in India is characterized by a high level of endemism. The diversity of insects is greater in the northeastern states, the Western Ghats and the Andaman and Nicobar Islands, and these areas also have a high level of endemism. A high percentage of endemism is noted in the primitive insect groups, viz., Protura (85%), Diplura (66%) and Thysanura (60%), followed by Collembola (15%). Among the exopterygotes, Thysanoptera has the highest percentage of endemism (75%), followed by Phasmida (68%), Ephemeroptera (58%), Plecoptera (57%), Orthoptera (54%), Embioptera (45%), and Isoptera (44%), and there is less than 40% endemism in the remaining orders. Among the endopterygotes, the endemism in species level is the highest in Mecoptera (86%), followed by Neuroptera (76%), Strepsiptera (71%), Hymenoptera (71%), Trichoptera (60%), Diptera (35%) and Coleoptera (17 %), while the order Lepidoptera shows only 10% endemism since the moth fauna is widely distributed in the Indo-Pacific region.

4.2.9   Analysis of Major Insect Orders

4.2.9.1   Orthoptera

Orthoptera comprise primitive group of insects includes short and long horned grass hoppers, crickets, mole crickets, katydids, grouse-locusts and pygmy mole-crickets. The number and diversity of the species (853 species belonging to 327 genera) was first recorded in the State of Art report of Zoological Survey of India, which was further reviewed in 1991. Presently, the country harbors 1033 species and subspecies belonging to 400 genera under 21 families (Shishodia et al., 2010). The family wise details of the group includes Acrididae (285), Dericorythidae (4), Pamphagidae (1), Chorotypidae (9), Eumastacidae (8), Mastacideidae (8), Pyrgomorhidae (47), Tetrigidae (137), Trigonidiidae (22), Gryllotapidae (8), Mogoplistidae (14), Myrmecophilidae (4), Prophalagapsidae (1), Raphidophoridae (14), Schyzodactylidae (3), Anostotomatidae (06), Gyllacrididae (49), Stenopelmatidae (3), and Tettigoniidae (160) species. An analysis of the species distribution indicates that Western Himalaya harbors 246 species, north eastern states with 380 species and more than 450 species are known from Western Ghats.

4.2.9.2   Lepidoptera

Lepidoptera includes moths, butterflies and skippers, largest group of insects next to Coleoptera. The family Gekuchiidae (Microlepidoptera) is the large family represented by 4530 species belonging to 507 genera worldwide and in India by 410 species belonging to 99 genera (Pathania and Kaur, 2010). Neuroptera: commonly known as 'Lace wings,' about 6,256 species are known from the world and Indian fauna accounts for 342 species and 15 subspecies belonging to 121 genera under 14 families (Oswald, 2007), which is about 5.5% of world occurrence (Chandra and Sharma, 2010).

4.2.9.3   Odonata

Globally 5,740 species of odonates are known, of this 470 species in 139 genera and 19 families exist in India. Extant Odonata is broadly divided into three suborders the Zygoptera or damselflies, Anisozygoptera and the Anisoptera or dragonflies. The Anisozygoptera with two relict species was earlier recognized as a third suborder of Odonata. However, recent studies groups Anisozygoptera with Anisoptera or some authors brings them together under a new name Epiprocta (Anisoptera + Anisozygoptera) (Lohmann et al., 1996; Kalkaman et al., 2008).

4.2.9.4 Ephemeroptera

Globally, about 3000 species in 400 genera and 42 families are currently known. Of this, 390 species in 84 genera and 20 families occur in the Oriental region. About 49% of the genera (41 genera) of this region are endemic. The Ephemeroptera fauna of India is represented by 4 suborders, 12 families, 46 genera, and 124 species (Sivaramakrishnan et al., 2011). Mantids were formerly placed under the Order Dictyoptera. Burmeister (1838) now placed them in a separate Order Mantodea, includes "Preying Mantids." Globally, 2300 species belonging to 434 genera and 15 families known to occur (Ehrmann, 2002), of which, 184 species in 73 genera and 11 families exist in India.

4.2.10 Animal Breeds in India

For century upon century domestic animals have been bred with specialized traits suited to particular tasks or to live and prosper in specific climates or regions. Arguably these domesticated breeds are of significance equal to their brethren who live in the wild and just like their wild counterparts. A "breed" is mostly the product of human intervention; it can only develop when a sub-population of a domestic animal species is separated and sexually isolated from the rest of the gene-pool. This can happen as a result of conscious selection for certain phenotypic or performance characteristics, as is typically the case in herd book registered livestock breeds. Local breeds play a multi-functional role in rural livelihoods, contributing not only cash products but also manure and traction. They are of social benefit as "insurance" against natural disasters or economic bottlenecks. Furthermore, indigenous breeds may have valuable genes with commercial potential. Especially the breeds kept by pastoralists are regularly exposed to stressful condition. The importance of livestock goes beyond its food production function. Livestock provides valuable draft power worth Rs. 33,792 crores, organic manure for agriculture worth Rs. 5,700 crores, dung as fuel for domestic purpose worth Rs. 4,482 crore and other by-products including leather, bones, and horns worth Rs. 100 crores.

In India, there are over 250 breeds of farm animals (30 of cattle, 10 of buffalo, 40 of sheep, 20 of goat, 18 of poultry, 9 of camel and 6 of horse). In addition, the other species found are pig, Mithun, yak, duck, goose, turkey, guinea fowl, pheasant, dog, and cat) are 'recognized,' though there are also a number of other breeds not yet described or recognized. Each of these breeds are indigenous to a specific region, have certain specializations, are hardy and adapted for conditions in their area of origin, require very little management when compared to exotic breeds and crossbreeds, especially veterinary costs; and can subsist on a far inferior diet during the leaner months. From the time of hunter gatherers, then to fishing to the latest development of agriculture and domestication of animal breeds or livestock, the role of biodiversity in feeding the ever-growing human population is significant.

Livestock production systems in India are mostly based on low-cost agro-byproducts and traditional technologies primarily for producing milk, draft power, meat, egg, fiber, etc. Out of over 40 animal species so for domesticated by man, 14 major species contribute 82% of global food and agricultural production. India is among the 19 mega biodiversity centers of the world, possessing all the major species of livestock and poultry. Cattle (37%), Buffalo (21%), goat (26%), sheep (13%), pig (2%), and the rest the others. As on date, 151 breeds so for registered have been characterized by performance recording through surveys and by microsatellite genotyping by the National Bureau of Animal Genetic Resources of India. Of the 151 registered breeds in the country, 39 are of cattle, 13 of buffalo, 24 of goat, 40 of sheep, 6 of horses and ponies, 9 of camel, 3 pig, 1 for donkey, 16 of chicken, besides yak, mithun, duck, quail, etc. According to the Livestock Census data, updated up to April, 2016, the country had 512.05 million livestock population and 729.2 million poultry population. The data pertaining to the year 2016 cattle (*Bos taurus* or *Bos indicus*) and water buffalo (*Bubalus bubalis*) population is estimated at 301.6 million head. Thereby, the genetic resources of domestic animals in India are represented by a broad spectrum of breeds, varieties, strains, and numbers. There is high biodiversity in cattle, goat, and sheep when compared to pig, fowl, quail, geese, ducks, mithun, yak and pet animals. Native breeds of these livestock, with remarkable ability to resist endemic diseases and to subsist on local feed and fodder resources, need to be conserved.

Acknowledgment

The authors wish to express deep-felt thanks to Dr. R. R. Rao, Director (Rtd.) CIMAP, Dr. M. Sanjappa, Director (Rtd.), BSI and Prof. M. Rajanna, University of Agricultural Sciences for their constant encouragement and help.

Keywords

- animal breeds
- biogeographic zones
- ecosystems
- Indo-Pacific region

References

Alfred, J. R. B., Das, A. K., & Sanyal, A. K., (1998). *Faunal Diversity in India*, ENVIS Center, ZSI, Calcutta.

Arora, R. K., (1990). *Plant Diversity in the Indian Gene Center.* http://www.bioversityinternational.org/ downloaded on 13.09.2016.

Arora, R. K., (1991). Plant diversity in the Indian gene center. In: *Plant Genetic Resources & Conservation and Management–Concepts & Approaches.* IBPGR Regional Office for South East Asia, pp. 25–54.

Beeson, C. F. C., (1941). *The Ecology and Control of the Forest Insects of India and the Neighboring Countries.* Dehra Dun, India.

Burmeister, H., (1838). *Handbuch Der Entomologie*, 2(1–8), 684.

Champion, H. G., & Seth, S. K., (1968). *A Revised Survey of the Forest Types of India.* Government of India, New Delhi, India.

Chandra, K., Sharma, R. M., & Ojha, P., (2010). *A Compendium on the Faunal Resources of Narmada River Basin in Madhya Pradesh.* Zoological Survey of India, Kolkata, India.

Chandra, S., (2000). *The Ferns of India (Enumeration, Synonyms and Distribution).* International Book Distributors, Dehradun, India.

Conservation International. www.ConservationInternational.org, (2011).

Crowther, T. W., Boddy, L., & Jones, T. H., (2011). Outcomes of fungal interactions are determined by soil invertebrate grazers. *Ecology Letters*, *14*, 1134–1142.

Das, I., (2003). Growth of knowledge on the reptiles of India with an introduction to systematics taxonomy and nomenclature. *J. Bombay Nat. Hist. Soc.*, *100*(2&3), 446–501.

Dixit, R. D., (1984). *A Census of the Indian Pteridophytes.* Botanical Survey of India, Kolkata, India.

Ehrmann, R., (2002). *Mantodea Gottesanbeterin De Welt.* Natur und Tier–Verlag, Münster.

Forest Survey of India, (2015). *India State of Forest Report*, www.fsi.nic.in.

George, A., Meadows, P., Metcalfe, H., & Rolfe, H., (2011). *Impact of Migration on the Consumption of Education and Children's Services and the Consumption of Health Services, Social Care and Social Services.* National Institute of Economic and Social Research, London, U.K.

Grewal, B., Sen, S., Singh, S., Devasar, S., & Bhatia, G., (2016). *A Pictorial Guide to Birds of India.* Om Books International.

Harlan, J., (1834). *Crops and Man.* American Society of Agronomy/ Crop Science Society of America, Madison.

Jefferson, T. A., Leatherwood, S., & Webber, M. A., (1993). *Marine Mammals of the World.* Food and Agriculture Organization of the United Nations.

Lohmann, D. R., Brandt, B., Höpping, W., Passarge, E., & Horsthemke, B., (1996). The spectrum of RB1 germ-line mutations in hereditary retinoblastoma. *Am. J. Hum. Genet.*, *58*(5), 940–949.

Martin, D., Pantoja, C., Fernández Miñán, A., Valdes-Quezada, C., et al. (2011). Genome-wide CTCF distribution in vertebrates defines equivalent sites that aid the identification of disease-associated genes. *Nature Structural & Molecular Biology, 18*, 708–714.

Maxwell-Lefroy, H., & Howlett, F. M., (1909). *Indian Insect Life: A Manual of the Insects of the Plains (Tropical India)*. W. Thacker & Co., Calcutta, India.

Menon., (1965). Taxonomy of Fishes of the Genus *Schizothorax* Heckel with the description of a new species from Kumaon Himalayas. *Records Zool. Surv. India, 63*, 1–4.

Nayar, M. P., (1996). *Hotspots of Endemic Plants of India*, Nepal and Bhutan. Tropical Botanic Garden and Research Institute, Thiruvananthapuram.

Nelson, J. S., (2006). *Fishes of the World*, 4th edn. John Wiley & Sons, New Jersey, USA.

Padmanabhan, P., & Dinesh, K. P., (2011). *A Checklist of Marine Mammals of India*. Zool. Surv. India, Kolkata, India.

Pathania, P., & Kaur, S., (2010). *A Checklist of Microlepidoptera of India (Part-II: Gelechiidae)*. Zool. Surv. India, Kolkata, India.

Ramakrishna, P., & Alfred, J. R. B., (2007). *Handbook on the Faunal Resources of India*. Zoological Survey of India.

Rao, P. S. N., & Gupta, S. L., (1997). Freshwater algae. In: *Floristic Diversity and Conservation Strategies in India, Vol. I,* Mudgal, V., & Hajra, P. K. (eds.). Botanical Survey of India, Calcutta, India, pp. 47–61.

Raven, P. H., Evert, R. F., & Eichhorn, S. E., (2004). *Biology of Plants,* 7th edition, McGraw-Hill, Boston, MA, U.S.A.

Rodgers, W. A., Panwar, H. S., & Mathur, V. B., (2002). *Wildlife Protected Area Network in India: A Review*. Wildlife Institute of India, Dehradun, India.

Shishodia, M. S., Chandra, K., & Gupta, S. K., (2010). An annotated checklist of Orthoptera (Insecta) from India. *Rec. Zool. Surv. India, Occ. Paper No., 314*, 1–366.

Singh, N. P., & Srivastava, R. C., (2013). *Gymnosperms of India, A Check List.* Botanical Survey of India, Kolkata, India.

Singh, P., &, Dash, S. S., (2014). *Plant Discoveries 2013–New Genera, Species and New Records.* Botanical Survey of India, Kolkata, India.

Sivaramakrishnan, K. G., Subramanian, K. A., Arunachalam, M., Selva kumar, C., & Sundar, S., (2011). Emerging trends in molecular systematics and molecular phylogeny of mayflies (Insecta: Ephemeroptera). *J. Threat. Taxa., 3*(8), 1972–1980.

Talmale, S. S., & Pradhan, M. S., (2009). Identification of some small Mammal species through owl pellet analysis. *Rec. Zool. Surv. India, Occ. Paper No., 294*.

Udvardy, M. D. F., (1975). A classification of the biogeographical provinces of the world – IUCN prepared as a contribution to UNESCO's. *Man and the Biosphere for Conservation of Nature and Natural Resources*. Morges, Switzerland.

Varshney, R. K., (1997). *Fauna of Delhi (Series 6) Protozoa to Mammalia*. Zoological Survey of India.

Venugopal, P. D., (2010). Population density estimates of agamid lizards in human- modified habitats of the Western Ghats, India. *Herpetological Journal, 20*(2), 69–76.

Willis, K. J., & McElwain, J. C., (2002). *The Evolution of Plants.* Oxford University Press, Oxford.

Wilson, D. E., & Reeder, D. M., (2005). *Mammal Species of the World. A Taxonomic and Geographic Reference.* Johns Hopkins University Press, U.S.A.

Zeven, A. C., & De Wet, J. M. J., (1982). *Dictionary of Cultivated Plants and Their Regions of Diversity: Excluding Most Ornamentals, Forest Trees and Lower Plants.* CAPD, Wageningen.

Zhang, Z. Q., (2011). Animal biodiversity: An introduction to higher-level classification and taxonomic richness. *Zootaxa, 3148,* 7–12.

ZSI, (2010). *Animal Discovery.* Zoological Survey of India. Kolkata, India.

ZSI, (2012). *Animal Discovery.* Zoological Survey of India. Kolkata, India.

Plate 4.1   A. *Ceropegia pullaiahii* Raja Kullayisw. et al., B. *Croton scabiosus* Bedd., C. *Cycas beddomei* Dyer, D. *Cycas indica* A. Lindstr. & K. D. Hill, E. *Elaeocarpus gaussenii* Weibel, F. *Euphorbia gokakensis* S. R. Yadav et al. (Photos by K. Raja Kullayi Swamy).

Plate 4.2   A. *Hildegardia populifolia* Schott & Endl., B. *Medinilla malabarica* Bedd., C. *Pimpinella tirupatiensis* N.P.Balakr. & Subram., D. *Shorea roxburghii* G. Don, E. *Sonerila grandiflora* R.Br. ex Wight & Arn., F. *Sonerila pulneyensis* Gamble (Photos by Raja Kullayi Swamy).

Batocera rufomaculata long horn beetle.

Common Barronet *Symphaedra nais.*

Blue Rock Pigeon.

Common Bush Brown *Mycalesis perseus.*

Commander *Moduza procris.*

Common Indian Crow *Euploea core.*

Plate 4.3

Junonia almana Peacock Pancy.

Common Leopard *Phalanta phalantha.*

Purple moorhen *Porphyrio porphyrio.*

Purple Moorhen.

Common Redshank – *Tringa totanus.*

Red Munia.

White-breasted Kingfisher *Halcyon smyrnensis* (Linnaeus).

Plate 4.4

CHAPTER 5

Biodiversity in Indonesia

IRAWATI and DIDIK WIDYATMOKO

Center for Plant Conservation, Botanic Gardens – Indonesian Institute of Sciences, Indonesia, E-mail: nfn.irawati@gmail.com

5.1 Geographical Status

Indonesia is the largest country in Southeast Asia, with a maximum dimension from east to west is about 5,100 km^2 and an extent from north to south of 1,800 km^2. Indonesia lies between 6° N to 11° S latitude and 95°–141° E longitude, a tropical archipelago between Asia and Australia continents and two oceans, Indian and Pacific. Indonesia gazetted 13,466 islands among 17,000 islands of the archipelago. The islands encompass an area of 1,919,440 km^2 between 3,257,483 km^2 of sea and 99,093 km^2 of coastline as a border.

Geological evidence involving the development of Indonesian archipelago is complex tectonic and glacioeustatic changes (Indonesia and Philippines archipelagos are a result of a combination of volcanic activity and tectonic plate movement) during the Pleistocene. The islands are varied with different altitude mountains and mountainous highlands. The highest of snowing mountainous and a range of active volcanoes known as part of Pacific Ring of Fire decorated the topography of the archipelago. The western part of the country is relatively wet while the southeastern part has long dry months.

Since the origin of the islands of Indonesia are from different parts of the continents, therefore, the biodiversity of the plant highly varied from the east to the west part of the country. The three main islands on the west, Sumatra, Jawa, and Kalimantan (Indonesian Borneo) were part of Asia, while the eastern part originated from Australian continent (Figure 5.1). The origin of Sulawesi Island was from different sources to make this island having a varied biodiversity.

However, the origin and the development of many "small islands" of Indonesia in between the two are varied. The belt of volcanoes that some have erupted destroyed many living organisms but on the other hand also give more minerals to the land. The biological diversity is strongly influenced by the last geological and climatic events, and geographic isolation and local adaptation. Furthermore, as the country stretches from the vast area, influences by different climates and many islands are isolated resulting

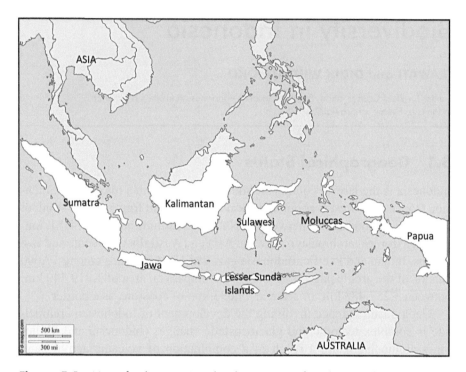

Figure 5.1 Map of Indonesia. (Used with permission from https://d-maps.com/
pays.php?num_pay=85&lang=en)

in a high diversity of land cover and rich in endemic flora. A study on the
relationship between geology and biodiversity was conducted by Cottle
(2004).

The richest zones in the world were classified by Barthlott et al. (2005)
with more than 5,000 species per 10,000 km² and Northern Borneo and New
Guinea are among five centers, which are "tropical moist broadleaf forest" biome according to WWF ecoregion scheme. The rich biodiversity in
Indonesia shows that this country occupying only 1.3% of the world's land
surface and possesses 10% of the world's flowering species according to
Conservation International (https://blogs.ntu.edu.sg/hp331-2014-03/?page_
id=27).

5.2 Ecosystem Diversity

Ecosystem types of Indonesia are related and dependent each other's and
consists of two main parts: natural and artificial/man-made ecosystems.

5.2.1 *Natural Ecosystems*

Marine ecosystem covers about 70% of the total Indonesian ecosystem and 30% is terrestrials ecosystem.

Indonesian waters are in the middle of the Coral Triangle, an epicenter of marine biodiversity in the world is home for both marine and coastal fauna and flora (plants, corals, sponges, Mollusks, Crustaceans, Echinoderms, fishes, reptiles, and mammals). Coastal ecosystem found in beaches, mud-flats, estuary, lagoon, mangroves, seagrass beds and coral reefs (Hutomo and Moosa, 2005). Indonesian coral reefs cover 32,000 km² (about 18% world coral reef area, about 25,000 km² have been mapped (Susanto et al., 2015). Unfortunately, there was a slight decline of coral reefs in the excellent shaped, from 6.9% in 2000 to 5.3% in 2012, 25.3% in good condition increased into 27.1% and fair from 33.1% slightly increased into 37.3%. Coral species that dominate are *Acropora* spp., *Montipora* spp. and *Porites* spp. (Susanto et al., 2015). The number of species according to Veron, (2000) is 590 belonging to 82 genera.

Seagrass ecosystem has an important role in supporting lives in shallow water. Indonesia has 31.000 km² seagrass, where food and spawning area for some marine biota are located. In this ecosystem, 16 species of seagrasses plants have been identified. Biota's associated in seagrass such as sea cucumber, sea urchin, shells, sea slugs, starfish and some fishes. Oceanic or pelagic zone of Indonesia has not been known in detail due to a limited expert working in this sector.

In freshwater ecosystem, the river ecosystem is under threat in many places in Indonesia due to encroachment, gold, nickel, copper mining, limestone quarry, sand, and chalk mining as well as industrial waste disposal. These activities affecting the lives of aquatic biota's such as fishes, shrimps, plankton, bentos, crabs and various mollusks and mussels.

Indonesia has 840 lakes and 735 small lakes, totally 550,000 ha. Biota that live in lakes varies, such as phytoplankton, zoo plankton, mollusk, crustaceans, insect, fishes, amphibians, reptile, such as turtle, snakes, ducks, goose, and mammals.

Semi-terrestrial ecosystem is an essential ecosystem both either in fresh or sea water areas. Mangrove ecosystem which is adapted to ocean tides fall in this category. Towards the sea is usually dominated by *Avicennia* and *Sonneratia* and behind them are *Avicennia* and *Rhizophora* with specific roots. After them are *Rhizophora*, *Bruguiera* growing on harder soil such as clay. Other vegetation bordering the mainland is *Ceriops* mixed with shrubs, *Bruguiera*, *Heritiera* and *Lumnitzera*. *Nypa fruticans* with strong roots withstand to stronger current wave at the estuarine area, sometimes growing together with *Avicennia*, *Sonneratia* or *Rhizophora*. Mangrove forest has an important role for the fishery in the tropic. However, mangrove areas have decreased over time from 4.25 million hectares in 1982 to 3.9 million hectares in 2013. The diversity of Indonesian mangrove is 243 species.

Conversion of mangrove area to agriculture, aquaculture, tourism, urban development, and over-exploitation are the main reasons for the decrease of mangrove areas rapidly.

The riparian ecosystem has important roles in maintaining the quality of surface as well as underground water. Plant species found in riparian zone are from Euphorbiaceae, Poaceae, Fagaceae, Melastomataceae, Dipterocarpaceae, Meliaceae, Myrtaceae or Rubiaceae.

Terrestrial ecosystem usually is classified according to the rainfall, elevation, water status, soil types, and vegetation types. However, Kartawinata (2013) classified (natural) vegetation types of Indonesia in more detail, and 74 vegetation types were distinguished based on composition, structure and the dominant plant species. Some of them are: Coastal forest, Dipterocarps forest, Heath forest (Kerangas forest), Swamp forest, Peat forest, Karst and Cave forest, Savanna, Lower mountain forest, Upper mountain forest, Subalpine forest, Alpine forest.

Indonesia's rainforests are home to some of the highest levels of biological diversity in the world. Indonesia contains the world's third largest area of rainforest after the Amazon and Africa's Congo Basin.

5.2.2 Artificial/Man-Made Ecosystem

There are types of artificial/man-made ecosystem. "Tegalan" is an arable land for "second crops" and horticultural crops, often with intercropping planting system to prevent the risk of crop failures due to climate as well as the pest. "Pekarangan" or garden surrounding the house with mixed planting, fruit trees, medicinal and ornamental plants, sometimes also for raising livestock or source of firewood, often on downhill sites and considered to be marginal land. Paddy fields are also known as an artificial ecosystem. There are different types of paddy field; irrigated paddy field, non-irrigated, the flat lowland or terraced land. In other wet areas, there are rice-field in flooded area, swamp rice field, and tidal rice fields.

Mix garden is another type of artificial ecosystem for annual plants as well as perennials. Plantation for estate crops is usually in monoculture in large areas. Another agro-ecosystem which developed near natural forest is shifting cultivation with low maintenance practice.

Fishpond is a dynamic ecosystem involving the interaction of biotic, abiotic and decomposer. And at the seaside another dynamic ecosystem called 'tambak" a fish/shrimp pond with brackish water is found.

5.3   Species Diversity

To describe the diversity of marine and terrestrial plant and animal in Indonesia facing challenges because the great size of the area. Furthermore, it requires many taxonomists for different biota.

5.3.1   Diversity of Marine Life

The number of marine life from Indonesia recorded is about 6,764 species (Table 5.1); consist of 5,459 fauna species and 1,305 algae and flora species. Marine microbes was estimated reach 38,000 species/liter.

Many sea algae are found in coral areas with different colors and the grouping is according to the color of the pigment in the alga. The green color is Chlorophyta containing green pigment, the brown color is Phaeophyta containing brown pigment and the red color is Rhodophyta containing red pigment. Sea grass in Indonesia waters are: *Halophila spinulosa, H. decipiens, H. minor, H, ovalis, H.sulawesii, Enhalus acoroide, Thalassia hemprichii, Cymodocea serrulata, C. rotundata, Halodule pinifolia, H. uninervis, Syringodium isoetifolium,* and *Ruppia maritima* (Romimohtarto and Juwana, 1999).

Sea flora found in many seashores are sea grass and mangrove. Sea grass found in relatively calm waters with fine sand and silt. Mangrove growing in saline water and produce a large number of stilt roots or pneumatophores.

Table 5.1   The Number of Recorded Marine Life in Indonesia

Fauna		Flora & Microbe	
Biota group	**Number of species**	**Biota group**	**Number of species**
Echinodermata	557	Sea grass	16
Polychaeta	527	Algae	971
Crustacea	309	Mangrove	243
Corals	590	Mangrove associate	75
Fishes	3,476	Total	1,305
Total	5,459	Microbe (estimated)	38,000/liter

Autotroph microbe and heterotroph microbes found in Indonesian water, however very few studies were conducted to explore the benefit of this group and some of them having a symbiotic relationship with corals.

5.4 Diversity of Terrestrial Biota

5.4.1 Fauna

The Indonesian archipelago's 17,000 islands are home to roughly 12% of the world's mammals, 16% of the world's reptiles and amphibians, 17% of the world's birds, and 25% of global fish populations.

The high diversity and number of endemic mammal's species in Indonesia were found by Wallace (1823–1913) on his famous exploration in this area and published "The Malay Archipelago." Since then imaginary biogeography: a Wallace line was postulated. Later, another lines; Lydekker, Weber's were also introduced. Indonesia has more species of mammal than any other nation, an incredible 515 species by most counts. Unfortunately, Indonesia also leads the world in the number of threatened mammals at 135 species, which is nearly a third of all of its native mammals. The number of mammals in Indonesia is: Kalimantan (268), Sumatra (257), Jawa (193), Sulawesi (207), Maluku (149), Lesser Sunda Island (125). Many of Indonesia's most iconic and well-known species are also its most endangered. With the extinction of the Balinese and Javan tigers, the Sumatran tiger is the only surviving species of Indonesian tiger. Its wild population is believed to total less than 500 animals, with an estimated 150 breeding pairs. One of the most immediate threats to their survival comes from the destruction of critical habitat by the pulp and paper industry as it converts high-value rainforests into monoculture pulp plantations. The Red Ape, the orangutan continues to suffer precipitous declines from deforestation. Sumatran orangutans are designated as Critically Endangered by the IUCN, with a population of just a few thousand, while the Borneo orangutan is considered Threatened. The island of Java is home to a small population of Javan Rhinoceros, the rarest of all rhinoceros, while Sumatra's swamp forests protect the last Sumatran Rhinoceros, second rarest of the rhinos. The Sumatran elephant, also listed as Endangered, is pre-eminently threatened by habitat loss, degradation, and fragmentation.

Indonesia is the largest host of bird species in the world, the number of bird species in Indonesia is 1,611 species including the migratory (Haryoko, 2014). Wallace area is the place where endemic birds are mostly found. The endemic bird species in Wallace area is 107 species, 21 of them are mono-typic species, some of them are critically endangered birds (http://www.worldatlas.com/articles/indonesia-s-critically-endangered-endemic-birds.html). Some of the most beautiful birds found in this country are: The Red Bird of Paradise (*Paradisaea rudolphi*), Lesser Bird of Paradise (*Paradisaea minor*), Ribbon-tailed Astrapia (*Astrapia mayeri*), Jewel-like Wilson's Bird of Paradise, Greater bird of Paradise (*Paradisaea apoda*), Silken Satin-bird (*Loboparadisea sericea*), and Bali Starling (*Leucopsar rothschidi*)

The last record of amphibians in Indonesia 385 species was found espe-cially from Papua, Kalimantan and Sumatra. Similar to amphibians, those three islands also harbor reptiles. A total number of Indonesian reptile is 723 species (Widjaja et al., 2014); among them is the famous Komodo dragon (*Varanus komodoensis*) the largest and heaviest lizard species in the world. This species inhabits five islands; Komodo, Flores, Padar, Rinca, and Gili Motang. The males can grow up to 3 meters and listed as vulnerable species in IUCN Redlist.

Indonesia has one of the world highest densities of fish species. The total number of fresh-water fish is 1218 species and 630 are endemic species. The diversity is due to the fragmentation of several rivers (Hutama et al., 2016).

A total number of Indonesian mollusks was estimated as 4,000 species, and 151,847 species of insects have been recorded, these are about 15% of the total world insect species.

Indonesian fauna has many endemic species such as: mammals (383 sp.); birds (323 sp.); amphibian (160 sp.); reptile (231 sp.); fresh-water fish (241 sp.); butterfly (38 sp.)

5.4.2 Wild Plants

Cryptogams (algae, fungi, moss, and fern) have an important role in the stability of an ecosystem. Algae, a diverse group of organisms from a single cell to large sizes and play important role in the aquatic ecosystem. Indo-nesia is estimated to have 1,500 species of algae among 72,500 species of world algae. Sometimes algae are not grouped as plants because they lack of vascular system, true root, stem, and leaves.

Fungi are found everywhere, macro and micro fungi both have been used by people for food and drink, the source of medicine and involve in improv-ing the quality of the product. The number of potential fungi are scattered in many institutions in Indonesia and more than 2,500 species of macro and micro fungi have been recorded.

Lichen as a bioindicator and also used as medicine in Indonesia. The number of lichen in Indonesia is 595 species; this is only 3% of the total world lichen.

Moss and liverworts play an important role as oxygen provider; improve the humidity as well as pollution absorber. Up to 2013, the number of moss species recorded is 1,510, while liverworts are 1,364 species.

The number of ferns in Indonesia is 2,197 species from about 10,000 species of world ferns, some of them are used as ornamental plants, medicine and growing media.

The number of Angiosperm in the world was estimated between 250,000–400,000 species. The number of species in Indonesia had been estimated between 30,000–40,000 species of plants including ferns and gymnosperms. They are source of timber, food including spice, medicine and cosmetics (Barthlott et al., 2005)

There are 120 species Gymnosperms in Indonesia and 72 species among them are endemics. They are not distributed evenly, Kalimantan, Sumatra and Papua Island have more species compared to other islands in Indonesia. As part of Kalimantan and Papua are under Malaysia and Papua New Guinea, therefore, information on the distribution of the species has to be verified.

In the lowland rainforest, some important families are: Dipterocarpaceae, Sapindaceae, Sterculiaceae, Leguminosae, Anacardiaceae, and Ebenaceae. Other important tree families are Lauraceae, Moraceae, Myrtaceae, Rubiaceae, and Gymnosperms. The complex topography and geology, the equatorial wet climate, and the island setting synergistically produced a species-rich and highly endemic flora. Indonesia is home to more than 100 Critically Endangered tree species, many of which require urgent conservation action to prevent their extinction.

Indonesia is the host of about 5,000 species of orchids, 161 bamboo species, more than 500 palm species (including rattan) (Personal communication, 2017) and many of them are endemic.

Two of the most charismatic plants are endemic to Indonesia, the largest flower in the world; *Rafflesia arnoldii* a true parasitic plants (Figure 5.2). The diameter of the flower is up to 100 cm. Unfortunately, these rare plants only can be seen in the wild, since up to now unable to be planted outside their habitat. The existence of *Rafflesia arnoldii* depends on its host, a woody climber. The big climber requires an external support (usually strong tree) to grow vertically and enhance light acquisition. When the habitat of the host deteriorated is affected the tree supporting the host, consequently interfere with the development of this obligate parasite.

Figure 5.2 *Rafflesia arnoldii.*

The second one is the Titan Arum, *Amorphophallus titanium* (Figure 5.3), the tallest flower clusters in the world, it reaches more than 3 meters, has a strange single leaf. This iconic species has been planted in many botanic gardens in the world and always attracts visitors during a short time of anthesis. Both species are protected since 1999. Recently these two charismatic species have special strategic and action plan to manage their existence in the future and have more benefit for people.

5.5 Genetic Diversity

Source of genetic diversity is total variation found in nature which is essential in the utilization of the species in creating new cultivar, varieties, species, family or race through conventional breeding or biotechnology.

5.5.1 Animal

Fish germplasm is found all over Indonesia, each location has their specific or endemic fish in their lakes or rivers and local people taking care to

Figure 5.3 *Amorphophallus titanium.*

maintain the availability in the future. Some important endemic fishes are: *Mystacoleucus padangensis* from Sumatra; *Leptobarbus hoevanii, Mystus planices, Chitala lopis* and *Barbodes schwanengeldii* from Kalimantan. Four important fish from Kapuas River are: *Tor tambra, T. souronensis, T. tambroides* and *Labeobarbus douronensis*. Introduction of new (invasive) species in freshwater fisheries would threaten the local fish species.

Three sources of cow germplasm in Indonesia are *Bos indicus* (fom India), *Bos taurus* (subtropical) and *Bos javanicus* (local cow). Long-time cow breeding produced important local races. Eight goat stocks in Indonesia originally were from different sources; therefore conservation of goat germplasm is important. For pig, it is not clear whether local wild pig was the parent of domestic pig. Local rabbit has come from Sumatra, *Nesolagus netscheri* but has not been used for breeding.

Domesticated Indonesian chicken has a wide diversity. Local chicken breeding was conducted to find out high yield for egg or meat producer. At least 18 local chickens have been identified genetically. Similarly, duck breeding is conducted to produce prime race and 16 local ducks have been characterized.

5.5.2 Plants

Important families in the surrounding lowland rainforests are especially Dipterocarpaceae, Sapindaceae, Sterculiaceae, Leguminosae, Anacardiaceae, and Ebenaceae. In the montane forests (1,400–3,400 m), temperate elements such as oaks and stone oaks (*Quercus/Lithocarpus*) and other Fagaceae show up. Other important tree families in this altitudinal zone include Lauraceae, Moraceae, Myrtaceae, Rubiaceae, and various gymnosperms.

Indo-Malayan region including Indonesia is the center of diversity for rice, Job's tears, wing bean, taro, Tacca, pamelo, banana, breadfruit, mangosteen, star-fruit, durian, rambutan, snake-fruit, langsat, mango, candle-nut tree, coconut, sugar cane, clove, nutmeg, pepper, abaca, sago, sandal wood and bamboo (Vavilov, 1926). Activities need to explore and conserve these valuable crops to prevent genetic erosion. Collection of these genetic materials is found in Tables 5.2 and 5.3.

5.5.3 Estate Crops

- Sugarcane, has 321 accessions and 70 varieties have been released, and 127 accessions of wild relatives are conserved at Sugar Research Center.
- Palm oil germplasm is found in some research institutes, seedlings producers, and corporate plantation gardens.
- Clonal rubber collections are planted in several institutes. From a previous collaboration with Rubber Research and Development Board, Indonesia received 7,778 genotype materials collected from Brazil to add the 583 collections at Rubber Research Center in North Sumatra.
- The origin of clove is Indonesia, cloves are also used for spices of local cigarette.

Table 5.2 Germplasm Collection in Research Centers and Institutes in Indonesia

No.	Germplasm	Σ accession	Remarks
1	Rice	6,179	Including landraces and cultivar
2	Corn	1,546	Including local variety (480), introduction (130) and introduction population (50)
3	Sorghum	255	
4	Job's tears	12	
5	*Setaria italica*	9	
6	*Sesamum indicum*	6	
7	*Vigna sinensis*	130	
8	*Vigna umbellata*	46	
9	*Vigna subterranea*	70	
10	*Cajanus cajan*	13	
11	*Lablab purpureus*	11	
12	*Mucuna pruriens*	9	
13	*Canavalia ensiformis*	7	
14	Wing bean	88	
15	Soybean	1,993	Variety (68), local variety (3)
16	Ground nut	1,194	Released varieties (8)
17	Green pea	1,492	
18	Cassava	706	Variety (11)
19	Sweet potato	1,702	Variety (17)
20	Taro	245	
21	*Xanthosoma* sp.	126	
22	*Maranta arundinacea*	34	
23	*Canna edulis*	63	
24	*Dioscorea esculenta*	17	
25	*Dioscorea hispida*	14	
26	*Dioscorea alata*	20	
27	*Amorphophallus campanulatus*	2	
28	Wheat	88	
29	Chilli	255	4 varieties, including released variety (7), breeding variety (79)
30	Shallot	55	Including released variety
31	Potato	95	Including released variety (25)

Table 5.2 (Continued)

No.	Germplasm	Σ accession	Remarks
32	Mango	208	298 clonal, 1,568 trees, released variety (14)
33	Citrus	213	Including local variety (22)
34	Salacca	615	Including breeding variety (7) and local variety
35	Durio	147	Including released variety (17), local variety (68)
36	Mangosteen	104	Including local variety (8), released variety (6)
37	Banana	921	Local variety (20), breeding variety (3), released variety (4)

Table 5.3 Other Minor Fruits Collection

No.	Germplasm	Σ accession	No.	Germplasm	Σ accession
1	Avocado	20	12	Persimon	6
2	Star fruit	17	13	Biwa	5
3	Water apple	16	14	Namnam	6
4	Guava	20	15	Watermelon	9
5	Pineapple	150	16	Duku	5
6	Papaya	25	17	Langsat	2
7	Rambutan	27	18	Matoa	3
8	Soursop	18	19	Longan	11
9	Jackfruit	12	20	Apple	72
10	Sapodilla	7	21	Grape	43
11	Melon	6	22	Strawberry	19

- It was estimated that Indonesia has about 500 coconut cultivars, and 115 cultivars has been collected.
- Pepper, 35 accession number has been collected including one registered variety.
- Cacao, originated from South America. 571 accession has been collected including 5 breeding varieties.
- Coffee and Cacao Research center has 36 coffee accession including registered one local variety.
- Five cashew breeding varieties has been registered and also five local varieties.

5.5.4 Forest Trees

Priority species for forest vegetation research Government of Indonesia have chosen 12 species for research and development; they are *Santalum album, Eusideroxylon zwageri, Araucaria cunninghamii, Melaleuca cajuputi, Acacia mangium, Alstonia* spp., *Intsia* spp., *Tectona grandis* and *Diospyros celebica*. Priority species for genetic conservation of forest tree also had been stated; they are *Santalum album, Araucaria cuninghamii, Intsia bijuga, Eusideroxylon zwageri, Shorea* spp, and *Diospyros celebica*.

5.6 Invasive Species

The biological diversity of Indonesia also threatened by the invasive species. These invasive species compete with local species. Invasive Species Specialist Group reported that 190 invasive species have been found in Indonesia, 53 plants, 43 animal and 2 microbes. Some invasive species were then used as handicraft material (*Eichhornia crassipes*), meat (*Rusa timorensis*) medicine or for insecticide (*Hyptis capitata, Piper aduncum*). A new regulation for invasive species management has been prepared.

5.7 Conservation of Biodiversity

The threat of the diversity occurred everywhere in the country, especially due to logging, conversion of forest area into plantation as well all illegal fishing. However, it is impossible to conserve all species under threat. Therefore, Myers et al. (2000) identified the biodiversity hotspots where exceptional concentrations of endemic species are undergoing exceptional loss of habitat. Indonesia is divided into two areas: Sundaland (west) and Wallacea, both areas are two among 34 hotspots in the world. Five factors (endemic plants, vertebratas and remaining primary vegetation per 100 km²) were applied to sort the hottest hotspots. *In situ* and *ex situ* conservation to save the remaining flora and fauna should be priorities.

In situ conservation is the most ideal way to conserve biodiversity. Natural habitat is protected through Ministry of Forestry regulation. The government also nominated some area to UNESCO Man and the Biosphere programme as Biosphere reserves.

Ex situ conservation as a complement of *in situ* conservation actively developed recently. Initially, only four Botanic Gardens were found in Indonesia, but now another 25 new Botanic Gardens were developed under local government.

Regulations to protect biodiversity have been started during Dutch occupation and now are improved through central government laws, local regulations as well as local wisdoms.

To conserve the biodiversity need immediate action involving political, socio-economic, multidisciplinary strategy as well as scientific input involving all stakeholder both government, non-government, national and international organizations) (Sodhi et al., 2004). Due to limited manpower and funds to manage the biodiversity, setting priority to manage the ecosystem and species have been conducted. Only decisive action and a paradigm shift towards meaningful conservation commitments by industry and the Indonesian government will prevent a catastrophic epidemic of extinctions in the coming decades.

Indonesia has ratified the Convention of Biological Diversity, therefore the government set a guideline document, Indonesian Biodiversity Strategy and Action Plan 2015–2020 (IBSAP 2015–2020) launched last year as an updated version of IBSAP 2003–2020 (Bappenas, 2016). Management of Indonesian biodiversity is in-line to three global agreements: Aichi Targets (Global target to prevent decreasing of biodiversity) (CBD, 2013), Nagoya Protocol (on access and benefit sharing) and identifying and Mobilizing Resources for biodiversity conservation.

Keywords

- conservation
- genetic diversity
- germplasm
- invasive species

References

Bappenas, (2016). Indonesian Biodiversity Strategy and Action Plan (2015–2020).

Barthlott, W., Mutke, J., Rafiqpoor, D., Kier, G., & Kreft, H., (2005). Global centers of vascular plant diversity. *Nov. Act. Leopoldina*, *92* (342), 61–83.

CBD, (2013). *Quick Guides to the Aichi Biodiversity Targets Version 2*, United Nations Decade for Biodiversity, 2011–2020.

Cottle, R., (2004). *Linking Geology and Biodiversity.* English Nature Research Reports. Report Number 562.

Haryoko, T., (2014). Persebaran dan habitat persinggahan burung migrant di kabupaten Natuna provinsi Kepulauan Riau (Distribution and Stopover habitat of migratory birds in Natuna, Riau Archipelago province). *Berita Biologi*, *13*(2), 222–230.

http://herbarium.desu.edu/downloads/files/Charismatic_Plants_Aug_2_.pdf (accessed, 10 February 2017).

http://indoplasma.or.id/index.php/en/artikel/264-kebun-plasma-nutfah-pisang-terlengkap-di-asia-tenggara-ada-di-yogyakarta (accessed, 10 February 2017).

http://www.worldatlas.com/articles/indonesia-s-critically-endangered-endemic-birds.html (accessed 9 February 2017).

https://blogs.ntu.edu.sg/hp331–2014–03/?page_id=27 (accessed 1 February 2017).

https://www.coralguardian.org/en/marine-ecosystem/ (accessed 2 February 2017).

Hutama, A. A., Hadiaty, R. K., & Hubert, N., (2016). Biogeography of Indonesian freshwater fishes: Current progress. *Treubia*, *43*, 17–30.

Hutomo, M., & Mosa, N. K., (2005). Indonesian marine and coastal biodiversity: Present status. *Indian J. Mar. Sci.*, *34*(1), 88–97.

Kartawinata, K., (2013). *Diversitas Ekosistem Alami Indonesia.* Ungkapan singkat dengan sajian foto dan gambar. Jakarta: LIPI Press bekerja sama dengan Yayasan Obor Indonesia.

Myers, N., Mittermeier, R. A., Mittermeier, C. G., Da Fonseca, G. A. B., & Kent, J., (2000). Biodiversity hotspots for conservation priorities. *Nature*, *403*, 853–858.

Romimohtarto, K., & Juwana, S., (1999). *Biologi Laut. Ilmu Pengetahuan Tentang Biota Laut.* Pusat Penelitian dan Pengembangan Oseanologi–LIPI, Jakarta.

Sodhi, N. S., Koh, L. P., Brook, B. W., & Ng, P. K. L., (2004). Southeast Asian biodiversity: An impending disaster. *Trends in Ecology and Evolution*, *19*(12), 654–660.

Susanto, H. A., & Tokeshi, S. M., (2015). Management of coral reef ecosystems in Indonesia: Past, present and the future. *Coastal Ecosys.*, *2*, 21–41.

Vavilov, N. I., (1926). Studies on the origin of cultivated plants. *Bull. Appl. Botany, Genetics and Plant Breeding*, *16*, 405–435.

Veron, J. E. N., (2000). *Corals of the World.* Australian Institute of Marine Science, Townsville, Queensland, Australia.

Widjaja, E. A., Rahayuningsih, Y., Rahajoe, J. S., Ubaidillah, R., Maryanto, I., Waluyo, E. B., & Semiadi, G., (2014). *Kekinian Keanekaragaman Hayati Indonesia.* Pusat Penelitian Biologi, LIPI.

CHAPTER 6

Biodiversity in Iran

M. M. DEHSHIRI

Department of Biology, Boroujerd Branch, Islamic Azad University, Boroujerd, Iran,
E-mail: dehsiri2005@yahoo.com

6.1 Introduction

The Islamic Republic of Iran is situated in the eastern portion of the northern hemisphere, in the southwest of Asia, and in the west of the Palaearctic region. With a total surface area of 1,648,000 km², Iran ranks 18th in size among the countries of the world, placed in the Middle East and surrounded by the Armenia, Azerbaijan, Caspian Sea, Turkmenistan on the north, Afghanistan and Pakistan on the east, Oman Sea and Persian Gulf on the south, and Iraq and Turkey on the west. The country is both a meeting point for many cultures as well as for many types of climate, land, water, and biodiversity of flora, fauna, and people (Collins, 2001). Population of Iran is about 73 million and the growth rate estimated as 1.3% and it has 31 provinces (FAO, 2005). Various environmental conditions with respect to wide latitude and longitude range; 44° 05' to 63° 18' E longitude, 25° 03' to 39° 47' N latitude, topographic diversity; altitude varies from –28 to 5774 m (the Caspian coastal area is below sea-level (–28 m) while Damavand peak in the Alborz reaches 5774 m above mean sea level) and high geological and geomorphologic diversity and also very variable from the stand point of climate producing variable ecological diversity and habitats (Sharifi, 2011). Iran is situated where three climatic zones meet the Mediterranean, the arid or semi-arid West Asian and the temperate humid/semi-humid Caspian zone. Of the total land area, about 55% are rangelands; 12% are forests and 33% are deserts including bare salty lands. Being dominantly in an arid environmental zone, approximately 85% of Iran's agricultural lands are located in arid and semi-arid areas (Misra, 2009). Iran is well-known as one of the world's major centers of biodiversity and natural heritage, because of the junction of major plant geographical regions. As a result, most rivers are seasonal and their flows depend heavily upon the amount of rainfall (Ansari, 2000; Calabrese et al., 2008). The climate is extremely continental with hot and dry summer and very cold winter particularly in inland areas. Apart from the coastal areas, the temperature in Iran is characterized by relatively large annual range about 22°C to 26°C. The rainy period in most of the country is from November to May followed by dry period between May and October with rare precipitation. The average annual rainfall of the country is about 240 mm (less than a third of world average precipitation) with maximum amounts in the Caspian Sea plains, Alborz and Zagros slopes with up to more than 2,000 and 480 mm, respectively. Going inland at the central and eastern plains, the ranges of precipitation decreases to less than 100 mm annually depending on the location. From the synoptic aspects, the climate of most part of Iran is dominated by subtropical high in most part of the year. This phenomenon causes hot and dry climate in summer. The rainfall in the country is produced by Mediterranean synoptic systems, which move eastward along with westerly winds in cold season. Synoptic systems and year-to-year variation in the number of passing cyclones cause high variability in annual rainfall. Frontal Mediterranean cyclones associated with the westerly air flows produce most of precipitation in the whole country in late autumn and particularly in winter. In addition to the frontal Mediterranean cyclones, rainfall bearing systems called Sudanian cyclones which come from the southwest make an important contribution to increase annual rainfall amount of the west and southwest of the country (Raziei et al., 2005). In northwest mountainous regions, convective and frontal thunderstorms are important atmospheric process responsible for rainfall in spring and early summer. These rainfalls bearing systems are restricted to the west portion of the country and have no more energy and moisture to pass far to the east. These systems sometimes may reach central and east dry regions of the country when there will be no potential to produce rainfall due to the long trajectory and loss of moisture. This region is the most drought-prone area in the country due to high inter- and intra-annual irregularity in rainfall and high coefficients of variation. This region that accounts for over half of Iran's land area is surrounded by Alborz mountain range from the north and Zagros range from the northwest to southwest. These mountains play an important role in determining the non-uniform spatial and temporal distribution of precipitation in the whole country. The eastern part consists mostly of desert basins such as the Dasht-e-Kavir, Iran's largest desert, in the north-central portion of the country, and the Dasht-e-Lut, in the east, as well as some salt lakes. This is because the mountain ranges are too high for rain clouds to reach these regions (Kazemi, 2014). The area within the mentioned mountain ranges is the high plateau with its own secondary ranges and gradually slopes down to become desert which continues into southern part of Afghanistan and near the Pakistan border. The Zagros range prevents Mediterranean moisture bearing systems to pass through to the east. This phenomenon gives rise to high irregularity in rainfall in the center of Iran. Lack of rainfall in May to October compounded with high temperature leads to high evapotranspiration and water deficit in this region. Iran is situated within the dry belt of Asia (Breckle, 2002). As a result, the central and eastern parts of Iran are arid and semi-arid regions. Since Iran is in the arid zone, approximately 85% has arid, semi-arid or hyper-arid environment. The specific features and location of Iran cause it to receive less than a third of the world average precipitation. Prolonged drought in this area and availability of moisture in other parts of Iran has led to the formation of different ecological zones (Heshmati, 2007). Iranians depends almost entirely on water resources in the form of spring rains or melting snow descending from the mountains or underground water resources. Traditionally, in most parts of the country, the rich underground water resources in the higher altitudes are transferred by the man-made underground water channels systems to the lower and dryer agricultural and settled areas which are known as Qanat or Karez. In contrast to Afghanistan, where the mountains form a 'backbone' of the country, in Iran, the mountains are surrounding the desert lowlands of the central part (Noroozi et al., 2008).

The complex and varied climates, topography, geological formations and anthropological management of natural resources have led to a varied and unique biological diversity. Only the Hyrcanian forests located in the South Caspian Sea characterized a humid climate (temperature rarely fall below freezing and the area remains humid for the rest of the year) with a rich vegetation of lowland and montane deciduous forests (Akhani, 1998; Noroozi et al., 2008). Forests cover about 12.4 million ha in Iran, including ca. 5.5 million hectares (about 40% of Iran's forests) in the mountainous Zagros region (Raunkiaer, 1934; Sabeti, 2002; Sagheb-Talebi et al., 2003; Haidari and Rezaei, 2013; Heidari et al., 2013). The northern forests of Iran cover an area of 1.9 million hectares (Pourbabaei and Poor-Rostam, 2009). Except for the interior deserts and the lowlands along the Caspian Sea, Persian Gulf and Gulf of Oman, ca. half of Iran is composed of high mountains with alpine life zones and a diverse alpine flora. It is a mountainous country of particular geopolitical significance owing to its location in three spheres of Asia and four-fifths of its surface lies at altitudes above 1000 m (Zohary, 1973). The mountains enclose several broad basins, or plateaus, on which major agricultural and urban settlements are located. The main mountain chains are Alborz, Zagros, Kopet Dagh, and Khorassan and Makran (Breckle, 2002).

6.2   Ecosystem Diversity

Extreme relief and climatic conditions have led to a great diversity of ecosystems and habitats over small geographical areas. In turn, this has created a home for a vast range of plant and animal species, especially endemic and endangered species (Olfat and Pourtahmasi, 2010; Heydari et al., 2013).

The Islamic Republic of Iran is an ecologically diverse country which includes rich agricultural areas, deserts, marshes, rivers, and mountain habitats (such as two important mountain ranges (Alborz and Zagros), 10 Biosphere Reserves, 24 Ramsar listed wetlands and unique drylands and forests). Woodlands cover 12.4 million hectares and there are 10,000 hectares of *Avicennia* mangroves along the Persian Gulf. In the Iranian ecosystems approximately 7,115 species of plants (of which 1,727 are endemic), 197 species of mammals, 541 species of birds, 241 species of reptiles, 23 species of amphibians, 160 species of freshwater fishes, and 1,080 species of marine fishes have been recorded. Iran is one of the most important countries in the Middle East and Western Asia for the conservation of biological diversity. Habitat diversity in Iran allows for a wide range of animals to inhibit in Iran. With regards to ecosystem diversity of marine and coastal zones in the North and South of the country, it consists of 25 ecological types and units, in which the most important are coral reefs, bays, and small islands.

Habitat diversity in Iran allows for a wide range of animals to inhabit in the country. Bears, Caucasian Black Grouse, Sacred Ibis, Eurasian Lynx, River Trout, and Crocodile can be found in these habitats. Red Deer and Roe Deer in Hyrcanian forests, Urial Sheep in steppe hills, Wild Goats, European Snow Vole, Caspian Snowcock and Caucasian Agama, and Leopards on cliffs and bare heights of dry areas, Wild Asses, Gazelles and Cheetahs in desert plains are found in Iran. The Pheasant, Partridge, Stork, Saker and Barbary Falcon, Golden and Imperial Eagle are also native to Iran. The Persian Leopard is said to be the largest of all the subspecies of Leopards in the world. The main range of this species in Iran closely overlaps with that of Bezoar Ibex. Hence, it is found throughout Alborz and Zagros mountain ranges, as well as smaller ranges within the Iranian plateau. Leopard population is very sparse, due to loss of habitat, loss of natural prey, and population fragmentation.

It is very difficult to assess biodiversity trends in Iran since the national biodiversity indicators are not fully developed. However, what is obvious from reports is that data and statistics in general biodiversity of Iran in different ecosystems and at different levels is degrading. Although the Iranian government has taken various measures to combat degradation and rehabilitate degraded natural resources, however, different factors such as prolonged and frequent drought cycles and maximized use of soil, water, and plant cover have aggravated the biodiversity decline in Iran. For example, the country's population is more than 73 million with an annual average growth rate of about 1.3% per annum. More than 21 million of the populations live in rural areas (IFNRCBD, 2010).

Zagros forests cover a vast area of the Zagros Mountain ranges. Geographically this important natural ecosystem has been divided into three parts: northern, central and southern Zagros (Valipour et al., 2009). It is globally significant for protection of ecosystems, species, and genetic biodiversity.

In order to conserve existing biodiversity of the country, representative samples of the nature of land have been selected and are being conserved under different categories such as Protected Areas, National Parks, Wildlife Refuges and National Natural Monuments (IFNRCBD, 2010).

6.3   Biomes and Vegetation Regions

Vegetation of each region indicates specific important features and phenomena of nature and is the best guide in judgments concerning ecological factors in the region. Plants are resistant organisms that tolerate all environmental conditions and occurrences over the long-term, including environmental stress (Nezhad, 1973). In any country, information on vegetation status—not just infrastructure development and scientific activities in the field—has commercial applications. It can also play an important role in restoration and use of natural resources (Shahsavari, 1994). Identification of plant vegetation and geography of each region based on regional ecological research and reviews provides for effective appraisal of current and anticipated future nature status, and in this context, proper management practices at a regional level play an important role (Shahsavari, 1994; Mataji et al., 2013).

Because of its large size and varied ecosystems, Iran is one of the most important countries for the conservation of biological diversity in the Middle East and West Asia. More basic technical works and field surveys in terms of eco-biology analyses are needed in order to fully understand the country's biodiversity characteristics.

The climate contrasts together with different geological substrates across the country providing various habitats and vegetation structures. By now, there is no comprehensive vegetation database dealing with all regions of

Iran. There is only scattered vegetation information mainly for the forest areas and the halophytic vegetation. The mountain steppes, wetlands, and the Zagros Mountains remain still unknown (Naqinezhad, 2012).

The following four biomes are considered as general vegetation regions across the country:

a) Irano-Turanian which covers an area of about 3,452,775 ha with arid and semi-arid deserts and plains of central Iran. Regarding topographical conditions and diversity of species, the region is divided into plain and mountainous sub-regions.

   Plain sub-region: This ecoregion is dominated by the central Iranian plateau, an immense area covering 1,648,000 km² in the center of Iran and encompassing a great variety of climates, soils, and topography. It is almost completely surrounded on all sides by mountain ranges. The area can be divided into two major sections: the Dasht-e-Kavir in the north, a vast saline desert, and the Dasht-e-Lut in the south, largely a sand and gravel desert and one of the hottest deserts in the world. The plateau is also partly covered with sand dunes. Adjacent regions, such as the Kavire-Namak (salt desert) and a series of marshes and lakes east of Qom, are also included in this ecoregion. In the northwestern corner of the central plateau, where the Kavir National Park is situated, habitat types range from desert and semi-desert to dry steppe.

   These biomes in Iran cover nearly 95% of the total area of the country (Zehzad et al., 2002). They differ particularly in their contemporary climatic conditions (with the steppe biome having a higher precipitation rate), vegetation complexity, topography (e.g., mean elevation), history (e.g., paleoclimate), geology, and flora (Zohary, 1973). From the conservation perspective, both steppe and desert biomes have been under disturbance pressure in different ways during recent decades. Desert areas have been used for military training exercises for many years, resulting in severe habitat disturbance. The steppe biome, by contrast, faces deforestation in huge areas, generating high rates of habitat loss and fragmentation (Paknia and Pfeiffer, 2011). In the northeastern reaches of the plateau, in the area of the Turan Biosphere Reserve, the variety of landforms includes extensive plains, a saline river system, alluvial fans, limestone outcrops, salt desert, and 200,000 ha of the northernmost sand dunes in Iran. While mean annual temperature ranges from 15 to 18°C, the extreme maximum temperature can reach 42°C and the extreme minimum temperature can fall to –20°C. In most of the region, annual rainfall does not exceed 200 mm and in much of it, rainfall is less than 100 mm. In the northwest corner of the region, precipitation is highly variable from year-to-year, ranging from less than 50 mm to over 300 mm and falling mostly as rain from November to May. As a result, most rivers are seasonal and their flows depend heavily upon the amount of rainfall. The driest parts of Iran are found in the central and eastern parts of the plateau, with the Dasht-e-Lut receiving only up to 50 mm per year. The low amount of rainfall is aggravated by high evaporation rates. The rivers descending into the central plateau from the surrounding mountain ranges carry high levels of soluble salts, and the ground in the plateau tends to be highly saline. On the margins of the plateau, and in a few patches in the interior where the topography is such that the soil is less saline, areas of piedmont fans and alluvial soils exist and can be farmed. The central plateau can be differentiated into a series of habitats; including poorly drained flats inhabited by halophytic communities and better-drained flats inhabited by a variety *Artemisia*.

Mountainous sub-region: In this sub-region, the *Juniperus polycarpus* species have been developed. It has the dry and cold climate, temperate summer, and the annual precipitation of about 400 mm. A variety of fruit trees, medicinal, industrial, and field crop plants are grown in the mountain ecoregions of the Irano-Turanian Zone. Some of the dominating plant species of these regions are among others: *Amygdalus scoparia, Onobrychis cornuta, Acantholimon* spp., *Astragalus* spp., *Artemisia aucheri, Allium* spp., *Bromus tomentellus*.

b) *Zagrosian*: This region with an area of about 5.5 million hectares covers semi-arid Zagros mountain ranges. This ecological zone extends throughout the Zagros Mountain in the west and south. This mountain range parallels the Persian Gulf and consists of numerous parallel ridges, with the highest peaks exceeding 4,000 m and maintaining permanent snow cover. Many large rivers, including Karun, Dez, and Kharkeh originate here, drain into the Persian Gulf or the Caspian Sea. Scenic waterfalls, pools, and lakes add to the beauty of the mountainous landscape. The forest and steppe forest areas of the Zagros Mountain ranges have a semi-arid temperate climate, with annual precipitation ranging from 400 m to 800 mm, falling mostly in winter and spring. Winters are severe, with winter minima often below 25°C, and extreme summer aridity prevails in the region.

The Kurdo-Zagrosian steppe-forest consists mainly of deciduous, broad-leaved trees or shrubs with a dense ground cover of steppe vegetation. The dominant species are oak (*Quercus* spp.), pistachio (*Pistacia* spp.) and a few others. In the northern reaches of the mountain range, lower altitudes (400 m to 500 m) host communities dominated by *Astragalus* spp., *Salvia* spp., or others while higher up (700 m to 800 m) forests or forest remnants of *Quercus brantii* and/or *Q. boissieri* occur up to an altitude of about 1,700 m. Above the timberline (1,900 m to 2,000 m) appears a relatively wide zone of subalpine vegetation. Further south along the range, the forest becomes more impoverished and a richer steppe flora develops among the trees. Forest remnants consist primarily of *Quercus persica* and, up to an elevation of 2,400 m, xerophilous forest of *Quercus* spp., hawthorn (*Crataegus*), almond (*Prunus amygdalus*), nettle tree (*Celtis*), and pear (*Pyrus* spp.) predominates. Below 1,400 m, the vegetation is steppic, with shrubs predominating.

The halophilous vegetation around Urmia Lake displays an interesting gradient of salinity-tolerance, ranging from annual obligatory hygro-halophytic communities on lake marshes dominated by *Salicornia* spp. up to hydrophytic communities dominated by *Alisma plantago-aquatica*. The latter species grows on the margins of salty and brackish water marshes where the fresh groundwater dilutes the salt contents of the soil (Asem et al., 2014).

c) Hyrcanian which covers semi-humid and humid Arasbaran and Hyrcanian mountains and Caspian plain. This region extends throughout the south coast of the Caspian Sea and the northern part of the country which is bordered by the largest lake in the world. Hyrcanian forest is natural refuges of West Eurasian temperate deciduous forest region, old growth, natural, self-regulating forest ecosystems exist up to now. Nemoral broad-leaved forest biomes are important, because of their history and the climate in the South of Caspian Sea (Sharifi, 2011). Mountains dominate the landscape of this ecoregion. Hyrcanian (Caspian) region could be divided into three subdivisions on the basis of geographical situations. These subdivisions are: (1) Alborz Range forest steppe, (2) Caspian Hyrcanian mixed forest, and (3) Caspian lowland desert.

At the first subdivision, the highest peak is Mount Damavand, a dormant volcano 5,774 m tall. Below Mount Damavand's crater are two small glaciers, as well as fumaroles, hot springs, and mineral deposits.

Only between 280 and 500 mm of precipitation fall annually on this high elevation of Alborz mountain. The dominant trees, *Juniperus sabina*, and *J. communis*, are resistant to summer drought and heat and can tolerate winter cold equally well. However, the tree grows so slowly that it is difficult to reestablish these forests once they are cut. Shrubs include pistachio (*Pistacia vera*), *Berberis integerrima*, *Acer* spp., and *Amygdalus*, spp., with such a plant species: *Onobrychis cornuta*, *Astragalus gossypinus*, *Agropyron* spp., *Bromus tomentellus* forming the ground cover. Overgrazing at higher elevations by sheep and the continued fragmentation of habitat, which is accelerated by road construction, are also of concern because so few natural areas remain.

The second subdivision is described in more details over a paragraph to introduce Iran's forests and rangelands.

Concerning the third subdivision (Caspian lowland desert), the ecoregion lies on the southern and eastern shores of the Caspian Sea at elevations between –28 to 100 m above the sea level. Average annual temperature is 17.1°C and annual precipitation is 187 mm. A long frostless period (271 days) encourages cultivation of crops such as olive, fig, pomegranate, and cotton. This ecoregion is covered by shrubs and grasses, which are used by livestock. The vegetation of the coastal Caspian desert within Iran is impoverished, it consists of highly specialized halophytes (salt-resistant plants) represented by shrubs and semi-shrubs such as various sagebrushes (*Artemisia*), Tetyr (*Salsola gemmascens*), Kevreik (*S. orientalis*), Boyalych (*S. arbuscula*), Biyurgun (*Anabasis salsa*, *A. ramosissimum*), Sarsazan (*Halocnemum strobilaceum*), *Halostachys*, *Ceratocarpus*, and *Nitraria*, *Kalidium*. Herbaceous vegetation is represented by species of *Aristida*, *Peganum*, *Agropyron*, *Anisantha*, and *Eremopyrum*. One of the most typical halophyte plant formations is dominated by Tetyr (*Salsola gemmascens*), a 30–50 cm shrub, associated with low species diversity and sparse coverage. Solonchaks are sometimes occupied exclusively by Sarsazan (*Halocnemum strobilaceum*). The European portions of the ecoregion consist of northern lowland dwarf semi-shrub deserts and small areas of floodplain vegetation and coastal and inland halophytic vegetation north of the Caspian Sea.

d) Khalijo-Ommanian which encompasses dry southern coastal plains with high humidity. The region with an area of 2,130,000 ha extends throughout southern parts of the country in Khuzestan, Bushehr,

Hormozgan, and Sistan-Baluchistan provinces. They are dominated by sub-equatorial climate. The main plant species of these regions are: *Acacia, Prosopis, Ziziphus, Avicennia, Rhizophora, Populus euphratica*, and *Prosopis stephaniana*.

The plant species of the above four ecological zones are classified on the basis of average rainfall and altitude (IFNRCBD, 2015).

6.4 Species Plant Diversity

Iran with approximately 1.65 million km² area is a large country and except for Turkey, it is the richest country in the Middle East in terms of plant diversity (White and Léonard, 1991). The country is one of the centers of plant diversity considered in the Old World with nearly 1,727 taxa (25%) endemic species of 7,115 plant species belong to 173 Families and 1206 Genera (Table 6.1). Out of 151 Angiosperms families, 9 families include c. 58% genera and c. 65% species. The remaining genera and species belong to 142 families (Table 6.2) (Ghahreman, 1994; Yousefi, 2007; Sharifi, 2011; Mataji et al., 2013; Ghelichnia, 2014).

A large portion of Iran's territory is located in the Palaearctic realm and is considered at the center of origin for many commercially valuable plant species, e.g., *Medicago sativa* or medicinal and aromatic herbs. The southwest has been characterized by Afro-tropical species, while the southeast has some species from the Indo-Malayan sub-tropical realm.

Wetlands and rivers are considered as inland water ecosystems. Investigations by the Department of Environment suggest that more than 100 big wetland sites are found nationwide. At this time 22 sites with a total area of 1,481,147 hectares are registered on the list of wetlands of international importance by the Ramsar Convention on Wetlands (Ramsar Convention Secretariat, 2013).

Table 6.1 Number of Species in Different Plant Groups in Comparison with World

Group	No. of Species in Iran	Endemic	%	No. of Species in World	% w.r.t. Iranian species
Angiosperms (151/173 families)	7,058 (1,175 genera)	1,726	24.45	352,000 (14,044 genera)	2
Gymnosperms	16	1	6.25	1,088	1.47
Pteridophytes	41	0	0	9,294	0.44

(adapted from Yousefi, 2007)

Table 6.2 Diversity of Angiosperms (151/1175/7058) from Iran

Families	No. of Genera & Species	Remarks
Asteraceae	148 & 1036	Out of 151 Angiosperms families, 9 families include c. 58% genera and c. 65% species. The remaining genera and species belong to 142 families.
Fabaceae	55 & 1385	
Poaceae	122 & 412	
Caryophyllaceae	38 & 350	
Boraginaceae	36 & 219	
Lamiaceae	44 & 348	
Apiaceae	112 & 320	
Brassicaceae	103 & 318	
Rosaceae	30 & 206	

(Yousefi, 2007)

Wetlands are the most valuable ecosystems since they offer an abundance of ecological goods and services. Iran's wetlands play an important role in water balance and well being given the fact that most of the climate is dry or semi-dry. Out of 22 registered internationally important wetland sites in Iran, seven are mismanaged and included in the Montreux Record that is "a record of Ramsar sites where changes in ecological character have occurred, are accruing or are likely to occur."

More than 3,450 permanent and seasonal rivers are extant in Iran. These rivers are categorized within six main watersheds and 37 sub-basins. Long-time measurements suggest that the largest portion of annual water discharge is into the Persian Gulf and the Caspian Sea, respectively, while other four watersheds produce less water in terms of quantity but not importance. Rivers are the natural habitat for specialized flora. They offer important and valuable ecological services as well.

There are some 12.4 million hectares of forests and 10,000 hectares of mangroves along the Persian Gulf coasts. Iran's forests are classified into five main categories as follows (Figure 6.1):

1. *Caspian broadleaf deciduous forests* consist of a narrow green belt in northern Iran with a current area of about 1.9 million hectares.

2. *Arasbaran broadleaf deciduous forests* are in the northwest of the country. They support many endemic species in an area of 120,000 hectares.

3. *Zagros broadleaf deciduous forests* consist mainly of oak trees from three main species, with an area of 5.5 million hectares. Zagros forests are located in the west of Iran.

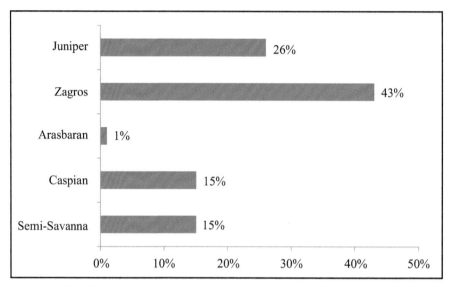

Figure 6.1   Classification of forests of Iran (adapted from IFNRCBD, 2010).

4. *Irano-Turanian evergreen Juniper forests* are almost all high moun-
    tains of the country outside the deciduous forests used to be covered
    by Persian Juniper (*Juniperus polycarpus*). Optimistically speaking,
    the area covered by this type of forest is about 500,000 hectares.
5. *Semi-Savanna thorn forests* with an area of about 2,130,000 hectares
    cover narrow bands in the west of the country and a wider belt in the
    south along the Persian Gulf and the Sea of Oman.

Rangelands cover some 55% of the total land area of the Country. While
8% out of 86.1 million hectares of Iran's rangelands are classified as excellent, 26% are favorable and 66% are classified as poor. Based on biomass
production and grazing season, rangelands can be categorized as: (i) summer
rangelands; (ii) winter rangelands; and (iii) arid rangelands (DOE, 2010).

Iran is a typical country of large Sabkhas, littoral, and inland salt marshes,
and diverse brackish and salty river ecosystems. A total of 365 species within
151 genera and 44 families of Iranian vascular plants are known to be true
halophytes, or species capable of successful growth on salty soils. The Chenopodiaceae family with 139 species ranks first, followed by the Poaceae
(35), Tamaricaceae (29), Asteraceae (23), and Plumbaginaceae (14). The
genus *Salsola* with 28 species, *Tamarix* with 25, *Atriplex* and *Suaeda* each
with 15 and *Limonium* with 10 species, respectively, are the largest halophytic genera in Iran. Diverse eco- and morphotypes, anatomical features,

life forms and photosynthetic pathways allow the halophytes of Iran to inhabit various ecological conditions from very high (an EC over 200 ds/m) to low salinity soils, and from seawater to very dry, salty and gypsum soils. In spite of a vast expanse of salty soils, the halophytic communities in Iran are in danger because of extensive damming, which has led to the drying out of many rivers and wetlands, deformation of salt depressions and overgrazing (Akhani, 2006).

Alpine regions are above timber-line, which is not easy to recognize, since aridity is prominent in most regions. The alpine zone in Alborz lies between 3,000 and 4,000 m, the nival zone is above 4,000 m, locally varying by some hundred meters. The first evaluation of vascular flora shows that 682 species belonging to 193 genera and 39 families are known from the alpine zone of Iran. The alpine zone is commonly characterized by many species of hemicryptophytes and thorny cushions. Species numbers decline very strongly with increasing altitude. New data indicate a transitional situation of the Iranian mountains between Anatolia/Caucasus and the Hindu Kush, but with a strong own element with high endemism and remarkable relict species. Ca. 58% of the alpine flora of Iran are endemic and subendemic. The Zagros Mountains harbor high endemism which justifies considering this area as a separate floristic province. Based on the evaluation of published data from 682 known alpine species ca. 160 species have been known only by one record, 110 species by 2–3 records and 87 endemic species have been known only based on the type location. These plants need a strong conservation and protection management since the fragile ecosystems are often very restricted, small and very isolated, nonetheless, grazing and overgrazing are still common threats (Noroozi et al., 2008).

The Hyrcanian forests stretch from Talish in Republic of Azerbaijan and cover the northern slopes of the Alborz Mountains in North Iran, in Gilan, Mazandaran and Golestan provinces. The vegetation is composed mostly of deciduous forests. In the lower altitudes it is represented by a number of relict Arcto-Tertiary thermophilous species such as *Parrotia persica, Gleditsia caspica, Zelkova carpinifolia,* and *Pterocarya fraxinifolia.* The diversity of tree species increases at higher elevations where the subalpine forests and scrubs of low shrubs of the timber-line are replaced by alpine grasslands in the northern slopes and the Irano-Turanian thorn-cushion steppe at the exposed summits and southern slopes. So far, 3,234 species belonging to 856 genera and 148 families of vascular plants have been reported from the northern provinces of Iran and Talish in the Republic of Azerbaijan. Main vegetation types of the Hyrcanian forest zone include: (i) sand dune vegetation along the Caspian Sea coasts; (ii) C4-dominated grass communities on rocky outcrops; (iii) aquatic vegetation on wetlands; (iv) riverine and valley forests; (v) alluvial and lowland deciduous forests; (vi) submontane and montane deciduous forests; (vii) subalpine deciduous forests (*Quercus macranthera*); (viii) successional and transitional scrub and woodlands; (ix) *Cupressus sempervirens* and *Thuja orientalis* woodlands; (x) juniper woodlands; (xi) subalpine and alpine meadows; (xii) montane steppe dominated by xerophytic and thorn-cushion species; (xiii) rock cliff communities; (xiv) halophytic communities; (xv) *Artemisia spicigera* steppe and desert like dunes; (xvi) ruderal habitats; and (xvii) cultural landscapes and artificial forests. Evidence from studies on loess/palaeosol sequences, long-term Caspian Sea-level fluctuations, and peat/lake deposits in northern Iran give some indication of the climate and vegetation history of the south Caspian region. Based on these investigations, during the early-Pleistocene, at least parts of the area were covered by steppe-like vegetation and the climate was slightly warmer than today. It is also postulated that northern Iran was an extensive area of increased dust accumulation and loess formation during the Pleistocene glaciations, which is contemporaneous with and similar to major climatic changes as in SE Central Europe and Central Asia. These studies further suggest pronounced climate changes for the north of the country in which a dry and cool climate changed to moist and warm conditions during the Pleistocene glaciations. Similarly, a markedly dry period occurred during the early Holocene for the south Caspian area, parallel to the climatic optimum in Europe. Palynological studies have also shown intensified human impact on the lowland forest composition and structure of the area over the last centuries. The forests of the south Caspian area are severely degraded and deforested; in particular, in the alluvial lowlands where only small remnants exist. There are several protected areas in the Alborz Mountains and south Caspian area which suffers from mismanagement. Therefore, improving their protection quality and increasing their area or addition of new sites are crucial to guarantee conservation of this very important natural heritage of SW Asia (Akhani et al., 2010).

6.5 Species Animal Diversity

Preservation of the biodiversity of Iran would benefit from the selection and priority conservation of flagship species, especially carnivores, which can provide habitat connectivity because of their relatively large home ranges. Rivers are considered as a natural habitat for aquatic species, small animals and birds (Linnell et al., 2000; Ghoddousi et al., 2010).

The complex and varied climates, topography, geological formations and anthropological management of natural resources have led to a varied and unique biological diversity. In the Iranian ecosystems over 197 species of mammals, 541 species of birds, 241 species of reptiles (26 endemic species), 1,240 species of fishes and 23 species of amphibians, 358 butterflies' species and 25000 insects have been recorded (Figure 6.2). Iranian fish resources include 900 species of Persian Gulf and Sea of Oman (with 9 endemic sp.) and 180 species of Caspian Sea (with 10 endemic sp.) and living in inland and fresh waters (160 species of freshwater fishes with 15 endemic sp.) (IFNRCBD, 2015). Because of its large size and varied ecosystems, Iran is one of the most important countries in the Middle East and Western Asia for the conservation of biological diversity. Habitat diversity in Iran allows for a wide range of animals to inhibit in Iran (Sharifi, 2011).

Marine living resources play an important role in the food security of the country. Many of the aquatic resources are exclusive to the region, and therefore are of great importance in the context of biological diversity. Seafood protein comprises the largest proportion of protein consumption in the world. In Iran, fish consumption has increased in the last two decades, but it is still below the average global consumption, at about one-third of international average. The marine environment of Iran comprises two distinct water

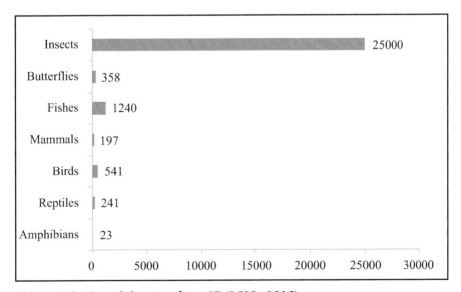

Figure 6.2 Faunal diversity of Iran (IFNRCBD, 2015).

bodies, namely, the Caspian to the north, and the Persian Gulf and the Sea of Oman to the south.

The decline of fish stocks has implications for food security in some coastal regions, as communities often rely on fish diet.

In the 1970s, increased coastal erosion in the delta of the Sefid Rud in the south Caspian was attributed to a reduction in the amount of silt reaching the delta following the construction of a large dam upstream in the Alborz Mountains. At some wetlands, especially in the Sistan Basin, heavy grazing of marsh vegetation by domestic livestock is inhibiting natural plant succession, and is causing permanent damage to aquatic plant communities as the highly palatable species are grazed to extinction. This degradation of wetland vegetation and the introduction of exotic fish species have had a detrimental effect on some of the native freshwater fishes.

In Iran, wetlands and marine ecosystems are especially important for the populations living in the north of Iran. They provide resources for food and livelihood of the people. Local communities have the right to hunt and trap non-protected bird and fish species.

The Caspian Sea is rich in biodiversity with about 315 species of zooplanktons, 1 species of mammal, 466 species of birds, 1394 species of invertebrates, and 2 species of reptiles.

There are 180 species of fish in the southern Caspian, which are commercially divided into sturgeons and bony fishes. The bony fishes are further divided into Kilka (small fish of the family Clupeidae) and other species. The main commercial species are as follows:

- **Sturgeons:** Beluga *Huso huso*, Russian sturgeon *Acipenser guldenstadti*, Iranian sturgeon *A. persicus*, and Sevruga *A. stellatus*. Iranian caviar, a famous and exclusive product worldwide, is produced by these species.
- **Kilkas:** *Clupeonella delicatula, C. engrauliformis, C. grimmi.*
- **Other bony fishes:** Kutum *Rutilus frisii kutum*, Mullets *Mugil auratus* and *M. saliens*, Carp *Cyprinus carpio*, Bream *Abramis brama*, Pikeperch *Lucioperca lucioperca*, Roach *Rutilus rutilus,* and Salmon *Salmo trutta caspius*.

A large variety of animal species are observed in the coastal ecosystems. Mangrove forests are unique coastal wetlands, important fish as habitats. Marine turtles, many on the endangered list, live in these ecosystems. The following marine turtles have been observed in Iranian waters: Green Turtle (*Chelonia mydas*), Leatherback Turtle (*Dermochelys coriacea*), Olive Ridley Turtle (*Lepidochelys olivacea*), Loggerhead Turtle (*Caretta caretta*),

Hawksbill Turtle (*Eretmochelys imbricata*), and Black Turtle (*Chelonia aqaziz*).

Different species of marine mammals are observed in the southern waters of Iran, which some of them include blue whale (*Balaenoptera musculus*), fin whale (*Balaenoptera physalus*), sperm whale (*Physeter musculus*), humpback whale (*Megaptera novaeangliae*), long-beaked common dolphin (*Delphinus capensis*), black finless porpoise (*Neophocaena phocaenoides*), and dugong (*Dugong dugon*).

There is no convincing report on the existence of the two big mammals, that is, Persian lion and Caspian tiger from their natural and endemic habitats during the last 50 years. Asiatic cheetah, Persian wild ass, Persian squirrel and three species of sea turtles are among critically endangered species and the population has rapidly declined during the past years (IFN-RCBD, 2010).

Birds are considered important markers of biodiversity; their abundance in an area is the indicator of diversity in other flora and fauna as well. As such, the study of diversity and abundance of birds in natural environments is considered an important line of enquiry for ecologists (Zarei et al., 2011). Although much of Iran is extremely dry; while deserts cover 33% and forests 12% of the country, the numerous wetlands stand out. The country possesses a great diversity of wetland ecosystems, most of which can be grouped into six major systems; large populations of migratory birds winter at these wetlands or use them on their way to and from wintering areas in Africa or the Indian sub-continent (including that of the Siberian Crane, flamingos, and pelicans, all listed as global heritage). The wetlands of Iran constitute vital staging and wintering areas for millions of migratory waterfowl using the West Siberian-Caspian-East African and Central Siberian-Indus South Asian flyways, and support large breeding populations of many species. Several million waterfowl utilize the wetlands as wintering habitat, while perhaps as many birds again use the wetlands as staging areas on their way to and from wintering areas further to the southwest or southeast. The wetlands of Iran are very important for seven species of birds listed as globally threatened in the 1994 IUCN List of Threatened Animals, namely Pygmy Cormorant (*Phalacrocorax pygmaeus*), Dalmatian Pelican (*Pelecanus crispus*), Lesser White-fronted Goose (*Anser erythropus*), Marbled Teal (*Marmaronetta angustirostris*), White-headed Duck (*Oxyura leucocephala*), White-tailed Eagle (*Haliaeetus albicilla*), and Siberian Crane (*Grus leucogeranus*). A further four threatened species formerly occurred in significant numbers, but are now only scarce passage migrants or vagrants, namely Red-breasted Goose

(*Branta ruficollis*), Pallas' Sea-Eagle (*Haliaeetus leucoryphus*), Sociable Plover (*Chettusia gregaria*) and Slender-billed Curlew (*Numenius tenuirostris*) (IFNRCBD, 2015).

6.6   Species Microbe Diversity

6.6.1   Urmia Lake

Halophiles are in all three domains of life: Archaea, Bacteria, and Eucarya. Halophilic microorganisms, in addition to forming a major part of life biodiversity, can have many biotechnological applications (Irannejad et al., 2015).

Urmia Lake harbors a diverse group of bacterial species. Two pathogenic bacteria *Clostridium perfringens* and *Enterococcus faecalis* have been identified in the lake water and particularly in the estuary sediments. As these bacteria constitute the natural flora of the human digestive tract, their presence in the lake water and sediments suggests that they have originated from the inflow of urban wastewater into Urmia Lake. In Urmia Lake, the mud contains green sulphur-bacteria, purple sulphur-bacteria, and ferrobacteria. Halophilic archaebacteria can usually synthesize red and pink pigments in response to environmental stress. During the summer of 2008, the water of Urmia Lake around Kaboudan Island changed from blue, its normal color, to red. This was the first report of this event that may be attributed to the bloom of *Archaebacteria* or of *Dunaliella* or both (Asem et al., 2014). Studied strains belonged to three genera: *Halomonas* 50% (including *H. andesensis* (12.5%), *H. gomseomensis* (12.5%), *H. hydrothermalis* (12.5%), *H. boliviensis* (6.25%) and *H. janggokensis* (6.25%)), *Salinivibrio* 25% (including *S. costicola* subsp. *alcaliphilus* (18.75%) and *S. sharmensis* (6.25%)) and *Idiomarina* 25% (including *I. loihiensis* (25%)) (Irannejad et al., 2015).

Fungi are a large group of eukaryotic organisms with worldwide distribution, inhabiting a diverse range of extreme habitats from deserts to hypersaline environments. Investigations of the distribution of fungi in extremely hypersaline environments are rare. Numerous fungi have been identified from *Artemia* cysts and hypersaline water. Fungal contamination of *Artemia* cysts is possibly one of the important reasons for their reduced hatchability, and therefore one of the main problems in cyst processing and culture. 12 fungi have been reported in some halophytes and glycophytes of islands and also from the western shores of Urmia Lake (Asem et al., 2014).

6.6.2 Hot Spring

Hot springs have been a subject of intense discussion for biologists in the last decades. Thermophilic Cyanobacteria are scientifically valuable for their analogy to the ancient life forms on earth and also as a source of thermostable biocompounds. Exploration of their biodiversity is an important step towards these goals. In total 43 species belonging to 20 genera, 11 families and 5 orders of the planktonic Cyanobacteria were identified from hot spring in Iran. Among these taxa *Chroococcus* with 7, *Oscillatoria* and *Phormidium* with 5 species found to be more predominant and noticeable among other genera (Heidari et al., 2013).

6.7 Genetic Diversity of Crop and Cultivated Plants

Knowledge of the genetic diversity in crops is important in crop breeding programs and in the conservation of primary gene pool (Beheshtizadeh et al., 2013). The findings of genetic diversity can be used in breeding programs for increasing the genetic variation in base populations by crossing cultivars with a high level of genetic distance as well as for the introgression of exotic germplasm (Faramarzi et al., 2014).

Iran is well known as one of the world's major centers of origin and diversity for many important crop plants. Activities related to conservation and utilization of plant genetic resources in Iran started nearly 70 years ago, mostly with cereals. There are considerable land areas under organic management but only a few registered organic farms (2%) (Ansari, 2000; Calabrese et al., 2008).

Approximately 33 million ha of land area has typical to good capacity for cultivation but just 18.5 million ha are cultivated of which 8.5 million ha are irrigated and 10 million ha are rain-fed. Major soil types of Iran are xerosols, arenosols, regosols, solonchaks, and lithosols. In other terms, roughly one-third of Iran's total surface area is suited for farmland, but because of soil type and lack of adequate water distribution, some of it is not under cultivation. Only 12% of the total land area is under cultivation (field crops and fruit trees) but less than half of the cultivated area is irrigated; the rest is devoted to dry farming. Some 92% of agro products depend on irrigation.

About half of the total land area (55%) is considered rangeland, used for grazing and small fodder production. Most of the grazing is done on mostly semi-dry rangeland in mountain areas and on areas surrounding the large deserts ("Dasht's") of Central Iran.

Approximately 51% of the total land area of Iran is not suited for typical agricultural activities and is characterized as follows:

Approximately 33% of the country is covered by deserts, salt flats ("Kavirs") and bare-rock mountains, not suited for agricultural purposes. An additional 11% of Iran's total surface is covered by woodlands. And 7% is covered by cities, towns, villages, industrial areas, and roads.

Iranian agriculture is thousands of years old and this reflects the length of time during which soil and water resources of the country have been utilized for crop production. Both systems of irrigated and rainfed farming (dry farming) are practiced in different parts of the country while the area devoted to each system varies considerably depending on the annual precipitation and agroclimatical conditions. Rainfed agriculture and dry farming are most successful in the north, west and northwestern of Iran. In other parts of the country, dry farming is also practiced in hilly areas, but the yield is very low. In the central plateau, as well as the southern plains and the southern coastal areas of Iran, crop production is mostly possible only under irrigation. This is because of low rainfall and high evaporation rates. In some low lying plains of the central plateau, the annual rainfall is about 50 mm while the annual evaporation may exceed 4000 mm.

In 2008, about 12.34 million hectares were under annual crops, out of which 59.21% was under cereals; wheat, rice, corn, and barley, producing 59.1%, 16.2%, 13.2%, and 11.4% of the cereals, respectively. All agricultural lands are owned by the people, except some portions of which are used by the government sector for particular (research, mechanized production, etc.) purposes.

About 90% of the irrigated land is under annual crops (The remaining 10% is used for the production of perennial crops, mostly fruit trees). In rained areas, annual crops constitute about 98% of the total production. During dry years, about 8 to 12% of the total production comes from dry land areas. However, in wet years this figure can rise to 35%.

Traditional small-scale farming was the main structure of farming communities for centuries and this has caused a tremendous accumulation of indigenous knowledge in farming practices and food production. Therefore, natural farming used to be a widespread practice not more than half a century ago and is still appreciated to some extent by the farmers. Although the traditional systems of food production are almost organic both in technical and social terms, however, they are in operation in small-scale farming or in remote areas. Production of some crops such as date, saffron, pistachio, and pomegranate is generally considered organic.

Extra water was used to irrigate the land around the water hole to culti-vate alfalfa as a supplement fodder for winters (Zakeri, 2011).

The cultivated spinach (*Spinacia oleracea* L.) belongs to the family Chenopodiaceae. Spinach has become an important vegetable crop in most regions of the world and remarkable changes in production amounts have occurred in the past decades due to demand increase in many countries. It is widely grown as a leafy vegetable both for fresh consumption and for indus-trial processing in many countries. Spinach is an excellent source of vitamin B and C in the human diet. It is a cool seasonal vegetable as an important source of minerals that produces a rosette during the vegetative stage. The production of spinach has been reported more than 105,000 tons in Iran in 2013 (Fahim et al., 2013).

A very interesting historical event in the domestication of crop plants happened in recent years in southern Iran. Traditionally, the young leaves of *Suaeda aegyptiaca* (their local name in Khuzestan is Googaleh and Kakol in Bushehr) have been used as a wild vegetable in many parts of the southern provinces. Recently, the people in Bushehr Province (near Borazjan) began cultivating *S. aegyptiaca* as a new vegetable crop, using saline water for irrigation. As a consequence, it is being sold for food by farmers in the local bazaars.

During the 1990's many studies have been aimed at evaluating *Salicor-nia* as a new oil seed crop plant (Glenn et al., 1991, 1998; Glenn and Watson, 1993). *S. persica* (Akhani, 2003) is one of the good candidates for use as a new crop plant, not only because of its high biomass (it grows up to one meter high in favorable conditions), but also due to the fact that it is nor-mally eaten by goats. The cultivation of halophytes and psammohalophytes from native and exotic species is a well funded and extensively expanded program in Iran. The two *Haloxylon* species (*Haloxylon ammodendron* for cultivation in saline and *H. persicum* for cultivation in sand dunes) and the exotic New World species *Atriplex canescens*, have been cultivated in many parts of Iran by the Forests and Rangeland Organization. Many of these pro-grams are not based on prior ecological studies. Therefore, in some places where the species were accidentally suitable for cultivation, the results have been good, but for the most part such activities have not been successful. Many *Haloxylon* plantations are being destroyed both because of unsuitable ecological habitats and due to diseases (Akhani, 2006).

The south Caspian area is an important center for the domestication of cultivated trees and shrubs (Khoshbakht and Hammer, 2006). The agricul-tural area of the south Caspian zone plays an important role in the agronomy of Iran. The very rich soils of the alluvial South Caspian lowlands provide extensive agronomic activities including rice, wheat, colza, citrus fruits, kiwi, peach, strawberry and tea cultivations. Olive is a major product of the local Mediterranean bioclimatic zones of the area which is concentrated mostly in Sefid-Rud valley and Gorgan. The history of re- and afforestation activities goes back to 1952. Thereafter, many exotic trees along with a few native species have been planted in the area either for commercial or ornamental purposes. Among them are: *Pinus brutia* subsp. *eldarica, Eucalyptus* spp., *Robinia pseudoacacia, Cryptomeria japonica, Taxodium distichum, Cupressus sempervirens, Populus nigra, Ailanthus altissima,* and *Picea abies* (Rezaiee, 2000; Bahri, 2004; Akhani et al., 2010).

Canola (*Brassica napus* L.) is the most important source of edible oil and the second most important oilseed crop in the international oilseed market after soybean. The interest of canola production has increased significantly worldwide in the last few decades. The seeds of modern varieties typically contain 40–45% oil, which provides a raw material for many other products ranging from rapeseed methyl ester (biodiesel) to industrial lubricants and hydraulic oils for detergent and soap production and biodegradable plastics (Sharafi et al., 2015). Canola has been widely grown in Iran in recent years for oil production and rarely livestock feed. High seed yield (over two tons per hectare) and high oil content (approximately 40%), in addition to low water needs in dry regions of Iran, have made it one of the most important crop plants that provide the oil needs of the country. Due to the economic importance of oilseed rape, several *B. napus* cultivars from diverse sources in the world have been introduced in Iran during the last decade and at present. Canola cultivars appear to be best adapted to the conditions of Iran. However, some cultivars are less tolerant to environmental conditions (Abtahi and Arzani, 2013).

Triticum aestivum L. and *T. turgidum* L. are important cereal grains that are mainly used for human consumption. Durum is the primary wheat for pasta and semolina production and the second most cultivated wheat after bread wheat (*T. aestivum*). Due to modern breeding, it has been suggested that genetic diversity in wheat and other crop species has been increasingly narrowed. A narrow genetic base is a problem in breeding for seed yield and for adaptation to biotic and abiotic stresses. Iran is one of the leading centers for domestication of cereals and according to genetic and archaeological evidence (Seyedimoradi et al., 2016).

Agropyron with high forage yield and wide stability in different climate especially drought and Salt tolerance is one of the most important forage crops. Since, there is high variation within and among different species of

Agropyron, selection response for improving important traits is high. *Agropyron* has been applied in wide hybridization specially to transfer alien genes into cultivated wheat. *A. trichophorum* is one of the most important pasture plants in Iran. Different species of this plant are growing in most of pastures. It is a perennial herbaceous plant with 19 different species distributed in Iran (Shirvani et al., 2013).

Maize (*Zea mays* L.) is a widely grown crop in most parts of the world due to its adaptability and productivity. It is one of the most important crops in Iran, accounting for over 4.13% of its total cropped area. Maize breeding faces unique challenges resulting from the narrow genetic background of commercial cultivars (Shiria et al., 2014).

Chickpea is an important self-pollinated grain legume crop, grown mainly in West Asia, North Africa, and the Indian subcontinent, where it is a basic component of the human diet. Iran is considered as a major center of diversity for chickpea. This species is an important food legume in several countries including Algeria, Ethiopia, India, Iran, Mexico, Morocco, Myanmar, Pakistan, Spain, Syria, Tanzania, Tunisia, and Turkey (Naghavi et al., 2012). Assessment of the extent of genetic variability within chickpea is fundamental for chickpea breeding and conservation of genetic resources, and is particularly useful as a general guide in the choice of parents for breeding hybrids (Talebi et al., 2008).

Safflower (*Carthamus tinctorius* L.) is one of the oldest oilseed crops and is widely grown under hot and dry climate of the Middle East. Seven diversity centers for safflower germplasm evolution as follows: the Far East, India-Pakistan, the Middle East, Egypt, Sudan, Ethiopia, and Europe. Safflower has some agronomic advantages such as drought resistance and adaptation to arid and semi-arid climatic conditions. Iran has one of the richest germplasm sources of safflower. For instance, out of the 2042 safflower genotypes deposited at the Western Regional Plant Introduction Station, Pullman, WA, USA, 199 are from Iran (Salamati et al., 2011).

Apple (*Malus × domestica* Borkh.) is one of the most widespread and popular fruit trees in the world. About 59 species and 7500 cultivars were identified in all over the world. Apple is an ancient fruit crop in Iran and there is an extremely abundant germplasm resource for it. The point in apple breeding is usually to make better-quality apples. There is a high level of genetic diversity in Iran's cultivated apple due to close distance to apple origin in Central Asia (Farrokhi et al., 2011).

Tobacco (*Nicotiana tabacum* L.) is a member of the nightshade (Solanaceae) family, the largest and most diverse family within the angiosperms, for it harbors 3,000–4,000 species, of which a considerable number are of major economic importance as crop, vegetable or ornamental species throughout the world. The genus *Nicotiana* consists of nearly 100 species mainly from tropical and subtropical America; almost all commercial tobacco belongs to the *N. tabacum* species. Several types of *N. tabacum* are defined to a large extent by region of production, projected use in manufacturing, method of curing (flue-, air-, sun- and fire-cured tobacco), as well as their morphological and biochemical characteristics (i.e., aromatic fire-cured, bright leaf tobacco, Burley tobacco, Turkish or oriental tobacco). Turkish or oriental tobacco is a sun-cured, highly aromatic, small-leafed type which is grown in Iran, Turkey, Greece, Bulgaria, Lebanon and the Republic of Macedonia. It can be grown on low fertility soils (Darvishzadeha and Hatami Malekib, 2012).

By 1975 potato (*Solanum tuberosum*) was planted in nearly all provinces of Iran and ranked nationally as the third most important crop, after wheat and rice. Although Iran is clearly one of the largest potato producers in the Middle East, reliable data on recent production and on the history of former and current varieties are rare or non-existent and production statistics published by different sources vary largely. According to the Iranian Ministry of Agriculture, the total potato production in the 2004/2005 and 2005/ 2006 growing seasons was 4,830,124 and 4,218,522 tons and the mean potato yield was 25.76 and 26.20 t/ha, respectively. Before 1986, no basic seed potato was produced in Iran and most farmers used to either save a portion of their harvest, buy uncertified seed tubers at local markets or trade seed potatoes with other farmers to provide seed for the following crop season. A program was established in the mid-1970 to multiply and distribute certified seed imported from Europe. Thereafter, despite occasional interruptions, almost all potatoes cultivated commercially and used in breeding and screening programs in Iran were introduced from European countries, particularly from the Netherlands and Germany. In recent years, some national research centers and producers began to introduce some North American and compare them with the traditional European varieties, to replace degenerating European by novel American varieties. The recently introduced North American varieties had been selected as commercially more profitable in their countries of origin. Actually, almost all studied varieties are globally known as outstanding varieties (Esfahani et al., 2009).

Citrus plants are cultivated in the North and South of Iran. Little is known about the genetic variability of Iranian cultivated *Citrus* germplasm collection. *Citrus* cultivated since ancient times in its center of origin in south eastern Asia, *Citrus* production has spread over the centuries into most areas that have a suitable climate. Today *Citrus* is one of the most widely cultivated fruit in the world, and most major production areas are far removed from the original areas. Different *Citrus* species are widely grown in more than 50 countries in the world. World production is increasing and has reached 70 million tons, according to FAO (Jannati et al., 2009).

6.8 Endangered Plants

The flora of Iran is a surprisingly rich one, comprising some 7,115 species of seed plants and 173 families (Rechinger, 1963–2015). 1,727 taxa are classed as endemics, belonging to 54 families. Phytogeographically, the large, central, Irano-Turanian region is by far the richest in flora and contains nearly 85% of the Iranian endemics. The much smaller, wetter and more northerly Euxino-Hyrcanian Province of the Euro-Siberian region, bordering the Caspian Sea, also contains a high number of endemics per unit area, giving the forest types a distinctive character. By contrast, the very dry and hot Saharo-Sindian region, adjoining the Persian Gulf, has a more limited flora and is relatively unimportant as a center for speciation.

Four categories of plants are recognized: endangered, vulnerable, lower-risk and data-deficient. It is estimated that 32% of the vulnerable and endangered species are grazed or harvested for medicinal or other purposes.

There being no fewer than 20 endemic monotypic genera in Iran, mainly in the Cruciferae (Brassicaceae) and the Umbelliferae (Apiaceae). Although about 20% of the Iranian flora are annuals (a high proportion in terms of the world flora) only 6.7% of the endemics are annuals, and there are only seven endemic tree species. The vast bulk of the endemics are herbaceous perennials, considered a relatively vulnerable life form. Outstanding in the flora is the genus *Astragalus*, with about 1000 species in the *Flora Iranica* region (Table 6.3) (Podlech, 1999). Almost 400 species found mostly in mountainous areas in the Irano-Turanian region, arguably the major center in the world for speciation of this huge genus. Other genera with many endemics in Iran include *Nepeta, Acantholimon* and *Cousinia*, this composite being represented by more than 200 spiny species, about three-quarters of them endemics, found mainly in the dry parts of Iran. Interestingly, plants familiar in the British Isles, listed here in the lower-risk category, include *Daphne mezereum*, several orchids (e.g., *Anacamptis pyramidalis, Epipactis palustris*), a number of grasses (e.g., *Milium effusum, Phleum bertolonii*), and the aquatic *Zostera noltii*, classed as vulnerable (Jalili and Jamzad, 1999; Willis, 2001). The lowland relict forests of the Caspian plain are certainly the most intensively suffering parts of the Hyrcanian forest. These forests host the main habitats for some

Table 6.3 Prolific Plant Genera in Iran

Dicotyledons		Monocotyledons	
Genus	**No. of species**	**Genus**	**No. of species**
Astragalus	1000	*Allium*	75
Cousinia	210		
Silene	100		
Acantholimon	83		
Centaurea	74		
Euphorbia	70		
Nepeta	67		
Veronica	58		
Taraxacum	58		

(Yousefi, 2007)

of the most endangered endemic plants such as *Gleditsia caspica* (Akhani et al., 2010).

Following is the list of 28 endangered halophytes in Iran: *Asparagus lycaonicus* P. H. Davis, *Piptoptera turkestana* Bunge, *Chenopodium chenopodioides* (L.) Aellen, *Psylliostachys beluchestanica*, *Diaphanoptera stenocalycina* Rech. f. & Schiman-Czeika, *Reaumuria fruticosa*, *Halanthium alaeflavum* Assadi, *Salsola abarghuensis* Assadi, *Halanthium mamamense* Bunge, *Salsola zehzadii* Akhani, *Halanthium purpureum* (Moq.) Bunge, *Suaeda dendroides* (C. A. Mey.) Moq., *Halimocnemis occulta* (Bunge) Hedge, *Suaeda linifolia* Pall., *Halopeplis perfoliata* (Forssk.) Bunge ex Aschers, *Suaeda monoica* J. F. Gmelin, *Helminthotheca echioides* (L.) Holub, *Suaeda physophora* Pall., *Hypericopsis persica* Boiss., *Tetradiclis tenella* (Ehrenbg.) Litw., *Malacocarpus crithmifolius* (Retz.) Fisch. & C. A. Mey., *Thellungiella parvula* (Schrenk) Al-Shehbaz & S. L. O'Kane, *Microcnemum coralloides* (Loscos & Pardo) Buen, *Thesium compressum* Boiss. & Heldr., *Nitraria retusa* (Forssk.) Aschers, *Trachyspermum* sp. (Akhani, 2006).

These 28 endangered halophytes represent 8% of all halophytic and salt tolerant plants of Iran. The main threat to these species is the habitat degradation caused either by changing hydrologic conditions or by overgrazing.

6.9 Endangered Animals

With an area of 1,648,000 km² Iran is a vast country with a diversity of landscapes, fauna (1,674 species of vertebrates) (Zehzad et al., 2002; Firouz,

2005; Darvishsefat, 2006). Iran has conducted research activities, enhanced legal protection of the habitats, increased public awareness and started captive breeding. Also, national targets relating to the endangered species; "supporting *in situ* and *ex situ* biodiversity emphasizing on endangered animal species" and "conservation of endangered species," have been included in National Biodiversity Strategy and Action Plan (Ansari, 2000; Calabrese et al., 2008).

Endangered species like Persian fallow deer (*Dama mesopotamica*), Goitered gazelle (*Gazella subgutturosa*), Jeeber gazelle (*Gazella dorcas*), Red deer (*Cervus elaphus*), Persian wild ass (*Equus hemionus*), Asiatic cheetah (*Acinonyx jubatus venaticus*) and Siberian Crane. Yellow deer, Persian wild ass, Asiatic cheetah and the crocodile are some of the endangered species inhabiting in arid region of the country.

The Asiatic cheetah (*Acinonyx jubatus venaticus*), also known as the Iranian cheetah, is a critically endangered cheetah surviving today only in Iran. It used to occur in India as well, where it is locally extinct (IFNRCBD, 2015). The endangered species of Asiatic cheetah that generally occurs in the east and central range of the country is faced by many threats, such as habitat disturbance and degradation, decline in prey and illegal hunting (IFNRCBD, 2010).

Since 2008 only one Siberian Crane arrived annually (excluding winter of 2009/10) in the wintering grounds of the Western Asian flock on the Caspian littoral, in Fereidoonkenar, Mazandaran Province, Iran (Vuosalo, 2013). The Siberian crane is a critically endangered species of crane. Three distinct populations of this magnificent bird wintered in Iran, India, and China. Of these, the Indian-central-group is extinct and none have been sighted in recent years. The eastern group wintering in China is relatively stable with approximately 3,000 birds remaining. The western population spends its winters in Iran, but of these birds only one has returned after two years of absence and has been nicknamed Omid-Hope-by the media (Yousefi, 2011).

Choghakhor supports 1% of the local population of Ferruginous Duck and two endangered species namely White-headed Duck and Marbled Teal have been observed in this wetland in winter. The area supports a large population of Eurasian Coot, especially during migration season. Choghakhor is particularly important for wintering birds, including White Egret and Mallard, and is also a natural summer habitat for White Stork (Zarei et al., 2011).

It is hopefully aimed that the knowledge could evaluate the efficacy of the current network of protected areas and national parks in Iran to save the endangered Persian leopards. Beside the leopard, several other endangered species occur within the leopards' habitats (e.g., Asiatic cheetah, Eurasian lynx, brown bear, caracal, Asiatic wild ass, etc.) which leopard's umbrella role is a guarantee for saving their habitats (Farhadinia et al., 2011).

The Persian leopard *Panthera pardus saxicolor* is a critically endangered flagship species of the Caucasus (Farhadinia et al., 2011) and, with the extinction of the Asiatic lion *Panthera leo persica* and Caspian tiger *Panthera tigris virgata*, is the only extant large felid in Iran. This subspecies also occurs in neighboring countries its stronghold is in Iran. After the disappearance of Asiatic lion and Caspian tiger prior to 1970s, the Persian leopard is the largest cat in western Asia with a global status category of Endangered in the IUCN red list of threatened species (IUCN, 2010). As an endangered subspecies, the Persian leopard was once abundant across most mountainous and forest habitats of Iran, but it is now one of the rarest species of carnivores in the country.

The leopard population in Iran is estimated to be 550–850 and its range extends over 850,000 km^2 wherever sufficient prey and protected habitat is present (Kiabi et al., 2002; Firouz, 2005). It is essential to count and determine the population structure of this predator so as to verify its status, monitor population viability, and identify the effects of natural and human factors on the species and to determine the impact of the decline of the leopard on the ecosystem.

As leopards are wide-ranging their occupancy, which is that part of the range (extent of occurrence) actually inhabited and used by the species, must be sufficiently large to full the species' ecological requirements. To assess the spatial distribution and viability of the species it is important to estimate population occupancy, study the relationship of the species with habitat fragmentation, examine the effects of study design on occupancy estimation, and to identify sites visited by leopards.

Bamu National Park is one of the most important habitats for the leopard in Iran. The Park has a long history of conservation, access for research is relatively easy compared to other leopard habitats in Iran, and sightings of leopards in the area are relatively common. However, fragmentation from human encroachment is ongoing and there is a high rate of poaching in the area (Ghoddousi et al., 2010).

6.10 Protected Areas

Relatively vast areas of high protection significance are selected with the purpose of preserving and restoring plants sites and animals habitats. Protected areas are appropriate places for the implementation of educational and research plans. Tourism and economic utilizations in proportion with each area under the comprehensive management plan of the area are allowed (Madjnoonian, 1993).

6.10.1   The History of Environmental Protection in Iran

The legislation of the Protection Bill and the establishment of the Iranian Center for Hunting in 1956 are considered as the first documented actions taken toward protecting the Iranian wildlife population and diversity. The main duties of the Center as an independent organization were defined to be the protection of game and monitoring the enforcement of the relevant regulations. These foremost experiences of the enforcement of the Bill revealed that the protection of the wildlife could only be achieved through protecting their habitats. This led to the formation of a new governmental organization in 1967 entitled the Hunting and Fishing Organization. This organization superseded the Center for Hunting in the same year serving as an independent governmental organization under the supervision of the Supreme Council of Hunting and Fishing. Laws pertaining to this organization, provisions were made to allow for the allocation of parts of the country for national parks (called wildlife parks at the time) and protected areas with predefined definitions. Therefore, the year 1967 marked the pioneer attempts for the foundation of the Iranian protected areas, 95 years after the establishment of the first national park in the world (Yellowstone National Park in the USA) and 19 years after the foundation of the World Conservation Union (IUCN, 2010). In this year, the proposal for the foundation of three national parks and 15 protected areas was approved by the Supreme Council of Hunting and Fishing as the first series of protected areas in Iran.

Two new classes of protected areas namely, wildlife refuge and national natural monument also came to existence. Based on what mentioned above, since 1974, the Iranian protected areas have been classified into four groups of national parks, national natural monuments, wildlife refuges and protected areas, according to the official definitions and criteria of the system of the classification of areas. In addition to these classes which conform to the IUCN classes and are known as the "four areas," other areas named no-hunt areas, wetlands, international wetlands and biosphere reserves are under management. While international wetlands and biosphere reserves are among the defined classes of international environmental protection societies, the no-hunt area is considered a national innovation. The present Atlas of protected areas of Iran contains only the national parks, national natural monument, wildlife refuges and protected areas (Darvishsefat et al., 2008).

6.10.2 The Development of the Iranian Protected Areas

The general acceptance of the concept of protected areas in Iran and the necessity to allocate areas to them which was materialized by the foundation of three national parks and 15 designation of 160 protected areas of the total area of 11824599 ha until 2005, covering 7.17% of the entire country area indicates an annual increase rate of 4.2 areas and 311174 ha (Darvishsefat et al., 2008).

In order to materialize one of the legislations of the Bali Conference in 1982 recommending that 10% of the total area of the world's forests be allocated to protected areas, an agreement was reached between the Forests, Rangelands and Watershed Management Organization and the Department of the Environment in 1999 to designate parts of the forests of the country as protected areas. According to this agreement, 22 areas in different floristic zones with the total area of 674181 ha were added to the Iranian protected areas network. This led to a 6.6% increase in the area of the Iranian protected areas.

Another action was to designate four Iranian summits of Damavand, Alamkuh, Sabalan, and Taftan all having unique characteristics as national natural monuments on the occasion of the International Year of Mountains in 2002. Although the selection of these summits as national natural monuments only marginally increased the protected areas of Iran due to their small areas, it is by far a clear sign indicating the awareness required to protect national reserves (Darvishsefat et al., 2008).

The country's protected areas have consistently been increased and from 194 sites in 2010 have established 274 sites in 2014, with 29 national parks, 44 wildlife refuges, 35 national natural monuments, and 166 protected areas (10.3% of the land area) (Table 6.4). Now the protected area of the country

Table 6.4 Protected Areas of Iran

Title	Number	Area in Hectares	Percent of the Land Area
Natural Parks	29	2,001,624	1.2
National Natural Monuments	35	372,253	0.2
Wildlife Refuges	44	5,595,746	3.4
Protected Areas	166	9,116,779	5.5
Number and Total Area	274	17,086,402	10.3

(adapted from IFNRCBD, 2015)

has increased to 184 sites. 34 are designated under international and regional agreements or conventions. 150 are national-level protected areas which IUCN Management Categories have been reported for 126 of these records and the remaining 24 features are recorded as Not Reported. Of the 184 protected areas, 6 (3%) include a marine component. These protected areas may be either partially or completely within the marine environment. About 81% of the data have boundaries in polygon format, with the remaining 19% depicted in point format (i.e., a single latitude and longitude point for the protected area); 24 are Ramsar Wetlands of International Importance; 10 are UNESCO Man and the Biosphere Reserves. To these areas, we should add the wetlands and forest reserves which also are managed as protected areas. There are more measures taken for the development and enhancement of aquatic and terrestrial areas and this goal is being done, but the priority for Environmental Protection Department is to improve the quality and reduce the destructed areas and also the restoration of existing protected areas. The protected area networks in Iran are dispersed and have spots. Some action has taken place for identifying the corridors in deserts and in the northeast of Iran (IFNRCBD, 2015).

6.10.3 Examples of Protected Areas of Iran

Some about 5,552,000 hectares are protected as hunting prohibited areas, and there are about 200 protected forest zones. Special projects are running to protect aquatic ecosystem (i.e., wetland; bog, bay, lake, river, estuarine etc.), and protection programs of marine ecosystems and species are under consideration (Ghoddousi et al., 2010). These areas have been selected on the basis of being representatives of the world's ecological regions. Conservation of threatened species and their habitats are one of the main concerns of the country (Ansari, 2000; Calabrese et al., 2008).

Many of the protected areas of Iran include high altitude mountains such as Central Alborz (Tehran and Mazandaran Provinces), Oshtorankuh area (Lorestan), Haftadgholeh (Arak), Korkhod and Golgulsarani (Khorassan) and Dena (Fars) protected areas and the national parks of Golestan (Golestan and Khorassan provinces) and Lar (Tehran) (Noroozi et al., 2008).

With regard to its ecological significance, unique biodiversity and the presence of indigenous communities, Urmia Lake has been recognized as a Protected Field since 1967 and was designated as a National Park in 1976 as one of 59 biosphere reserves by UNESCO. In 1975, it was also registered in the Ramsar Convention on Wetlands as a wetland of international importance (Asem et al., 2014).

6.11 Concluding Remarks

Conservation measures on biodiversity and ecosystems in some sectors of Iran have made progress but there are still many challenges ahead. In general, it appears that the rate of decline in many species and ecosystems has increased during the past few years. The reasons for the current major challenges on biodiversity conservation are as follows:

1. Ecosystems are disturbed by the dramatic increase in human activities and naturally occurring events. As a result of the rapid population growth and more pressure on utilization of natural resources, the area and biodiversity of different ecosystems including forest, rangeland, wetland, mountain, and marine and coastal ecosystems are decreasing.
2. Integrating biodiversity conservation into other sectors and integrated management of biodiversity in Iran is a new concept. Because of being new and challenging, it is very important to consider the time and the process of integrated management. It should not be expected to see the final results and achievements very soon but it is very important to plan the right process and move forward gradually.
3. A great deal of progress has to be made in the areas of education and public awareness in order to achieve an integrated biodiversity management at the national, provincial and local levels. Iran is a huge country with very diverse people and different cultures, livelihood and behaviors. With this socio-economic complexity, community participation and stakeholders' involvement may be very welcome in some areas but there may be resistance to it at other sites.
4. Administrative and technical capacity at different levels as well as funding policies of government to conservation activities need to be improved. To date the government has provided funding to support biodiversity conservation of the nation; however, this is not enough to address all the challenges on the way (IFNRCBD, 2010).

Keywords

- biodiversity
- conservation
- protected areas
- Iran

References

Abtahi, M., & Arzani, A., (2013). Molecular and morphological assessment of genetic variability induced by gamma radiation in canola. *J. Plant Molecular Breeding, 1*(2), 69–84.

Akhani, H., (1998). Plant biodiversity of Golestan national park. *Stapfia, 53*, 1–411.

Akhani, H., (2006). Biodiversity of halophytic and Sabkha ecosystems in Iran. In: *Tasks for Vegetation Science-42*: *Sabkha Ecosystems: West and Central Asia*. Springer, *vol. 2*, pp. 71–78.

Akhani, H., (2003). *Salicornia persica* Akhani (Chenopodiaceae), A remarkable new species from Central Iran. *Linzer Biol. Beitr.*, *35*, 607–612.

Akhani, H., Djamali, M., & Ramezani, E., (2010). Plant biodiversity of hyrcanian relict forests, N Iran: An overview of the flora, vegetation, paleoecology and conservation. *Pak. J. Bot.*, *42*, 231–258.

Ansari, M. T., (2000). *Environmental Policies in Iran*. Samt Publication.

Asem, A., Eimanifer, A., Djamali, M., De los Rios, P., & Wink, M., (2014). Biodiversity of the hypersaline Urmia lake National Park (NW Iran). *Diversity*, *6*, 102–132.

Bahri, Y. G., (2004). Results of loblolly pine (*Pinus taeda* L.) and Caucasian alder (*Alnus subcordata* C. A. M.) silvicultural operations in the Caspian low land regions of Iran. *Pajouhesh va Sazandegi*, *17*(2), 2–9.

Beheshti, Z. H., Ressale, A., & Ghandi, A., (2013). Genetic variability assessment in bread wheat (*Triticum aestivum* L.) cultivars using multivariate statistical analysis. *Intern. J. Farming and Allied Sci.*, *2*(16), 520–523.

Breckle, S. W., (2002). *Walter's Vegetation of the Earth*. The ecological systems of the geobiosphere. Springer, Heidelberg.

Calabrese, D., Kalantari, K., Santucci, F. M., & Stanghellini, E., (2008). *Environmental Policies and Strategic Communication in Iran, The International Bank for Reconstruction and Development, No. 132*, The World Bank, Washington DC.

Collins, B. J., (2001). A history of the animal world in the ancient Near East. *Handbook of Oriental Studies, Sect. 1: The Near and Middle East. vol. 64*. Brill Academic Pub, Boston.

Darvishsefat, A. A., (2006). *Atlas of Protected Areas of Iran*. University of Tehran Press, Tehran, Iran.

Darvishsefat, A. A., Khosravi, A., & Borzuii, A., (2008). Concept of the national atlas of protected areas of Iran and its realization. *Proceeding Conference of TS 8H-GIS in Environmental Management*, Integrating Generations in Stockholm, Sweden.

Darvishzadeha, R., & Hatami, M. H., (2012). Analysis of genetic variation for morphological and agronomic traits in Iranian oriental tobacco (*Nicotiana tabacum* L.) genotypes.*Crop Breeding Journal*, *2*(1), 57–61.

DOE, (2010). *Iran's Second National Communication to the UNFCCC*. National Climate Change Office (NCCO). Department of Environment, Tehran.

Esfahani, S. T., Shiran, B., & Balali, G., (2009). AFLP markers for the assessment of genetic diversity in European and North American potato varieties cultivated in Iran. *Crop Breeding and Applied Biotechnology*, *9*, 75–86.

Fahim, S., Shojaeiyam, A., Mehrabi, A. A., & Ghanbari, E., (2013). Analysis of genetic diversity between and within Iranian accessions of spinach (*Spinacia oleracea* L.) by SRAP markers. *Iranian J. Genetics and Plant Breeding*, *2*(2), 21–27.

FAO, (2005). *Agricultural Statistics of Iran*. Rome, Italy.

Faramarzi, S., Yadollahi, A., & Soltani, B. M., (2014). Preliminary evaluation of genetic diversity among Iranian red fleshed apples using microsatellite markers. *J. Agr. Sci. Tech.*, *16*, 373–384.

Farhadinia, M. S., Jafarzadeh, F., Sharbafi, E., & Moqanaki, E. M., (2011). *Conservation Education to Save the Endangered Persian Leopard in Iran*. Report submitted to people's trust for endangered species, UK.

Farrokhi, J., Darvishzadeh, R., Naseri, L., Azar, M. M., & Maleki, H. H., (2011). Evaluation of genetic diversity among Iranian apple (*Malus* × *domestica* Borkh.) cultivars and landraces using simple sequence repeat markers. *Australian J. Crop Sci. (AJCS)*, *5*(7), 815–821.

Firouz, E., (2005). *The Complete Fauna of Iran*. I. B. Tauris & Co. Ltd, London, UK.

Ghahreman, A., (1994). *Iran Chromophytes, vol. 4*. University of Tehran Press, Tehran, Iran.

Ghelichnia, H., (2014). Flora and vegetation of Mt Damavand in Iran. *Phytologia Balcanica*, *20*(2–3), 257–265.

Ghoddousi, A., Hamidi, A. M., Ghadirian, T., & Khorozyan, I., (2010). The status of the endangered Persian leopard (*Panthera pardus saxicolor*) in Bamu national park, Iran. *Fauna & Flora International, Oryx, 44*(4), 551–557.

Glenn, E. P., Brown, J. J., & O'Leary, J. W., (1998). Irrigation crops with seawater. *Scientific American, 8*, 76–81.

Glenn, E. P., & Watson, C., (1993). Halophyte crops for direct salt water irrigation. In: *Towards the Rational Use of High Salinity Tolerant Plants*. Kluwer Academic Publisher, Dordrecht, pp. 379–385.

Glenn, E. P., O'Leary, J. W., Watson, M. C., Thompson, T. L., & Kuehl, R. O., (1991). *Salicornia bigelovii* Torr: An oilseed halophyte for seawater irrigation. *Science, 251*, 1065–1067.

Haidari, M., & Rezaei, D., (2013). Study of plant diversity in the northern Zagros forest (Case study: Marivan region). *Intern. J. Advanced Biol. Biomed. Res. (IJABBR), 1*(1), 1–10.

Heidari, F., Riahi, H., Yousefzadi, M., & Shariatmadari, Z., (2013). Morphological and phylogenetic diversity of Cyanobacteria in four hot springs of Iran. *Iran. J. Bot., 19*(2), 162–172.

Heshmati, G. A., (2007). Vegetation characteristics of four ecological zones of Iran. *Intern. J. Plant Prod., 1*(2), 215–224.

Heydari, M., Poorbabaei, H., Hatami, K., Salehi, A., & Faghir, M. B., (2013). Floristic study of Dalab woodlands, northeast of Ilam province, west Iran. *Iranian J. Sci. & Tech., 37*(A3), 301–308.

[IFNRCBD], (2010). Department of Environment, *Iran's Fourth National Report to the Convention of Biological Diversity*. Department of the Environment of Iran.

[IFNRCBD], (2015). Department of Environment, *Iran's Fifth National Report to the Convention of Biological Diversity*. Department of the Environment of Iran.

Irannejad, S., Sepahi, A. A., Amoozegar, M. A., Tukmechi, A., & Moghanjoghi, A. A. M., (2015). Isolation and identification of halophilic bacteria from Urmia lake in Iran. *Iranian J. Fisheries Sci., 14*(1), 45–59.

IUCN, (2010). *IUCN Red List of Threatened Species*. Available from http://www.redlist.org.

Jalili, A., & Jamzad, Z., (1999). *Endangered Plants in Iran: Red Data Book of Iran: A Preliminary Survey of Endemic, Rare and Endangered Plant Species in Iran*. Research Institute of Forests and Rangelands, Tehran, Iran.

Jannati, M., Fotouhi, R., Abad, A. P., & Salehi, Z., (2009). Genetic diversity analysis of Iranian citrus varieties using micro satellite (SSR) based marker. *J. Horticulture and Forestry, 1*(7), 120–125.

Kazemi, E., (2014). Biodiversity of the subfamily catocalinae boisduval, [1818] (Lepidoptera, Noctuidae) in Iran. *Intern. J. Advanced Biol. Biomed. Res., 2*(1), 25–33.

Khoshbakht, K., & Hammer, K., (2006). Savadkouh (Iran): An evolutionary center for fruit trees and shrubs. *Genet. Resour. Crop Evol., 53*(3), 641–651.

Kiabi, B. H., Dareshouri, B. F., Ghaemi, R. A., & Jahanshahi, M., (2002). Population status of the Persian leopard (*Panthera pardus saxicolor* Pocock, 1927) in Iran. *Zoology in the Middle East*, *26*, 41–47.

Linnell, J. D. C., Swenson, J. E., & Andersen, R., (2000). Conservation of biodiversity in Scandinavian boreal forests: Large carnivores as flagships, umbrellas, indicators or keystones? *Biodiversity Conservation*, *9*, 857–868.

Madjnoonian, H., (1993). *National Parks and Protected Areas, Values, Functions & Characteristics*. Department of the Environment, Tehran.

Mataji, A., Kia, H., Babale, S., & Roshan, S. A., (2013). Flora diversity in burned forest areas in Dehdez, Iran. *Folia Forestalia Polonica, Series A.*, *55*(1), 33–41.

Misra, M., (2009). Management of Iran's arid lands. *International Journal of Environmental Studies*, *66*(3), 287–288.

Naghavi, M. R., Monfared, S. R., & Humberto, G., (2012). Genetic diversity in Iranian chickpea (*Cicer arietinum* L.) landraces as revealed by microsatellite markers. *Czech J. Genet. Plant Breed.*, *48*(3), 131–138.

Naqinezhad, A., (2012). Vegetation database of Iran. In: *Vegetation Databases for the 21st Century. Biodiversity & Ecology*, *4*, 305–305.

Nezhad, M. J. M., (1973). Zagros vegetation disturbance. *Mohit Shenasi*, *1*, 97–107.

Noroozi, J., Akhani, H., & Breckle, S. W., (2008). Biodiversity and phytogeography of the alpine flora of Iran. *Biodiversity and Conservation*, *17*(3), 493–521.

Olfat, A. M., & Pourtahmasi, K., (2010). Anatomical characters in three oak species (*Q. libani*, *Q. brantii* and *Q. infectoria*) from Iranian Zagros mountains. *Australian J. Basic and Applied Sci.*, *4*, 3230–3237.

Paknia, O., & Pfeiffer, M., (2011). Steppe versus desert: Multi-scale spatial patterns in diversity of ant communities in Iran. *Insect Conservation and Diversity*, *4*, 297–306.

Podlech, D., (1999). Papilionaceae III: *Astragalus*. In: *Flora Iranica, vol. 174*. Akademische Druck-u. Verlagsanstalt, Graz, Austria.

Pourbabaei, H., & Poor-Rostam, A., (2009). The effect of shelterwood silvicultural method on the plant species diversity in a beech (*Fagus orientalis* Lipsky) forest in the north of Iran. *J. Forest Science*, *55*(8), 387–394.

Ramsar Convention Secretariat, (2013). *The Ramsar Convention Manual: A Guide to the Convention on Wetlands (Ramsar, Iran, 1971)*, 6th edition, Ramsar Convention Secretariat, Gland, Switzerland.

Raunkiaer, C., (1934). *The Life Forms of Plant and Statistical Plant Geography*. Clarendon Press, Oxford.

Raziei, T., Daneshkar, A. P., & Saghfian, B., (2005). *Annual Rainfall Trend in Arid and Semi-Arid Regions of Iran*. In: *ICID 21st European Regional Conference*, Frankfurt (Oder) and Slubice-Germany and Poland.

Rechinger, K. H., (1963–2015). *Flora Iranica, vols. 1–181*. Akademische Druck-u Verlagsanstalt, Graz, Austria.

Rezaiee, A. A., (2000). An investigation on growth and yield of Norway spruce in Ladjim forest region. *Pajouhesh va Sazandegi*, *13*(3), 56–59.

Sabeti, H., (2002). *Forests, Trees and Shrubs of Iran*. 3rd edition. Iran University of Science and Technology Press, Yazd.

Sagheb-Talebi, K., Sajedi, T., & Yazdian, F., (2003). *Forests of Iran*. Research Institute of Forests and Rangelands, Tehran, Iran.

Salamati, M. S., Zeinali, H., & Yousefi, E., (2011). Investigation of genetic variation in *Carthamus tinctorius* L. genotypes using agro-morphological traits. *J. Res. Agri. Sci.*, *7*(2), 101–108.

Seyedimoradi, H., Talebi, R., & Fayaz, F., (2016). Geographical diversity pattern in Iranian landrace durum wheat (*Triticum turgidum*) accessions using start codon targeted polymorphism and conserved DNA-derived polymorphism markers. *Environ. Experim. Biol.*, *14*, 63–68.

Shahsavari, A., (1994). *Natural Forests and Woody Plants of Iran.* Research Institute of Forests and Rangelands, Tehran, Iran.

Sharafi, Y., Majidi, M. M., Jafarzadeh, M., & Mirlohi, A., (2015). Multivariate analysis of genetic variation in winter rapeseed (*Brassica napus* L.) cultivars. *J. Agr. Sci. Tech.*, *17*, 1319–1331.

Sharifi, M., (2011). An overview of ecological potential and the outstanding universal value of forests resources of I. R. Iran with respect to climate change. In: *Regional Workshop "Forests, Rangelands and Climate Change in the Near East Region,"* Cairo, Egypt.

Shiria, M. R., Choukanb, R., & Aliyevc, R. T., (2014). Study of genetic diversity among maize hybrids using SSR markers and morphological traits under two different irrigation conditions. *Crop Breeding Journal*, *4*(1), 65–72.

Shirvani, H., Etminan, A. R., & Safari, H., (2013). Evaluation of genetic diversity within and between populations for *Agropyron trichophorum* by ISSR marker. *Intern. J. Farming and Allied Sci.*, *2*(21), 949–954.

Talebi, R., Naji, A. M., & Fayaz, F., (2008). Geographical patterns of genetic diversity in cultivated chickpea (*Cicer arietinum* L.) characterized by amplified fragment length polymorphism. *Plant Soil Environ.*, *54*(10), 447–452.

Valipour, A., Namiraninan, M., Etemad, V., & Ghazanfari, H., (2009). Relationships between diameter, height and geographical aspects with bark thickness of Lebanon oak tree (*Quercus libani* Oliv.) in Armardeh, Baneh, (Northern Zagros of Iran). *Research J. Forestry*, *3*, 1–7.

Vuosalo, E., (2013). Siberian crane wintering in Iran in 2011/12 and 2012/13. *Siberian Crane Flyway News*, *12*, 11–12.

White, F., & Léonard, J., (1991). Phytogeographical links between Africa and southwest Asia. *Flora Veg. Mundi*, *9*, 229–246.

Willis, A. J., (2001). Endangered plants in Iran: Red data book of Iran: A preliminary survey of endemic, rare and endangered plant species in Iran. *New Phytologist*, *149*, 165.

Yousefi, M., (2007). *Flora of Iran.* Payame Noor University Press, Tehran, Iran.

Yousefi, S. B., (2011). The role of wildlife documentaries in nature and wildlife conservation. *Persian Wildlife Heritage Foundation Newsletter*, *1*(2), 8–9.

Zakeri, H., (2011). Improving wild sheep habitat in Varjin protected area. *Persian Wildlife Heritage Foundation Newsletter*, *1*(2), 5.

Zarei, A., Cheraghi, S., & Pourhedayat, M., (2011). A study of birds of Chughakhor Wetland. *Persian Wildlife Heritage Foundation Newsletter*, *1*(2), 7.

Zehzad, B., Kiabi, B. H., & Madjnoonian, H., (2002). The natural areas and landscape of Iran: An overview. *Zoology in the Middle East*, *26*, 7–10.

Zohary, M., (1973). *Geobotanical Foundations of the Middle East, vols. 1–2.* Gustav Fischer Verlag, Stuttgart, Germany.

Plate 6.1   Plant diversity in Iran; from top left: (a) *Astragalus leucargyreus*, (b) *Vicia garinensis*, (c) *Hedysarum gypsophilum*, (d) *Hedysarum kalatense*, (e) *Cerasus paradoxa*, and (f) *Lagochilus lorestanicus* (Photo credits: M. M. Dehshiri).

CHAPTER 7

Biodiversity in Iraq

NATURE IRAQ

Pak City Apartments, Block A2, Floor 1, Flat No. 8, Near City Center Mall, Kurdistan Region, Iraq, E-mail: info@natureiraq.org

7.1 Overview of the Country

Iraq covers an area of 434,320 km², with the Gulf coastline comprising 58 km in length. The land upon which the modern state of Iraq has been founded came to be called 'Mesopotamia' (from the Greek meaning 'land between the rivers). This region is known as the cradle of western civilization. The modern boundaries of Iraq were drawn in 1920 to 1925 by the League of Nations in the Treaty of Sèvres upon the collapse of the Ottoman Empire after the First World War. The Republic of Iraq was created with the overthrow of the royalty in a coup d'état on the 14th of July 1958 but this led to a period of instability and a series of military coups until the Ba'athist Party took control from 1968 to 2003. Saddam Hussein, 5th president, ruled the country from July 1979 to April 2003. During his presidency Iraq fought three major and very destructive wars (the Iran–Iraq War in 1980–88, first Gulf War in 1990–91, and the second Gulf War in 2003); conducted a genocidal campaign against its Kurdish population (Anfal in 1986–89); drained the Mesopotamian Marshlands displacing and repressing people of southern Iraq (early 1990s), and suffered through 13 years of economic sanctions (1990–2003). The Ba'ath Party was removed from power in 2003 after a United States-led invasion of the country followed by an occupation that lasted until 2011. Iraq is now governed as a federated Republic with a president, a Prime Minster heading the Council of Ministers and an elected Parliament. There is a federal regional government in the northern part of the country governing the Kurdish mountains and plains. Unfortunately, there remains considerable unrest with the remnant of the Islamic State of the Levant being removed from Mosul, the country's second largest city.

As of 2012, the World Bank estimates Iraq's population as 32.6 million people, living in 18 provinces in Iraq and since 1970 three of the northern governorates (Dohuk, Erbil, and Sulaymaniyah) have been officially designated as a Kurdish autonomous region, with a separate elected legislature, the Kurdish Regional Government (KRG). In 2016, the district of Halabja was designated as a separate governorate, making the KRG into four governorates.

7.2 Ecosystem Diversity in Iraq

Iraq is part of the Palearctic Realm, the largest of the eight terrestrial biogeographic areas or ecozones that have been defined for the Earth. The World Wildlife Fund (2014) defines five terrestrial biomes and nine ecoregions in Iraq. These are listed in Table 7.1 with their conservation status, area and percentage within Iraq. Of these nine main ecoregions, five accounts for 96% of the total area of Iraq and these are discussed in more detail below. Note, that much of the information in the list below comes directly from the WWF's (2014) descriptions of these specific ecoregions but it should also be noted that in the Iraq Inventory 'biome-restricted bird species' were evaluated based on a slightly different list of biomes used by BirdLife International. For Iraq, the BirdLife list of biomes includes the Mediterranean (ME12), Eurasian High-Montane (ME05), Irano-Turanian (ME06), Eurasian Steppe and Desert (ME04) and Sahara-Sindian Desert (ME13) Biomes. Except for the Mediterranean biome (ME12), which does not constitute a major part of Iraq, these biomes are identified in the following list:

1. **Zagros Mountains Forest-steppe:** BirdLife International includes this region in the Irano-Turanian Biome (ME06), with higher elevations part of the Eurasian High-Montane Biome (ME05). Climate is semi-arid, with warm summers and cold winter and annual precipitations (falling mainly in winter and spring) averaging from 400 to 800 mm.

2. **Middle East Steppe:** BirdLife International includes this area within the Eurasian Steppe and Desert Biome (ME04).

3. **Mesopotamian Shrub Desert:** BirdLife International incorporates this area within the Sahara-Sindian Desert Biome (ME13).

4. **Tigris-Euphrates alluvial salt marsh (Southern Iraq):** The Mesopotamian Marshlands of southern Iraq, the largest wetland ecosystem in western Eurasia, are located in this ecoregion [BirdLife International incorporates this area within the Sahara-Sindian Desert Biome (ME13) and considers the area an important Endemic Bird Area (BirdLife International, 2014)].

5. **Arabian Desert and East Sahero-Arabian Xeric Shrublands** (Southwestern Asia: Most of Saudi Arabia, extending into Oman, United Arab Emirates, Yemen, Egypt, Iraq, Jordan, and Syria). This is the largest ecoregion of the Arabian Peninsula, stretching from the Yemeni border to the Arabian Gulf and from Oman to Jordan and Iraq [BirdLife International includes this within the Sahara-Sindian Desert Biome (ME13)].

Table 7.1 WWF Biomes and Terrestrial Ecoregions of Iraq

WWF Biome	Terrestrial Ecoregion*,**	Conservation Status	Total Area (km²)	Area in Iraq (km²)	% in Iraq
Temperate Broadleaf and Mixed Forests	Zagros Mountains Forest Steppe (PA0446)	Critical	397,800	29,376	7%
Mediterranean Forest, Woodland and shrubs	Eastern Mediterranean conifer-sclerophyllous-broadleaf forest (PA1207)	Critical	143,800	1,475	1%
Temperate Grasslands. Savannas and Shrublands	Middle East Steppe (PA0812)	Vulnerable	132,300	37,598	28%
	Arabian Desert and East Sahero-Arabian Xeric Shrublands (PA1303)	Critical	1,851,300	192,853	10%
	Mesopotamian Shrub Desert (PA1320)	Vulnerable	211,000	129,995	62%
	Red Sea Nubo-Sindian Tropical Desert and Semi-Desert (PA1325)	Critical	651,300	5,189	1%
Deserts and Xeric Shrublands	South Iran Nubo-Sindian Desert and Semi-Desert (PA1328)	Critical	351,500	7,993	2%
	Gulf Desert and Semi-Desert (PA1323)	Critical	72,600	1,480	2%
Flooded Grasslands and Savannas	Tigris-Euphrates alluvial salt marsh (PA0906)	Critical	35,600	28,795	81%
	Total		3,847,200	434,753	

* Eastern Anatolian Montane Steppe (PA0805) is an ecoregion with only 3 hectares considered part of Iraq.

** Each terrestrial ecoregion has a specific ID code in the format XXnnNN (where XX is the ecozone, nn is the biome number, and NN is the individual ecoregion number).

Iraq comprises three different freshwater ecoregions: Arabian Interior (440), Lower Tigris and Euphrates (441), Upper Tigris and Euphrates (442), and one marine ecoregion (Arabian Gulf (90) (Table 7.2, Figure 7.1).

7.3 Landscape and Species Diversity in Iraq

Due to the variety of habitats in Iraq from the snow-capped mountains to the hot deserts, the country is biologically diverse. In terms of birds, more

Table 7.2 WWF Freshwater and Marine Ecoregions of Iraq

Aquatic Ecoregions	Total Area (km²)	Area in Iraq (km²)	% in Iraq
441 Lower Tigris and Euphrates River Basin	340,633	227,497	67%
442 Upper Tigris and Euphrates River Basin	507,236	64,745	13%
440 Arabian Interior	2,334,454	142,494	6%
90 The Arabian Gulf	251,000	Territorial Sea: 4,910	2%

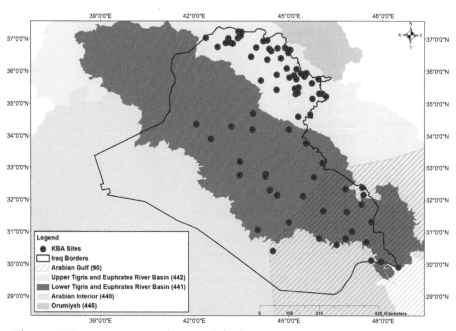

Figure 7.1 Iraq KBA Sites shown with freshwater & marine ecoregions (WWF 2014).

than 400 species have been recorded (Table 7.8); among them several that are Critically Endangered, such as Northern Bald Ibis *Geronticus eremita,* Slender-billed Curlew *Numenius tenuirostris,* and Sociable Lapwing *Vanellus gregarius.* There are a total of 37 globally threatened bird species (3 – Critically Endangered, 5 – Endangered, 11 – Vulnerable, and 18 – Near Threatened species). Several migration routes cross Iraq, including the African–Eurasian flyways. Millions of birds depend on the wide variety of habitats that Iraq has to offer as well as vital stopover sites in their annual migrations.

This diversity extends to plants and other fauna as well, with over 2500 plant species (the largest percentage of which are found in the mountain-forest region in the north) and 90 mammals, 98 reptiles, 10 amphibians, and about 100 freshwater and marine fish species. It is important to note that most species checklists for Iraq are still incomplete and this chapter only lists species that have recent hard evidence for their presence in the country. The botanical diversity of the Zagros Mountains Forest-steppe is quite rich due to the variety of landscapes and terrains within the Zagros range that extends through northern Iraq. The dominant habitat in this region is oak woodlands and important genera representative of the Irano-Turanian Steppe including Tragacanth dominated by species of *Astragalus, Cousinia* and many endemic species. Low slopes are clothed in open, oak forests again home to a good assemblage of endemic and rare species. Key threatened fauna species of this region include Persian Leopard *Panthera pardus saxicolor* (EN), Wild Goat *Capra aegagrus* (VU), Spur-thighed Tortoise *Testudo graeca* (VU), Azerbaijan Newt *Neurergus crocatus* (VU), Kurdistan Newt *Neurergus derjugini* (CR), and Spotted Belly Salamander *Salamandra infraimmaculata* (NT). Recent fish surveys in the Kurdistan region of Iraq found a number of new, undescribed species (Freyhof, 2012, 2016) and there are likely other new species to be discovered in this region.

Desert landscapes in Iraq, which comprise much of Iraq's west and southwest are extensions of the deserts of Syria and Saudi Arabia in the west and south. Additional desert landscapes lie between the Tigris and Euphrates extending from lower Babil to Nasiria, and areas east of the Tigris from Wasit southward. Average annual rainfall in this area is less than 100 mm but increases towards the north and east to between 100 and 240 mm (Batanouny, 2001). The majority of the deserts in Iraq lie within the Sahara-Sindian Desert biome/Arabian Desert and East Sahero-Arabian Xeric Shrublands Ecoregion. The genera *Artemisia* and *Haloxylon* are important here and there are relic areas of tropical species that date back to a time of climatic optimum, e.g., *Acacia gerrardii* and *Razya stricta.* In addition to being relatively rich with unique plant life, these areas are also important with considerable fauna taxa (mammals, reptiles, and birds) of which many are threatened or vulnerable.

Major river ecosystems of Iraq include the Tigris and Euphrates Rivers but a number of other important tributaries to these two main systems such as the Khaboor, Greater Zab, Lesser Zab, Diyala, Kharun Rivers, Gharraf, Shatt Al-Arab, Shatt Al-Hilla and others. Whereas the majority of the Euphrates River is sourced in the upstream, riparian countries of Turkey and Syria, approximately 40% of the waters of the Tigris River arise from within the borders of Iraq (the remainder coming from Turkey and Iran). Lake eco-systems in Iraq are largely formed by dam/reservoir systems such as Mosul, Dukan, Darbandikhan, Habbaniya, Himreen, Samara, and Qadassiya Lakes as well as depressions that have been used for water storage such as Tharthar and Razzaza Lakes (in fact, these lakes have their origins in the flood control era of the 1950's through 1980's. However since that time major dams have been built upstream making the flood control function secondary to water storage). There are some natural lakes within Iraq including Sawa Lake in southern Iraq, small mineral ponds such as Genaw Lake near Rania in the north and many desert oases. Often water bodies have been created by water management systems in Iraq that subsequently became biologically rich and important natural resources such as Dalmaj, Suweibaat, Al-Rashed Lake, and the marshes of the southern part of East Hammar.

Many of these systems now contain introduced, sometime invasive spe-cies, but the important native fish species include: *Acanthopagrus cf. arabi-cus, Alburnus mossulensis, Arabibarbus grypus* (Vulnerable), *Luciobarbus esocinus* (Vulnerable) and *Mesopotamichthys sharpeyi* (Vulnerable). There are also two important cave-dwelling fish reported in Iraq *Caecocypris basimi* and *Garra widdowsoni* (both Critically Endangered). Much field-work remains to be done on fish in Iraq. The same is true for herpetology though some important species are known such as the endangered Euphrates Softshell Turtle *Rafetus euphraticus* associated with the Tigris-Euphrates River Basins.

One of the most notable landscapes in Iraq is the Lower Mesopotamian Marshlands, once the 3rd largest wetlands in the world (the largest in west-ern Asia). This extensive network of marshes formed by the lower Tigris and Euphrates River formerly fluctuated seasonally up to 12,000–15,000 km². These marshlands were subject to a major drainage campaign in the 1990s. Though not botanically rich, the thick beds of *Phragmites australis* and *Typha domingensis* occasionally shelter water lilies *Nymphoides indica* and other locally unique species and this area remains an important habitat for mammals, amphibians, fish, and of particular importance for migrating birds making the long journeys between Eurasia and Africa.

Other key species found in the Mesopotamian marshlands include the vulnerable Smooth- coated Otter (*Lutrogale perspicillata*), which has a sub-species *L. p. maxwelli* and the Endangered Bunn's Short-tailed Bandicoot Rat (*Nesokia bunnii*) (both associated with this area); the endangered Euphrates Softshell Turtle (*Rafetus euphraticus*), and many important fish species such as those previously mentioned. Many threatened bird species such as the vulnerable Marbled Teal (*Marmaronetta angustirostris*) and Iraq's two endemic bird species the Basra Reed Warbler (*Acrocephalus griseldis*) (EN) and Iraq Babbler (*Turdoides altirostris*) are also found in the marshlands and this area is the main breeding ground for the latter species.

The Lower Mesopotamian Marshlands are also of global cultural significance. It was in this region that humans first developed writing and agriculture. The marshlands are the legendary location of the biblical Garden of Eden and the indigenous Ma'dan Marsh Arabs can trace their lineage directly to the ancient Sumerians whose lands were centered in the vast wetlands formed around the confluence of the Tigris and Euphrates River.

7.4 Protected Areas Network in Iraq

Under the Aichi Biodiversity Targets of the CBD, a goal has been set to conserve at least 17% of terrestrial and inland water, and 10% of coastal and marine areas by 2020 (CBD, 2014, Strategic Goal C, Target 11). Iraq has initiated this process with the designation of the Hawizeh Marshes, East and West Hammar Marshes, and Sawa Lake Ramsar sites; the Central Marshes National Park, and the recent inclusion of the Mesopotamian marshlands as a World Heritage Site. The first official protected area designation in Iraq was assigned to the Hawizeh Marshes Ramsar site in 2007 when Iraq became signatory to the Ramsar International Convention on Wetlands. Hawizeh Marshes are unique in Iraq as a transboundary wetland with Iran (the Iranian portion is called Hor Al-Azim) and were the only part of the Mesopotamian Marshlands that were never completely drained in the 1990s. Unfortunately, these wetlands have faced severe reductions due to drought as well as hydro/agricultural irrigation projects both inside Iraq and in upstream states. In addition, in 2011 Iran completed a 90 km embankment through the marsh along the border restricting the flow of water from Hor Al-Azim into Hawizeh. These threats led Iraq to include this Ramsar site on the Convention's Montreux List of ecologically threatened sites in April 2010.

A total of 137,700 hectares (1,377 km^2) in Hawizeh is covered by its Ramsar designation as shown in the Map below, but while management plans have been developed for Hawizeh implementation of these plans has lagged behind. Additionally in 2014, 500 hectares of the Sawa Lake were listed as a Ramsar Site as well as the 219,700 hectares of the Central Marshes and 180,000 hectares of the Hammar Marshes (including both East and West Hammar).

Additionally, it should be noted that the Kurdistan region is also interested in developing regional parks and an initiative to develop a park encompassing Halgurd and Sakran Mountains has started with support from the Protected Areas High Committee of the Kurdistan Regional Government. A park committee was formed and an office for the park was developed in Choman, Erbil.

Moreover, one site in Iraqi Kurdistan, the Barzan Area, is noteworthy for having, until recently, only informal environmental protection. Protected since the early 20th century through tribal prohibition and controls on hunting and habitat uses, this KBA site has an area of over 110,000 hectares representing approximately 4% of the Zagros Mountains Forest Steppe (PA0446) ecoregion in Iraq. The benefit of long-term protection for this extensive area can be seen in the fact that this is one of the few sites in Iraq where a large population of the globally Vulnerable Wild Goat *Capra aegagrus* can easily be seen. It is a noteworthy lesson that the Barzan Protected Area, which has achieved a relatively high level of protection since the beginning of the last century, was created only by local community agreement.

Iraq has no Man and Biosphere Reserves presently but has been a signatory nation since 1974 to the UNESCO World Heritage Convention (WHC). In July of 2016, UNESCO formally accepted the Ahwar of Southern Iraq as a cultural and natural World Heritage Site made up of seven components: three archaeological sites and four wetland marsh areas in southern Iraq (2016). These marsh sites include the Central Marshes, Hawizeh, and East and West Hammar. Under the WHC, Iraq needs to implement management plans to protect these sites and report regularly on the state of conservation of its World Heritage sites. Combined the four marsh areas of the Ahwar of Southern Iraq covers an area of 210,899 hectares with an additional 207,643 hectares of buffer.

There are a total of 82 sites that meet one or more of the KBA Criteria systems that were applied in Nature Iraq's KBA program: KBA for non-avian fauna, IBA for avian fauna, and IPA for plants and habitats. The rigorous application of quantitative KBA criteria assures that the network of these 82 identified sites is of global importance. While some sites may have met only one criterion within one system (KBA for non-avian fauna, IBA, or IPA), the majority met several criteria often within one or more of the systems (Figures 7.2 and 7.3).

7.5 Floral Diversity in Iraq

The flora of Iraq is composed of five phytogeographical groups: Mediterranean; Irano-Turanian; Saharo-Sindian; Eurosiberian-Boreoamerican; and bi- and pluri-regional. The Irano-Turanian region includes two sub regions, which are Irano-Anatolian and Mesopotamian sub-regions. Characteristically, the vegetation in the Irano-Anatolian sub-region is mainly mountainous forests and foothills with grass, herbs and small shrubs. The vegetation of Mesopotamian sub-region comprises mainly steppe or sub-desert communities on grey calcareous desert soil, often gravelly or stony; there is no mountain vegetation in the whole of this sub-region (Guest, 1966). The annual rainfall 350–500 mm where winter cultivation, without irrigation, normally succeeds in this region; but much of the land here has been cultivated for millennia, so that the soil is generally impoverished and has often suffered from erosion and leaching. Consequently, even the smaller woody shrublets of the original vegetation have long been eradicated by

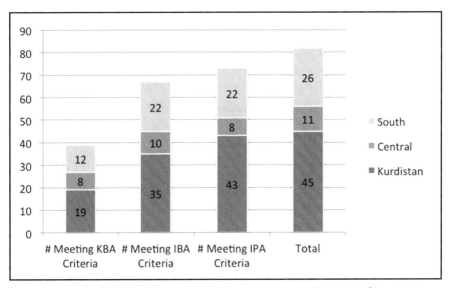

Figure 7.2 Number of sites meeting global criteria in each region of Iraq (Nature Iraq, 2017).

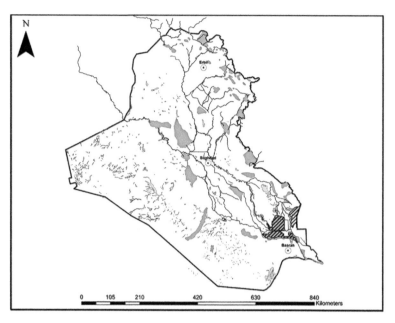

Figure 7.3 The potential for KBAs to extend and develop the existing network in
Iraq. (blue polygons = KBAs and hatched areas = Current Protected
Areas). (Nature Iraq, 2017).

the plough, the woodcutters and the fuel gatherer, while the more palatable
perennials have been greatly reduced or eliminated by overgrazing.

The Middle Saharo-Sindian sub-region (The Saharo-Sindian region is
the great desert belt that extends from the Atlantic coast in Africa to the Thar
desert in India) is bordered by the Mediterranean and Irano-Turanian regions
in the north and the Sudano-Deccanian region in the south. It is divided into
three sub-regions: the Western, the Middle, and the Eastern Saharo-Sindian
sub-region. The Middle Saharo-Sindian sub-region includes the northern
part of the Arabian Peninsula, Egypt's Sinai, parts of lower Palestine, southern and central Jordan, and lower Iraq (Figure 7.4).

The map (Figure 7.5), also from Guest (1966, p.64), shows the zones of
vegetation in Iraq that was believed to be the natural "climax" states of these
areas. The spring aspect of uncultivated and protected places in the steppe
zone today is luxuriant grassland dominated by *Poa bulbosa*, with *Hordeum
bulbosum*, other grasses and herbs (including *Anemone coronaria* in spring
and the early summer grass *Aegilops speltoides*), and clumps of semi-woody
or herbaceous survivors of the original perennial vegetation: usually unpalatable or obnoxious species of *Phlomis, Gundelia* (*G. tournefortii*), *Cousinia,*

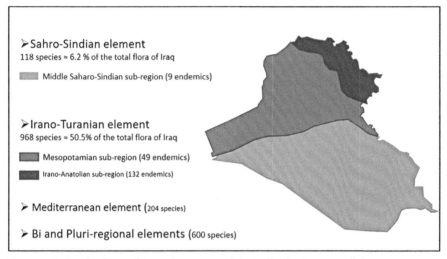

Figure 7.4 The flora of Iraq elements and their distribution overall the country (Nature Iraq).

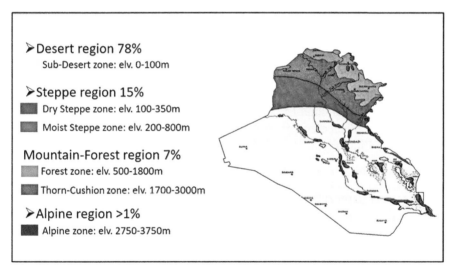

Figure 7.5 The vegetation types in Iraq and their distribution overall the country (Reprinted with permission from Guest, 1966. © Board of Trustees of the Royal Botanic Gardens, Kew.)

Hypericum (*H. triquetrifolium*), etc. Though the season of rains is somewhat longer than the desert further south, on summer cultivation is possible without irrigation. However, there is large part of the cultivated areas uses irrigation from both surface and ground water. As was stated in the caption of the original map, "in the Desert region, most of the alluvial floodplain below the 33rd to 34th parallel and the delta consist of irrigated, arable land, ancient abandoned cultivation and marshlands. The principal areas of date cultivation are indicated."

The flora of Iraq in summary:

- There are 151 plant families in Iraq (128 – Dicotyledons and 24 – Monocotyledons).
- Major plant families are: Poaceae – 412 species; Compositae ≈ 363; Fabaceae – 290; Lamiaceae ≈ 160; Apiaceae ≈ 148; Scrophulariaceae ≈ 100.
- Minor families are: Acanthaceae – 1; Equisetaceae – 2; Asclepediaceae – 3 species.
- The Flora of Iraq book is to be published in 9 volumes (Published: 6 volumes, from 1966–1985; Species published: 1783. The rest of volumes are in publishing after resuming the work by Iraqi National Herbarium and Kew Royal Botanic Garden.).

Plant species distribution in country as the following:

- The total number of plant species in Iraq is ± 3303.
- Sahro-Sindian element: 118 species ≈ 6.2% of the total flora of Iraq (Middle Saharo-Sindian sub-region contains 9 endemics).
- Irano-Turanian element: 968 species ≈ 50.5% of the total flora of Iraq (Mesopotamian sub-region contains 49 endemics and Irano-Anatolian sub-region contains 132 endemics).
- Mediterranean element: 204 species.
- Bi and Pluri-regional elements: 600 species.

Vegetation type overall the country is the following:

- Desert region 78% of Iraq's area (Sub-Desert zone: elevation 0–100 m).
- Steppe region 15% of Iraq's area (Dry Steppe zone: elevation 100–350 m and Moist Steppe zone: elevation 200–800 m).
- Mountain-Forest region 7% of Iraq's area (Forest zone: elevation. 500–1800 m; Thorn-Cushion zone: elevation 1700–3000 m.
- Alpine region >1% of Iraq's area (Alpine zone: elevation 2750–3750 m).

Plantlife International has developed criteria for the designation of Important Plant Areas (IPA) throughout the globe. Also there are criteria developed by the IUCN Arabian Plant Specialist Group to assess the IPAs in Arabian Peninsula. Nature Iraq adopted the Arabian Peninsula guideline with some modifications to make these criteria more applicable for Iraqi plants and habitats. Iraq is only in the initial stages of assessing sites based on these criteria. Comprehensive plant lists for species in Iraq do not yet exist as well as the list of threatened plants.

More than 200 sites throughout Iraq were surveyed seasonally from 2005 to 2010 for plants and other biodiversity components during the KBA program. In 2012, a provisional list of the IPAs in Iraq has been developed after the assessment of all the surveyed areas in Iraq. This list includes more than 80 IPAs (some of them are cluster of more than one site), which contain threatened plant species, particular example of species richness, and/or outstanding example of threatened habitats (see Figure 7.6).

7.6 Faunal Resource

Iraq is located in both the northern and eastern hemisphere between 29° to 38° N latitudes and 39° to 49° E longitudes. Situated in the southeast corner of the Palearctic biogeographical realm, covering an area of 438,317 km², Iraq holds a rich biological diversity due to its geographical diversity from mountainous areas reaching an altitude of 3607 m in Halgurd Mountain from the north, to plains and desert habitat in the west, and marshlands in the south. Due to the fact that conservation in Iraq has been undermined by extensive political and social upheaval within the country since the 1970s it is difficult to obtain robust data on species population estimates, the information in this chapter is based on the data collected by Nature Iraq as part of the Key Biodiversity Areas of Iraq program and the remaining gaps in our knowledge of the

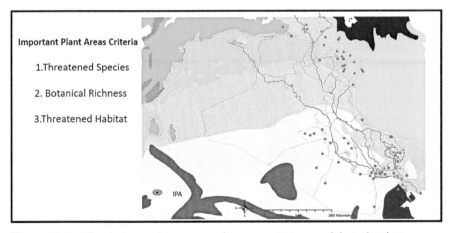

Figure 7.6 The designated important plant areas in Iraq and their distribution based on the ecoregions of country (Source: Nature Iraq, and the source of map is WWF- terrestrial ecoregions).

fauna species diversity is derived from historical data available for the country. The Iraq checklists currently have 9 threatened mammals (2 – Endangered, 5 – Vulnerable, and 2 – Near Threatened), 4 reptiles (1 – Endangered, 2 – Vulnerable, and 1 – Near Threatened), and 3 amphibians (1 – Critically Endangered, 1 – Vulnerable, and 1 – Near Threatened).

Most of the Iraqi non-avian fauna checklists are not yet complete and the lists provided below only include species seen during Nature Iraq Key Biodiversity Areas (KBA) surveys or on other surveys [Conservation Leadership Program through Conserving the Wild Goats on Peramagroon and Qara Dagh Mountains, 2011)]. Also included in the list are species that were camera trapped or reliably reported as well as unconfirmed reports by locals through the "Rehabilitating the Transboundary Habitat of the Persian Leopard through the Creation of a Peace Park" project Funded by Prince Bernhard Nature Fund, the Goldman Environmental Foundation, and the Federal Foreign Office, 2016-2018

- *Mammalia:* The mammalian fauna in Iraq (Table 7.3) comprises about 90 species, belonging to eight orders, 28 families and 65 genera. The first checklist of the mammals of Iraq since Hatt (1959) had seen little change, until Harrison's review of Hatt's observations and his additional information which he later compiled in three volumes of his book "The Mammals of Arabia" (1964, 1968, and 1972) and his second edition of the book in 1991. Recent additions to the checklist and more hard evidence to unconfirmed species have been provided by Nature Iraq, such as hard evidence for the Least Weasel *Mustela nivalis*, the Persian Leopard *Panthera pardus saxicolor* and the Mouflon *Ovis orientalis*. More recent photographic records of species such as the Near Threatened Striped Hyena *Hyaena hyaena* and the Grey Wolf *Canis lupus* has been recorded through camera traps set from June to October 2016 in Darbandikhan area of Iraqi Kurdistan.

- *Chiroptera:* Bats are the most understudied order of Iraq's mammals so far. According to the preliminary checklists of Nature Iraq, Iraq has seven Bat families, 11 genera, and 18 individual species. Most of these Bat species are listed as LC on IUCN except the Mehely's Horseshoe Bat *Rhinolophus mehelyi* and the Long-Fingered Bat *Myotis capaccinii*, which are listed as Vulnerable.

- *Rodentia:* Like Bats, Rodents are not well studied in Iraq. Among 28 families, 19 genera, and 27 species historically recorded, only 12 species have been confirmed over the last decade.

- *Reptilia:* A large number of reptiles occur in Iraq (Table 7.4), but information on their distribution and conservation status is limited. The

Table 7.3 Mammals List Found During Nature Iraq Surveys from 2007 to 2016

Scientific Name	Common Name	IUCN Status
Canis aureus	Golden Jackal	LC
Canis lupus	Grey Wolf	LC
Capra aegagrus	Wild Goat	VU
Capreolus capreolus	Roe Deer	LC
Caracal caracal	Caracal	LC
Crocidura suaveolens	Lesser Shrew	LC
Felis chaus	Jungle Cat	LC
Felis silvestris ornata	Wild Cat	LC
Gazella subgutturosa	Goitered Gazelle	VU
Herpestes edwardsii	Indian Grey Mongoose	LC
Herpestes javanicus	Small Asian Mongoose	LC
Hyaena hyaena	Striped Hyena	NT
Hystrix indica	Indian Crested Porcupine	LC
Lutrogale perspicillata	Smooth-coated Otter	VU
Lynx lynx	Eurasian Lynx	LC
Martes foina	Beech Marten	LC
Meles canescens	Eurasian Badger	LC
Mellivora capensis	Honey Badger	LC
Meriones persicus	Persian Jird	LC
Microtus socialis	Social Vole	LC
Microtus irani	Persian Vole	DD
Panthera pardus saxicolor	Persian Leopard	EN
Sciurus anomalus	Caucasian Squirrel	LC
Sus scrofa	Wild Boar	LC
Ursus arctos syriacus	Syrian Brown Bear	Not evaluated
Vulpes vulpes	Red Fox	LC
Vulpes rueppelli	Rüppell's Fox	LC

2016 IUCN Red List ranks the Euphrates Softshell Turtle *Rafetus euphraticus*, an endemic species, as 'endangered,' and the Spur-thighed Tortoise *Testudo graeca* as 'vulnerable.' The Desert Monitor *Varanus griseus* is listed on CITES Appendix I.

- *Amphibia:* Very few studies have been undertaken on the Amphibians of Iraq and the status and distribution of most, if not all species

Table 7.4 Reptile Species Found During Nature Iraq's Surveys from 2007 to 2016

Scientific Name	Common Name	IUCN Status
Acanthodactylus opheodurus	Arnold's Fringe-fingered Lizard	LC
Apathya cappadocica urmiana	Urmia Rock Lizard	LC
Asaccus griseonotus	Gray-spotted Leaf-toed Gecko	LC
Cerastes gasperettii	Arabian Horned Viper	LC
Dolichophis jugularis	Large Whip Snake	LC
Eirenis collaris	Collared Dwarf Snake	LC
Eumeces schneideri princeps	Schneideri's Red-marked Skink	
Hemidactylus turcicus	Turkish Gecko	LC
Lacerta media	Three-lined Lizard	LC
Laudakia nupta	Large-scaled Rock Agama	
Macrovipera lebetina	Levantine Viper	
Mediodactylus heteropholis	Iraqi Keel-scaled Gecko	DD
Mesalina brevirostris	Blanford's Short-nosed Desert Lizard	LC
Montivipera raddei kurdistanica	Kurdistan Viper	NT
Natrix tessellata	Dice Snake	LC
Ophisops elegans	Snake Eyed Lizard	
Paralaudakia caucasia	Caucasian Agama	LC
Platyceps najadum	Dahls Wipe Snake	LC
Pseudopus apodus	European Glass Lizard	
Rafetus euphraticus	Euphrates Softshell Turtle	EN
Saara loricata	Mesopotamian Spiny-tailed Lizard	LC
Spalerosophis diadema	Diadem Snake	
Testudo graeca	Spur-thighed Tortoise	VU
Timon princeps kurdistanica	Zagrosian Lizard	LC
Trachylepis aurata	Levant Skink	LC
Trapelus lessonae		LC
Uromastyx aegyptia	Egyptian Spiny-tailed Lizard	VU
Varanus griseus	Desert Monitor	
Walterinnesia morgani	Desert Cobra	

Those species that have either their common names or IUCN status or both missing, means they are unknown or not evaluated by IUCN in the country are poorly known. The current draft checklist of Iraq's Amphibians which is extracted from historical information as well as present observations includes 10 species (three of which are threatened according to IUCN) belonging to 5 families and 7 genera (Table 7.5).

• *Insects:* Insects are not fully evaluated, although extensive collections of benthic macro-invertebrates were undertaken during the Nature Iraq surveys from 2007 to 2010. Information collected by Nature Iraq about them is largely anecdotal and opportunistic, but where photographs were obtained later identification was often possible with the assistance of outside expertise. Currently efforts are underway to describe and assign names for species found in the Nature Iraq KBA surveys but this may take several years. Nevertheless preliminary information has indicated hotspot areas of conservation concern for aquatic insects in Kurdistan, northern Iraq based on the presence of rare and sensitive water quality bio-indicators (Mohammed Al-Saffar, Personal communication, 6 November 2013).

• *Odonata:* This biological group includes the Dragonflies and Butterflies. Dragonflies (and Damselflies) are amongst the biggest of insects, which are almost 6000 species in the world but we know very little about these insect groups in Iraq. There are 47 species of dragonfly known in Iraq and probably more than 200 species of butterflies. Recent fieldwork in Kurdistan has discovered three species of dragonfly new to Iraq (Porter, 2016).

7.6.1 Freshwater and Marine Fish Species Found During Nature Iraq's Surveys From 2007 to 2014

Eighty-six fish species [60 freshwater (Table 7.6) and 26 marine species (Table 7.7)] have been seen during Nature Iraq surveys and by subsequent surveys including those conducted by Dr. Jörg Freyhof of the Leibniz-Institute

Table 7.5 Amphibians Found During Nature Iraq Surveys

Scientific Name	Common Name	IUCN status
Bufotes variabilis	Variable Toad	DD
Hyla savignyi	Lemon-yellow Tree Frog	LC
Neurergus crocatus	Azerbaijan Newt	VU
Neurergus derjugini	Kurdistan Newt	CR
Pelophylax ridibundus	Marsh Frog	LC
Salamandra infraimmaculata	Spotted Belly Salamander	NT

Table 7.6 The Alien and Native Freshwater Fish Species Found During Nature
Iraq Surveys from 2007–2016

Scientific Name	Common Name	IUCN Status
Carassius auratus	Goldfish	LC
Ctenopharyngodon idella	Grass Carp	Not Evaluated
Cyprinus carpio	Wild Common Carp	VU
Gambusia holbrooki	Eastern Mosquitofish	LC
Heteropneustes fossilis	Stinging Catfsih	LC
Hypophthalmichthys molitrix		NT
Hypophthalmichthys nobilis	Bighead Carp	DD
Hemiculter leucisculus	Sharpbelly	LC
Pseudorasbora parva	Topmouth Gudgeon	LC
Tilapia/Coptodon zillii		LC
Native Species		
Acanthobrama marmid	Mesopotamian Bream	LC
Acanthopagrus arabicus	Arabian Yellowfin Seabream	LC
Alburnoides velioglui	Velioglu's Chub	Not Evaluated
Alburnus caeruleus	Black Spotted Bleak	LC
Alburnus mossulensis	Mossul Bleak	LC
Arabibarbus grypus	Shabout	VU
Leuciscus vorax	Mesopotamian Asp	LC
Barbus lacerta	Lizard Barbel	LC
Bathygobius fuscus	Brown Frillfin	LC
Caecocypris basimi	Haditha Cavefish	CR
Capoeta trutta	Longspine Scraper	LC
Capoeta damascina	Levantine Scraper	LC
Capoeta umbla	Tigris Scraper	LC
Carasobarbus luteus	Mesopotamian Himri	LC
Chondrostoma regium	Mesopotamian Nase	LC
Cobitis taenia	Spined Loach. The name for this species is to be updated to *Cobitis avicennae.*	LC
Cyprinion kais	Smallmouth Lotak	LC
Cyprinion macrostomum	Largemouth Lotak	LC
Eidinemacheilus proudlovei	New species, described from subterranean waters in the Little Zab River drainage in Iraqi Kurdistan. (Freyhof et al., 2016)	Not Evaluated

Table 7.6 (Continued)

Scientific Name	Common Name	IUCN Status
Garra elegans	Mesopotamian Garra	LC
Garra rufa	Red Garra	LC
Garra widdowsoni	Haditha Cave Garra	CR
Glyptothorax kurdistanicus	Mesopotamian Sucking Catfish	DD
Liza abu	Abu Mullet. All *Liza* species will soon be updated to the genus *Planiliza.*	LC
Liza carinata	Keeled Mullet	Not Evaluated
Liza klunzingeri	Klunzinger's Mullet	Not Evaluated
Liza subviridis		Not Evaluated
Luciobarbus esocinus	Pike Barbel	VU
Luciobarbus subquincunciatus	Leopard Barbel	CR
Luciobarbus xanthopterus	Gattan	VU
Mastacembelus mastacembelus	Mesopotamian Spiny Eel	LC
Mesopotamichthys sharpeyi	Binni	VU
Mystus pelusius	Zugzug Catfish	LC
Oxynoemacheilus bergianus	Kura Sportive Loach	LC
Oxynoemacheilus frenatus	Tigris Loach	LC
Oxynoemacheilus argyrogramma	Two Spot Loach	LC
Oxynoemacheilus chomanicus		Not Evaluated
Oxynoemacheilus kurdistanicus		Not Evaluated
Oxynoemacheilus zagrosensis		Not Evaluated
Paracobitis molavii	New species from the Sirwan and Lesser Zab drainages in the Iranian and Iraqi Tigris catchments.	Not Evaluated
Paracobitis zabgawraensis	New species from the Greater Zab drainage in the Iraqi Tigris catchment. Found in the Rezan River in the Barzan area.	Not Evaluated
Silurus triostegus	Mesopotamian Catfish	LC
Squalius berak	Mesopotamian Chub	LC
Squalius cephalus	Chub	LC
Squalius lepidus	Mesopotamian Pike Chub	LC
Tenualosa ilisha	Hilsa	LC
Thryssa hamiltonii	Hamilton's Thryssa	Not Evaluated
Turcinoemacheilus kosswigi	Zagroz Dwarf Loach	LC

Table 7.7   Marine Fish Species in Iraq

Scientific Name	Common Name	IUCN Status
Ablennes hians	Flat Needlefish	LC
Acanthopagrus berda	Goldsilk Seabream	Not Evaluated
Argyrosomus aeneus		
Bothus pantherinus	Leopard Flounder	Not Evaluated
Brachirus orientalis	Oriental Sole	Not Evaluated
Caranx malabaricus	Malabar Trevally	Not Evaluated
Chirocentrus dorab	Dorab Wolf-herring	Not Evaluated
Cynoglossus arel	Largescale Tonguesole	Not Evaluated
Eleutheronema tetradactylum	Fourfinger Threadfin	Not Evaluated
Epinephelus tauvina	Greasy Grouper	DD
Ilisha elongata	Elongate Ilisha	Not Evaluated
Johnius belangerii	Belanger's Croaker	Not Evaluated
Lutjanus rivulatus	Blubberlip Snapper	Not Evaluated
Mugil cephalus	Flathead Mullet	LC
Nemipterus bleekeri	Delagoa Threadfin Bream	Not Evaluated
Otolithes ruber	Tigertooth Croaker	Not Evaluated
Pampus argenteus	Silver Pomfret	Not Evaluated
Platycephalus indicus	Bartail Flathead	DD
Pseudorhombus arsius	Largetooth Flounder	Not Evaluated
Sarda orientalis	Oriental Bonito	LC
Sciaena dussumieri	Sin Croaker	Not Evaluated
Scomberomorous commerson	Narrow-barred Spanish Mackerel	NT
Scomberomorous guttatus	Indo-Pacific King Mackerel	DD
Siganus canaliculatus	White-spotted Spinefoot	Not Evaluated
Sillago sihama	Silver Sillago	Not Evaluated

of Freshwater Ecology and Inland Fisheries in Germany, which included 10 threatened fish species (3 – Critically Endangered (CR), 5 – Vulnerable (VU), and 1 – Near Threatened (NT) freshwater, and 1 – Near Threatened (NT) marine species). Additional effort is needed to complete the Iraqi freshwater and marine fish checklists and recent work indicates that Iraq has 100 freshwater and marine fish species with 4 – Critically Endangered (CR), 2 – Endangered (EN), 11 – Vulnerable (VU), and 10 – Near Threatened (NT) species (Tables 7.6 and 7.7).

Birds: Birds reported from Iraq are listed in Table 7.8.

Table 7.8 Complete checklist of Birds of Iraq

Scientific Name	Common name	IUCN status
Accipiter badius	Shikra	LC
Accipiter brevipes	Levant Sparrowhawk	LC
Accipiter gentilis	Northern Goshawk	LC
Accipiter nisus	Eurasian Sparrowhawk	LC
Acrocephalus arundinaceus	Great Reed-warbler	LC
Acrocephalus griseldis	Basra Reed-warbler	EN
Acrocephalus melanopogon	Moustached Warbler	LC
Acrocephalus palustris	Marsh Warbler	LC
Acrocephalus schoenobaenus	Sedge Warbler	LC
Acrocephalus scirpaceus	Eurasian Reed-warbler	LC
Acrocephalus stentoreus	Clamorous Reed-warbler	LC
Actitis hypoleucos	Common Sandpiper	LC
Aegithalos caudatus	Long-tailed Tit	LC
Aegypius monachus	Cinereous Vulture	NT
Alaemon alaudipes	Greater Hoopoe-lark	LC
Alauda arvensis	Eurasian Skylark	LC
Alcedo atthis	Common Kingfisher	LC
Alectoris chukar	Chukar	LC
Ammomanes cinctura	Bar-tailed Lark	LC
Ammomanes deserti	Desert Lark	LC
Ammoperdix griseogularis	See-see Partridge	LC
Anas acuta	Northern Pintail	LC
Anas crecca	Common Teal	LC
Anas platyrhynchos	Mallard	LC
Anhinga rufa	African Darter	LC
Anser albifrons	Greater White-fronted Goose	LC
Anser anser	Greylag Goose	LC
Anser erythropus	Lesser White-fronted Goose	VU
Anthropoides virgo	Demoiselle Crane	LC
Anthus campestris	Tawny Pipit	LC
Anthus cervinus	Red-throated Pipit	LC
Anthus pratensis	Meadow Pipit	NT
Anthus richardi	Richard's Pipit	LC
Anthus similis	Long-billed Pipit	LC
Anthus spinoletta	Water Pipit	LC

Table 7.8 (Continued)

Scientific Name	Common name	IUCN status
Anthus trivialis	Tree Pipit	LC
Apus affinis	Little Swift	LC
Apus apus	Common Swift	LC
Apus pallidus	Pallid Swift	LC
Aquila chrysaetos	Golden Eagle	LC
Aquila fasciata	Bonelli's Eagle	LC
Aquila heliaca	Eastern Imperial Eagle	VU
Aquila nipalensis	Steppe Eagle	EN
Ardea alba	Great White Egret	LC
Ardea cinerea	Grey Heron	LC
Ardea goliath	Goliath Heron	LC
Ardea purpurea	Purple Heron	LC
Ardeola ralloides	Squacco Heron	LC
Arenaria interpres	Ruddy Turnstone	LC
Asio flammeus	Short-eared Owl	LC
Asio otus	Northern Long-eared Owl	LC
Athene noctua	Little Owl	LC
Aythya ferina	Common Pochard	VU
Aythya fuligula	Tufted Duck	LC
Aythya marila	Greater Scaup	LC
Aythya nyroca	Ferruginous Duck	NT
Botaurus stellaris	Eurasian Bittern	LC
Branta ruficollis	Red-breasted Goose	VU
Bubo ascalaphus	Pharaoh Eagle-owl	LC
Bubo bubo	Eurasian Eagle-owl	LC
Bubulcus ibis	Cattle Egret	LC
Bucanetes githagineus	Trumpeter Finch	LC
Bucephala clangula	Common Goldeneye	LC
Burhinus oedicnemus	Eurasian Thick-knee	LC
Buteo buteo	Eurasian Buzzard	LC
Buteo rufinus	Long-legged Buzzard	LC
Calandrella brachydactyla	Greater Short-toed Lark	LC
Calandrella cheleensis	Asian Short-toed Lark	LC
Calandrella rufescens	Lesser Short-toed Lark	LC

Table 7.8 (Continued)

Scientific Name	Common name	IUCN status
Calidris alba	Sanderling	LC
Calidris alpina	Dunlin	LC
Calidris falcinellus	Broad-billed Sandpiper	LC
Calidris ferruginea	Curlew Sandpiper	NT
Calidris minuta	Little Stint	LC
Calidris pugnax	Ruff	LC
Calidris temminckii	Temminck's Stint	LC
Caprimulgus aegyptius	Egyptian Nightjar	LC
Caprimulgus europaeus	European Nightjar	LC
Carduelis cannabina	Eurasian Linnet	LC
Carduelis carduelis	European Goldfinch	LC
Carduelis chloris	European Greenfinch	LC
Carduelis flavirostris	Twite	LC
Carduelis spinus	Eurasian Siskin	LC
Ceryle rudis	Pied Kingfisher	LC
Cettia cetti	Cetti's Warbler	LC
Charadrius alexandrinus	Kentish Plover	LC
Charadrius asiaticus	Caspian Plover	LC
Charadrius dubius	Little Ringed Plover	LC
Charadrius hiaticula	Common Ringed Plover	LC
Charadrius leschenaultii	Greater Sandplover	LC
Charadrius mongolus	Lesser Sandplover	LC
Chlamydotis macqueenii	Asian Houbara	VU
Chlidonias hybrida	Whiskered Tern	LC
Chlidonias leucopterus	White-winged Tern	LC
Ciconia ciconia	White Stork	LC
Ciconia nigra	Black Stork	LC
Cinclus cinclus	White-throated Dipper	LC
Circaetus gallicus	Short-toed Snake-eagle	LC
Circus aeruginosus	Western Marsh-harrier	LC
Circus cyaneus	Hen Harrier	LC
Circus macrourus	Pallid Harrier	NT
Circus pygargus	Montagu's Harrier	LC
Cisticola juncidis	Zitting Cisticola	LC

Table 7.8 (Continued)

Scientific Name	Common name	IUCN status
Clamator glandarius	Great Spotted Cuckoo	LC
Clanga clanga	Greater Spotted Eagle	VU
Coccothraustes coccothraustes	Hawfinch	LC
Columba livia	Rock Dove	LC
Columba oenas	Stock Dove	LC
Columba palumbus	Common Woodpigeon	LC
Coracias benghalensis	Indian Roller	LC
Coracias garrulus	European Roller	LC
Corvus corax	Common Raven	LC
Corvus corone	Carrion Crow	LC
Corvus frugilegus	Rook	LC
Corvus monedula	Eurasian Jackdaw	LC
Corvus ruficollis	Brown-necked Raven	LC
Coturnix coturnix	Common Quail	LC
Crex crex	Corncrake	LC
Cuculus canorus	Common Cuckoo	LC
Cursorius cursor	Cream-colored Courser	LC
Cygnus columbianus	Tundra Swan	LC
Cygnus cygnus	Whooper Swan	LC
Cygnus olor	Mute Swan	LC
Delichon urbicum	Northern House-martin	LC
Dendrocopos syriacus	Syrian Woodpecker	LC
Dromas ardeola	Crab-plover	LC
Dryobates minor	Lesser Spotted Woodpecker	LC
Egretta garzetta	Little Egret	LC
Egretta gularis	Western Reef-egret	LC
Elanus caeruleus	Black-winged Kite	LC
Emberiza buchanani	Grey-necked Bunting	LC
Emberiza cia	Rock Bunting	LC
Emberiza cineracea	Cinereous Bunting	NT
Emberiza citrinella	Yellowhammer	LC
Emberiza hortulana	Ortolan Bunting	LC
Emberiza leucocephalos	Pine Bunting	LC
Emberiza melanocephala	Black-headed Bunting	LC

Table 7.8 (Continued)

Scientific Name	Common name	IUCN status
Emberiza schoeniclus	Reed Bunting	LC
Eremophila alpestris	Horned Lark	LC
Eremophila bilopha	Temminck's Lark	LC
Eremopterix nigriceps	Black-crowned Sparrow-lark	LC
Erithacus rubecula	European Robin	LC
Erythropygia galactotes	Rufous-tailed Scrub-robin	LC
Eudromias morinellus	Eurasian Dotterel	LC
Falco biarmicus	Lanner Falcon	LC
Falco cherrug	Saker Falcon	EN
Falco columbarius	Merlin	LC
Falco naumanni	Lesser Kestrel	LC
Falco peregrinus	Peregrine Falcon	LC
Falco subbuteo	Eurasian Hobby	LC
Falco tinnunculus	Common Kestrel	LC
Falco vespertinus	Red-footed Falcon	NT
Ficedula albicollis	Collared Flycatcher	LC
Ficedula hypoleuca	European Pied Flycatcher	LC
Ficedula parva	Red-breasted Flycatcher	LC
Ficedula semitorquata	Semi-collared Flycatcher	LC
Francolinus francolinus	Black Francolin	LC
Fringilla coelebs	Eurasian Chaffinch	LC
Fringilla montifringilla	Brambling	LC
Fulica atra	Common Coot	LC
Galerida cristata	Crested Lark	LC
Gallinago gallinago	Common Snipe	LC
Gallinago media	Great Snipe	NT
Gallinula chloropus	Common Moorhen	LC
Garrulus glandarius	Eurasian Jay	LC
Gelochelidon nilotica	Common Gull-billed Tern	LC
Geronticus eremita	Northern Bald Ibis	CR
Glareola nordmanni	Black-winged Pratincole	NT
Glareola pratincola	Collared Pratincole	LC
Grus grus	Common Crane	LC
Gypaetus barbatus	Bearded Vulture	NT

Table 7.8 (Continued)

Scientific Name	Common name	IUCN status
Gyps fulvus	Griffon Vulture	LC
Haematopus ostralegus	Eurasian Oystercatcher	NT
Halcyon smyrnensis	White-breasted Kingfisher	LC
Haliaeetus albicilla	White-tailed Sea-eagle	LC
Hieraaetus pennatus	Booted Eagle	LC
Himantopus himantopus	Black-winged Stilt	LC
Hippolais icterina	Icterine Warbler	LC
Hippolais languida	Upcher's Warbler	LC
Hippolais pallida	Eastern Olivaceous Warbler	LC
Hirundo daurica	Red-rumped Swallow	LC
Hirundo obsoleta	Pale Crag-martin	LC
Hirundo rupestris	Eurasian Crag-martin	LC
Hirundo rustica	Barn Swallow	LC
Hydroprogne caspia	Caspian Tern	LC
Hypocolius ampelinus	Grey Hypocolius	LC
Irania gutturalis	White-throated Robin	LC
Ixobrychus minutus	Common Little Bittern	LC
Jynx torquilla	Eurasian Wryneck	LC
Lanius collurio	Red-backed Shrike	LC
Lanius excubitor	Great Grey Shrike	LC
Lanius isabellinus	Rufous-tailed Shrike	LC
Lanius minor	Lesser Grey Shrike	LC
Lanius nubicus	Masked Shrike	LC
Lanius senator	Woodchat Shrike	LC
Larus armenicus	Armenian Gull	NT
Larus cachinnans	Caspian Gull	LC
Larus canus	Mew Gull	LC
Larus fuscus	Lesser Black-backed Gull	LC
Larus genei	Slender-billed Gull	LC
Larus ichthyaetus	Pallas's Gull	LC
Larus ridibundus	Black-headed Gull	LC
Leiopicus medius	Middle Spotted Woodpecker	LC
Limosa lapponica	Bar-tailed Godwit	NT
Limosa limosa	Black-tailed Godwit	NT

Table 7.8   (Continued)

Scientific Name	Common name	IUCN status
Locustella fluviatilis	Eurasian River Warbler	LC
Locustella luscinioides	Savi's Warbler	LC
Locustella naevia	Common Grasshopper-warbler	LC
Lullula arborea	Wood Lark	LC
Luscinia luscinia	Thrush Nightingale	LC
Luscinia megarhynchos	Common Nightingale	LC
Luscinia svecica	Bluethroat	LC
Lymnocryptes minimus	Jack Snipe	LC
Mareca penelope	Eurasian Wigeon	LC
Mareca strepera	Gadwall	LC
Marmaronetta angustirostris	Marbled Teal	VU
Melanocorypha bimaculata	Bimaculated Lark	LC
Melanocorypha calandra	Calandra Lark	LC
Mergellus albellus	Smew	LC
Mergus merganser	Goosander	LC
Mergus serrator	Red-breasted Merganser	LC
Merops apiaster	European Bee-eater	LC
Merops persicus	Blue-cheeked Bee-eater	LC
Microcarbo pygmaeus	Pygmy Cormorant	LC
Miliaria calandra	Corn Bunting	LC
Milvus migrans	Black Kite	LC
Monticola saxatilis	Rufous-tailed Rock-thrush	LC
Monticola solitarius	Blue Rock-thrush	LC
Montifringilla nivalis	White-winged Snowfinch	LC
Motacilla alba	White Wagtail	LC
Motacilla cinerea	Grey Wagtail	LC
Motacilla citreola	Citrine Wagtail	LC
Motacilla flava	Yellow Wagtail	LC
Muscicapa striata	Spotted Flycatcher	LC
Neophron percnopterus	Egyptian Vulture	EN
Netta rufina	Red-crested Pochard	LC
Numenius arquata	Eurasian Curlew	NT
Numenius phaeopus	Whimbrel	LC
Numenius tenuirostris	Slender-billed Curlew	CR

Table 7.8 (Continued)

Scientific Name	Common name	IUCN status
Nycticorax nycticorax	Black-crowned Night-heron	LC
Oena capensis	Namaqua Dove	LC
Oenanthe albonigra	Hume's Wheatear	LC
Oenanthe chrysopygia	Red-tailed Wheatear	LC
Oenanthe deserti	Desert Wheatear	LC
Oenanthe finschii	Finsch's Wheatear	LC
Oenanthe hispanica	Black-eared Wheatear	LC
Oenanthe isabellina	Isabelline Wheatear	LC
Oenanthe lugens	Mourning Wheatear	LC
Oenanthe moesta	Red-rumped Wheatear	LC
Oenanthe oenanthe	Northern Wheatear	LC
Oenanthe pleschanka	Pied Wheatear	LC
Oenanthe xanthoprymna	Kurdish Wheatear	LC
Onychoprion anaethetus	Bridled Tern	LC
Oriolus oriolus	Eurasian Golden Oriole	LC
Otis tarda	Great Bustard	VU
Otus brucei	Pallid Scops-owl	LC
Otus scops	Eurasian Scops-owl	LC
Oxyura leucocephala	White-headed Duck	EN
Pandion haliaetus	Osprey	LC
Parus caeruleus	Blue Tit	LC
Parus lugubris	Sombre Tit	LC
Parus major	Great Tit	LC
Passer domesticus	House Sparrow	LC
Passer hispaniolensis	Spanish Sparrow	LC
Passer moabiticus	Dead Sea Sparrow	LC
Passer montanus	Eurasian Tree Sparrow	LC
Pelecanus crispus	Dalmatian Pelican	VU
Pelecanus onocrotalus	Great White Pelican	LC
Pernis apivorus	European Honey-buzzard	LC
Petronia brachydactyla	Pale Rock Sparrow	LC
Petronia petronia	Rock Sparrow	LC
Petronia xanthocollis	Chestnut-shouldered Petronia	LC
Phalacrocorax carbo	Great Cormorant	LC

Table 7.8   (Continued)

Scientific Name	Common name	IUCN status
Phalaropus lobatus	Red-necked Phalarope	LC
Phoenicopterus roseus	Greater Flamingo	LC
Phoenicurus erythronotus	Rufous-backed Redstart	LC
Phoenicurus ochruros	Black Redstart	LC
Phoenicurus phoenicurus	Common Redstart	LC
Phylloscopus collybita	Common Chiffchaff	LC
Phylloscopus humei	Hume's Leaf-warbler	LC
Phylloscopus sibilatrix	Wood Warbler	LC
Phylloscopus sindianus	Mountain Chiffchaff	LC
Phylloscopus trochilus	Willow Warbler	LC
Pica pica	Black-billed Magpie	LC
Picus viridis	Eurasian Green Woodpecker	LC
Platalea leucorodia	Eurasian Spoonbill	LC
Plegadis falcinellus	Glossy Ibis	LC
Pluvialis squatarola	Grey Plover	LC
Podiceps cristatus	Great Crested Grebe	LC
Podiceps nigricollis	Black-necked Grebe	LC
Porphyrio porphyrio	Purple Swamphen	LC
Porzana porzana	Spotted Crake	LC
Prinia gracilis	Graceful Prinia	LC
Prunella collaris	Alpine Accentor	LC
Prunella modularis	Hedge Accentor	LC
Prunella ocularis	Radde's Accentor	LC
Pterocles alchata	Pin-tailed Sandgrouse	LC
Pterocles orientalis	Black-bellied Sandgrouse	LC
Pterocles senegallus	Spotted Sandgrouse	LC
Pycnonotus leucotis	White-eared Bulbul	LC
Pyrrhocorax graculus	Yellow-billed Chough	LC
Pyrrhocorax pyrrhocorax	Red-billed Chough	LC
Rallus aquaticus	Western Water Rail	LC
Recurvirostra avosetta	Pied Avocet	LC
Remiz pendulinus	Eurasian Penduline-tit	LC
Rhodopechys obsoletus	Desert Finch	LC
Rhodopechys sanguineus	Asian Crimson-winged Finch	LC

Table 7.8   (Continued)

Scientific Name	Common name	IUCN status
Riparia riparia	Sand Martin	LC
Saxicola rubetra	Whinchat	LC
Saxicola torquatus	Common Stonechat	LC
Scolopax rusticola	Eurasian Woodcock	LC
Scotocerca inquieta	Streaked Scrub-warbler	LC
Serinus pusillus	Fire-fronted Serin	LC
Serinus syriacus	Syrian Serin	VU
Sitta europaea	Wood Nuthatch	LC
Sitta neumayer	Western Rock-nuthatch	LC
Sitta tephronota	Eastern Rock-nuthatch	LC
Spatula clypeata	Northern Shoveler	LC
Spatula querquedula	Garganey	LC
Spilopelia senegalensis	Laughing Dove	LC
Stercorarius parasiticus	Arctic Jaeger	LC
Sterna hirundo	Common Tern	LC
Sterna repressa	White-cheeked Tern	LC
Sternula albifrons	Little Tern	LC
Streptopelia decaocto	Eurasian Collared-dove	LC
Streptopelia turtur	European Turtle-dove	VU
Strix aluco	Tawny Owl	LC
Sturnus roseus	Rosy Starling	LC
Sturnus vulgaris	Common Starling	LC
Sylvia atricapilla	Blackcap	LC
Sylvia borin	Garden Warbler	LC
Sylvia communis	Common Whitethroat	LC
Sylvia curruca	Lesser Whitethroat	LC
Sylvia hortensis	Orphean Warbler	LC
Sylvia mystacea	Menetries's Warbler	LC
Sylvia nana	Desert Warbler	LC
Sylvia nisoria	Barred Warbler	LC
Tachybaptus ruficollis	Little Grebe	LC
Tachymarptis melba	Alpine Swift	LC
Tadorna ferruginea	Ruddy Shelduck	LC
Tadorna tadorna	Common Shelduck	LC

Table 7.8   (Continued)

Scientific Name	Common name	IUCN status
Tetraogallus caspius	Caspian Snowcock	LC
Tetrax tetrax	Little Bustard	NT
Thalasseus bengalensis	Lesser Crested Tern	LC
Thalasseus bergii	Greater Crested Tern	LC
Thalasseus sandvicensis	Sandwich Tern	LC
Threskiornis aethiopicus	African Sacred Ibis	LC
Tichodroma muraria	Wallcreeper	LC
Tringa erythropus	Spotted Redshank	LC
Tringa glareola	Wood Sandpiper	LC
Tringa nebularia	Common Greenshank	LC
Tringa ochropus	Green Sandpiper	LC
Tringa stagnatilis	Marsh Sandpiper	LC
Tringa totanus	Common Redshank	LC
Troglodytes troglodytes	Winter Wren	LC
Turdoides altirostris	Iraq Babbler	LC
Turdoides caudata	Common Babbler	LC
Turdus iliacus	Redwing	NT
Turdus merula	Eurasian Blackbird	LC
Turdus philomelos	Song Thrush	LC
Turdus pilaris	Fieldfare	LC
Turdus ruficollis	Dark-throated Thrush	LC
Turdus torquatus	Ring Ouzel	LC
Turdus viscivorus	Mistle Thrush	LC
Tyto alba	Common Barn-owl	LC
Upupa epops	Common Hoopoe	LC
Vanellus gregarius	Sociable Lapwing	CR
Vanellus indicus	Red-wattled Lapwing	LC
Vanellus leucurus	White-tailed Lapwing	LC
Vanellus spinosus	Spur-winged Lapwing	LC
Vanellus vanellus	Northern Lapwing	NT
Xenus cinereus	Terek Sandpiper	LC
Zapornia parva	Little Crake	LC
Zapornia pusilla	Baillon's Crake	LC

(Salim et al., 2016)

Web Resources

Amphibia Web: This website provides information on amphibian declines, natural history, conservation, and taxonomy. www.amphibiaweb. org.

FishBase: FishBase is a relational database with information to cater to different professionals such as research scientists, fisheries managers, zoologists and many more. FishBase on the web contains practically all fish species known to science. www.fishbase.org.

Freshwater Fishes of Iraq: Brian W. Coad's Personal Website. Brian is an ichthyologist (student of fishes), working for the Canadian Museum of Nature in Ottawa, Canada. His main research interests are on Canadian freshwater and marine fishes and Southwest Asian, principally Iranian, freshwater fishes. He also has an inordinate fondness for words. www.briancoad.com.

Nature Iraq: The largest Iraqi conservation, non-governmental organization in Iraq. Nature Iraq is an affiliate of BirdLife International. www. natureiraq.org.

The Reptile Database: This database, edited by P. Uetz, provides a catalog of all living reptile species and their classification. The database covers all living snakes, lizards, turtles, amphisbaenians, tuataras, and crocodiles. Currently there are about 10,000 species including another 2,800 subspecies. The database focuses on taxonomic data, that is, names and synonyms, distribution and type data and literature references. www.reptile-database.org.

Keywords

- birds checklist
- freshwater fish species
- marine fish species
- mesopotamia

References

Batanouny, K. H., (2001). *Plants in the Deserts of the Middle East-Adaptations of Desert Organisms.* Springer-Verlag Berlin Heidelberg, New York.

Coad, B. W., (2010). *Freshwater Fishes of Iraq.* Pensoft Series Faunistica #93. Sofia-Moscow: Pensoft Publishers.

Convention on Biological Diversity (CBD), (2014). *Aichi Biodiversity Targets.* CBD Secretariat. Retrieved from http://www.cbd.int/sp/targets/ on 13 November.

Freyhof, J., (2013). *Fieldwork in Iraqi Kurdistan in June 2012.* Retrieved from http://joergfreyhof.de/fieldwork/278-fieldwork-in-iraqi-kurdistan-in-june-2012.

Freyhof, J., Abdullah, Y., Ararat, K., & Ibrahim, H., (2016). *Eidinemacheilus proudlovei, a new subterranean loach from Iraqi Kurdistan* (Teleostei, Nemacheilidae). *Zootaxa., 4173*(3), 466–476.

Guest, E., (1966). *Flora of Iraq, vol. 1.* Baghdad: Ministry of Agriculture of the Republic of Iraq.

Harrison, D. L., & Bates, P. J. J., (1991). *The Mammals of Arabia* (2nd ed.). Tonbridge, UK: Harrison Zoological Museum.

Harrison, D. L., (1964–1972). *The Mammals of Arabia, vols. 1–3.* London: Ernest Benn Limited.

Hatt, R. T., (1959). *The Mammals of Iraq.* Museum of Zoology, Univ. of Michigan.

Nature Iraq, (2017). *Key Biodiversity Areas of Iraq.* Sulaimaniyah, Iraq: Tablet House Publishing.

Porter, R. F., (2016). Odonata from Iraq, with three new records. *Notulae Odonatologicae, 8*(8), 247–318.

United Nations Environment Programme (UNEP), (2003). *Desk Study on the Environment in Iraq.* Switzerland: UNEP. Retrieved from http://postconflict.unep.ch/publications/ Iraq_DS. pdfon 26 October 2014.

United Nations Environment Programme (UNEP), (2006). *Global Desert Outlook.* Nairobi Kenya: UNEP. Retrieved from http://www.unep.org/geo/gdoutlook/.

Wikipedia, (2012). *Cheekha Dar.* Retrieved from http://en.wikipedia.org/w/index. php?title=Cheekha_Dar&oldid=490631565.

World Wide Fund for Nature (WWF), (2014). *Ecoregions.* Retrieved from www.worldwildlife.org/biome-categories/terrestrial-ecoregions.

Plate 7.1 Qara Dagh Mountain, Kurdistan Region (Photo by Hana Raza, 2010).

Plate 7.2 The Mesopotamian Marshlands (Photo by Jassim Al-Asadi).

Plate 7.3 Persian Leopard *Panthera pardus saxicolor* (EN) (Photo credit: Nature Iraq/CLP, 2012).

Plate 7.4 Wild Goat *Capra aegagrus*(VU) (Photo credit: Korsh Ararat/Prince Bernhard Nature Fund PBNF, 2016).

Plate 7.5 Striped Hyena *Hyaena hyaena* (NT) (Photo credit: Nature Iraq/PBNF, 2016).

Plate 7.6 Egyptian Vulture *Neophron percnopterus* (EN) (Photo credit: Korsh Ararat/Prince Bernhard Nature Fund PBNF, 2016).

Plate 7.7 Spur-thighed Turtoise *Testudo graeca* (VU) (Photo credit: Korsh Ararat/ Prince Bernhard Nature Fund PBNF, 2016).

CHAPTER 8

Biodiversity in Japan

**TAKAFUMI OHSAWA,[1] FUMIKO NAKAO,[1] and
TOHRU NAKASHIZUKA[2]**

[1]Nature Conservation Bureau, Ministry of the Environment, Japan, 1-2-2, Kasumigaseki,
Chiyoda-ku, Tokyo, 100-8975, Japan, E-mail: takafumi_osawa@env.go.jp
[2]Research Institute for Humanity and Nature, 457-4 Kamigamo-Motoyama, Kitaku, Kyoto,
603-8047, Japan

8.1   Introduction

More than 90,000 species of wild flora and fauna have been found in Japan (Ministry of the Environment of Japan, 2002) (Table 8.1).

Thus, in spite of its limited territory area (380,000 km^2 land area), Japan holds affluent biodiversity. Besides, high percentage of endemic species is noteworthy: around 40% of land mammals and vascular plants, around 60% of reptiles and around 80% of amphibians are endemic to Japan. Among all developed countries, only Japan harbors wild monkey populations. Other wildlife, including large mammals such as Sika deer and bears, also inhabits Japan.

According to Ministry of the Environment of Japan (2010), Flora in Japan belongs to the Paleotropical floristic region and the Holarctic region among the six floristic regions in the world. Pandanacae and palm trees are peculiar to the Paleotropical region, while Japanese chestnuts and willows are characteristic species of the Holarctic region. Fauna in Japan belongs to the Palearctic region as well as the Indomalaya region among six zoogeographic zones. The Watase Line, which corresponds to the Tokara Strait, is the boundary between the two regions.

In addition to the Watase L., there are also several biogeographical boundaries in and around the Japanese archipelago (Hatta L., Ishikari lowland L., Kuromatsunai lowland L., Blakiston L., *Crinum asiaticum* L., Lewis L., Makino L., Korean Strait L., Tsushima Strait L., and Miyake L.). Most of them, running from west to east, have been shaped due to temperature gradients. The northern temperate region of the archipelago is mostly covered with deciduous forests (mainly *Fagus* and *Quercus* species), while evergreen broadleaved forests (mainly *Quercus* and *Castanopsis* species) dominate in the southern warmer part of the archipelago (Hotta, 1974). For instance, Japanese beech (*F. crenata*) and Japanese oak (*Q. crispula*) are typical climax species in the cool-temperate region, though many beech

Table 8.1   Species Number of Japanese Fauna and Flora

Mammals (except cetaceans and sirenians)	109
Insectivores	20
Chiropterans (bats)	35
Primate	1
Carnivoras	23
Artiodactyls	3
Rodents	23
Lagomorphs	4
Birds	ca. 540
Reptiles	98
Amphibians	64
Freshwater fish	ca. 200
Insects	> ca. 30,000
Vascular plants	ca. 6,000*
Bryophytes	ca. 1,600
Algae	ca. 5,500
Lichen	ca. 1,800
Fungi	ca. 16,500

*, the number increases up to > 8,000, when including subspecies and variant species.
The information in the table is cited from Biodiversity Center of Japan (2010).

forests have been replaced by secondary forests. Differences in the amount of snowfall may also result in biogeographical boundaries. Specifically, the main mountain range separates the Japan Sea side from the Pacific Ocean side of the country. In winter, cold Siberian air masses blow over the warmer Japan Sea, resulting in the deep snowfalls (up to 10 m) falling on this side of the country; the Pacific Ocean side, in contrast, receives little precipitation (Uemura, 1989; Gansert, 2004). Such environmental clines in the Japanese Archipelago have resulted differentiations of plants at the species and intra-species levels (Hotta, 1974; Gansert, 2004; Ohsawa and Ide, 2011). The Japan Sea side is characterized with extensive beech forests together with deciduous broadleaved woodlands (i.e., quasi-alpine zone).

According to Ministry of the Environment of Japan (2010), fauna in the southern side of the Watase L. includes many species that are taxonomically related (close) to species in Taiwan and Southeastern Asia. In the northern sides of the same line, fauna is similar to that of the Eurasian continent. Furthermore, the Blakiston L., which corresponds to the Tsugaru Strait, divides the northern side into two sub-regions (the Siberian sub-region and the

Chinese sub-region). The Siberian sub-region harbors many common and related species in relation to Siberia and the Russian Far East, such as Blakiston's fish owl, Brown bear, and Pika. The Chinese sub-region harbors many common species with the Korean Peninsula, such as the Asian black bear.

Historically, the Japanese archipelago was connected to the Eurasian continent during the coldest period of the Last Glacial Maximum (LGM), around 18,000–20,000 years before the present (Minato, 1966). During such periods, many plant and animal species have immigrated into the archipelago through multiple land connection routes. Subsequently, the species' populations were shaped with re-isolation events from the continent (Ministry of the Environment of Japan, 2010). Hence, rich biodiversity in Japan can be attributed to topographical complexity and geohistorical repetition of land connection with and separation from the Eurasian continent (Biodiversity Center of Japan, 2010).

Microbes include viruses, bacterium, fungi, protozoans, and algae, and each taxon comprises a number of species. Around 5,500 species of algae and 16,500 species of fungi in Figure 8.1 are also microbes, while each of lichen species (around 1,800 spp.) is a symbiont of fungi (mostly ascomycetes) and algae respectively. Among Japanese microbes, some basidiomycetes, which are categorized in fungi, are known as mushrooms. Japanese beech (*Fagus*) forests harbor diverse mushrooms, including some edible species, such as *Pholiota nameko* and *Mycoleptodonoides aitchisonii* (Kaneko and Sahashi, 1998). Such basidiomycetes (e.g., Mycena) can decompose lignin as well as

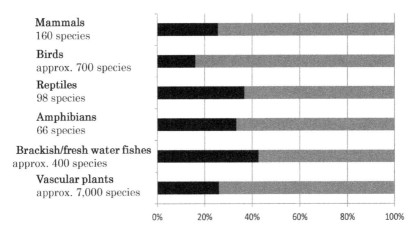

Figure 8.1 The ratio of extinct/extinct in the wild/endangered species (all put together shown in black color) and total number of species of each taxonomic group (shown as black + grey) reported from Japan (Ministry of the Environment of Japan, 2016a).

holocellulose, while ascomycetes can decompose holocellulose (Japanese Ecological Society, 2011). As such, more than hundred fungi species were found even in a three-year decomposition experiment of beech litter (Osono and Takeda, 2001), and these species are crucial to sustain forest ecosystems. Besides, some Japanese drinks and foods are produced through fermentation processes. For instance, Japanese miso (fermented soybean paste) is made by bacteria (*Tetragenococcus halophilus*), yeast (*Zygosaccharomyces rouxii, Candida versatilis*, and *Pichia miso*), and fungi (*Aspergillus oryzae*) (Sakamoto, 2008). Japanese sake (Japanese traditional rice wine) is made by bacteria (*Lactobacillus sakei* and *Leuconostoc mesenteroides*), yeast (*Saccharomyces cerevisiae*), and fungi (*Aspergillus oryzae*) (Sakamoto, 2008).

In contrast, some microbes can influence ecosystems and human beings negatively. For instance, avian influenza virus, which belongs to Orthomyxoviridae, can bring about mass death of wild birds, and wild waterfowls can spread the virus infection (Kajihara et al., 2011). The subtype of H5N1 of the virus is the most pathogenic, and it could be also fatal for humans occasionally. Chytrid fungus, *Batrachochytrium dendrobatidis*, can cause chytridiomycosis. However, in Japan, there are at least 26 types of the fungus, some of which seem to have commensal relationships with their host species (the Japanese giant salamander) in Japan (Goka et al., 2010). Therefore, such Japanese amphibians have acquired resistance to the sublethal impacts of the fungus, though oversea amphibians could be devastated by the same fungus (Goka et al., 2010). As such, Japanese microbes are diverse, including the species that are beneficial and those that are problematic.

The outline of ecosystems and species diversity (except microbes) in Japan is described in National Biodiversity Strategy of Japan 2012–2020 (Ministry of the Environment of Japan, 2012a). As well, Ministry of the Environment of Japan (2016a) published Japan Biodiversity Outlook 2 (JBO2), which was an assessment on biodiversity and ecosystem services in this country. Thus, hereafter we describe "Ecosystem diversity/Vegetation/Biomes" based on National Biodiversity Strategy of Japan 2012–2020 and the following five sections based on JBO2. Lastly, a few additional publications are introduced in the section of "Supplemental information" for reference.

8.2   Ecosystem Diversity/Vegetation/Biomes

Japanese land consists of arc islands located at the edge of the Eurasian continent on multiple plate borders. The land stretches over around 3,000 km from north (45°N) to south (20°N). There are several thousand islands of different size with high variation in altitude (up to 3,765 m of Mt. Fuji). This country has geological history to be connected to Asian continent for several times in the past million years, which relates to the uniqueness of flora and fauna of each island and district.

The land covers from sub-tropical to boreal climate zones with monsoons, generally humid enough to develop forest vegetation (1,600 mm annual rainfall on the average). As topographies of Japanese land areas are steep, rain water can flow downwards through rivers rapidly. Water discharge in rivers fluctuates due to seasonal rains and typhoons. Within each river, deep pools and rapids are shaped, while floodplains are also found in downstream areas. In addition, disturbances such as volcanic eruptions, earthquakes, and tsunamis, have contributed to create diversity of habitats.

Vegetation maps (1:50,000) cover the whole area of this country through the National Survey on the Natural Environment by Ministry of the Environment. They show that 67% of the total land area is covered by forests. This percentage is as high as that of Sweden (70%), and rather higher than those of some other developed countries, such as the United Kingdom (12%) and the United States (33%). In contrast, only 17.9% of the total area is covered by natural (primeval) forests. Including natural grasslands, totally 19.0% of land area is covered by natural vegetation, on which human activities are unlikely to have influenced mostly because of some geographical reasons, such as remote areas, steep mountains, peninsulas, and isolated islands. On flat sites or gentle hills, substitutional vegetation (e.g., secondary forests and secondary grasslands), artificial forests, and croplands are common. Natural succession proceeds swiftly due to ample precipitation and relatively warm climate.

Human beings have reached the islands about 20,000 years ago, and their activities like agriculture and forestry provided open habitats which are disturbed periodically. Some species have adapted to such human induced habitats, typically seen as *Pasque* plant and *Shijimiaeoides divina* butterfly. Yet, such environments are now being lost widely, because of the changes in agricultural and forestry systems and human life style in recent decades. These diverse types of vegetation originated from the variations in climate, topography, human disturbances, and their combinations, have created diverse habitats for organisms in this country.

There are six major types of ecosystems in Japan. They are discussed in the following subsections.

8.2.1 Forest Ecosystems

Forests are distributed broadly in Japanese archipelago due to humid climates. From south to north or from lowlands to highlands, ever-green broad-leaved forests (sub-tropical or warm temperate zones), deciduous broad-leaved forests (cool temperate), and coniferous forests (boreal) associated with climate zones. Upper limit of forest ecosystems (tree line) locates at about 2,500 m a.s.l. in Honshu Island, but at about 1,500 m a.s.l. in Hokkaido. Such natural forests are important habitats for fauna and flora unique to Japan.

Historically, forests have been utilized for slash-and-burn agriculture in ancient days, then coppice forest for fuels (firewood and charcoal) until the 1970s. Secondary forests were also utilized for collections of non-timber forest products including food. Plantation forests have been increasing since more than 1,000 years ago until the 1980s. Under periodical disturbances, some of these forests have shaped unique landscape as mosaics of secondary forests, plantations, and remnant mature forests.

8.2.2 Farmland Ecosystems

Since rice agriculture was inherited from the Eurasian continent, farmlands (e.g., rice paddy fields, and slash-and-burn fields), waterways/artificial ponds, forests for agricultural use, and secondary grasslands for stock-farming have been created in mosaic manners. Besides, water use for rice agriculture has resulted in unique types of rice paddy fields, such as Yatsuda (paddy fields at valley bottom) and Tanada (Terrace paddy fields). Such mosaic landscapes around residential areas (Satoyama) developed during thousands years of history have become important habitats for wild species that have adapted to human disturbances.

8.2.3 Urban Ecosystems

Farmlands and residential areas have developed on flat sites, such as flood plains, river mouths and alluvial fans, and tablelands. Reed beds and mudflats were dominant vegetation in river mouths, but utilized as basis of major urban areas by the 1700s or the 1900s. Urban areas had increased from 3% to 6% of total land area in Japan between 1850 and 1950, and still increasing in slower rate.

8.2.4 Inland Water Ecosystems

Catchment sizes of rivers are usually small in Japan, and flashy streams are common. Seasonal rains and typhoons sometimes cause extreme precipitation temporarily, and geological complexity increases sediment discharge from these rivers. As a consequence, riverside boulders have been common and large floodplains are easily shaped. There are many species that migrate between sea and rivers and also those have specific to brackish water. Besides, there are many endemic species among freshwater fish in Japan. Wetlands and riparian areas are key habitats for terrestrial wildlife, such as large cranes, storks, many migratory birds, amphibians, and insects.

8.2.5 Coastal and Marine Ecosystems

Japan has a number of islands that are scattered across 3,000 km from north to south, being surrounded by four marine areas, such as the Sea of Okhotsk, the Japan Sea, the East China Sea, and the Pacific Ocean. Steep topographies are found between continental shelves and deep oceans. Cold Oyashio Current flows southwards, while warm Kuroshio Current flows northwards. These intricate environmental factors contribute to growing fish populations of more than 3,500 species.

Such seafoods have been utilized as main source of proteins for Japanese people, while sea algae have been also used as food and green manure. As such, coastal and marine ecosystems have been utilized in many ways. Coastal and shallow water areas (e.g., tidal flats, seaweed beds, coral reefs, sandy beaches, dunes, and rocky shores) have been crucial habitats for wildlife but also important for human activities for a long time. Many tidal flats and seaweed beds were retained in good conditions though were rapidly landfilled during the period of rapid economic growth.

8.2.6 Island Ecosystems

Japan comprises of 6,800 small islands not only four main islands. Among them, there are islands which were isolated for a long time such as Ogasawara Archipelago and Nansei Archipelago. Unique biota is observed in these islands and are important as stopping points (or breeding grounds in uninhabited islands) of migratory bird species.

8.3 Species Diversity

The 4th Version of the Japanese Red Lists by Ministry of the Environment Japan reported that 26% of mammals, 16% of birds, 37% of reptiles, 33% of amphibians, 43% of brackish water/freshwater fish and 26% of vascular plants were extinct or endangered in Japan (Figure 8.2). Among mammals, birds, amphibians, reptiles, brackish water/freshwater fish, and coleopteran insects, 30 species (including the species that were extinct in wild) have

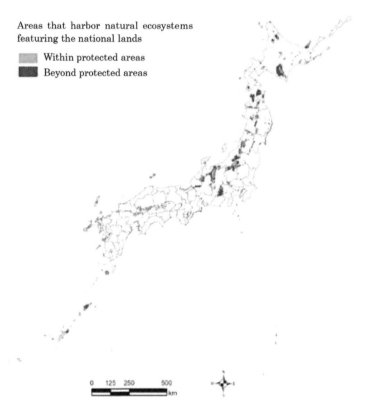

Areas that harbor natural ecosystems
featuring the national lands

☐ Within protected areas
■ Beyond protected areas

0   125 250   500
▬▬▬▬▬▬▬ km

Figure 8.2 The gap between protected areas and "areas that harbor natural ecosystems featuring the national lands" (Ministry of the Environment of Japan, 2012b).

been extinct since early 19th century, while 12 species of them disappeared since the late 1950s.

Besides, 40 species of vascular plants have been extinct completely or extinct in the wild since late 1920s, and 22 species are also about to be extinct (almost extinct). The average extinction rate during the past 50 years was 8.6 species/10 years. The number of extinct species gradually decreased in the later assessment period. However, if "almost extinct" species are included, the number of extinction has not been decreased yet. Regarding endangered vascular plants whose distribution data are available, the number of such species is large in Kagoshima, Okinawa, Hokkaido prefectures, where many endemic species are reported.

Limited information is available as to endangered species in coastal/marine areas. Yet, the data book by Fisheries Agency of Japan (1998) listed

118 aquatic species as endangered/vulnerable species, which include 6 marine shells, 15 saltwater fish and 8 marine algae, and the red data book by the Japanese Association of Benthology (2012) listed 651 invertebrates that are living in tidal flats area in Japan as the species could be possibly extinct.

8.4 Genetic Diversity

Variety of cultivars has lost by pursing productivity in Japanese agriculture. Number of district-specific cultivars that used to be cultivated by farmers in each district over a long time (> approx. 100 years) have been decreasing.

Regarding rice, around 4,000 cultivars used to be cultivated at the beginning of the Meiji Era (1868), though only 88 varieties are cultivated in 2005, suggesting a significant reduction in the number of cultivated varieties. Besides, along with changes in dietary life and improvement of incomes, the cultivated area of miscellaneous millets (e.g., *Setaria italica* and *Echinochloa esculenta*), which used to be several 100 km^2 in the 1950s, has drastically decreased.

District-specific dietary culture has been shaped by local cuisines. Nonetheless, dietary culture became much more homogenized in 1990 than that in 1963, and also coefficient of variation in expenditure on each type of food among districts was still declining in 2014. On the other hand, various crops (e.g., western vegetables) are now introduced in response to diversification of dietary life. About 2.2 billion plant genetic resources are stored as of March in 2012 mainly by National Institute of Agrobiological Sciences, which is contributing to not only developing new cultivars but also reviving past cultivars. For instance, Shirome-mai rice, which used to be cultivated during the Edo and Meiji eras, was becoming unpopular after the World War II in spite of its delicious taste, since it cannot be harvested efficiently. However, this variety was reintroduced for cultivation by using the seeds stored at Gene Bank of Agrobiological Sciences in order to revive towns as an unique local product.

8.5 Protected Area Network in Japan

During period of rapid economic growth, when lands were developed drastically, the basis of the current distribution of protected areas was shaped. The total area of national/quasi-national parks and wildlife protected areas was also increased rapidly during the same period. With regard to terrestrial protected areas managed by Ministry of the Environment of Japan, the total area of national parks (under the Natural Parks Act), quasi-national parks (under the Natural Parks Act), and wildlife protected areas (under the Wildlife Protection and Hunting Management Law) was around 32,000 km^2 in around 1960. Thereafter, including additional institutions of protected areas, such as prefectural nature parks (under the Natural Parks Act), wilderness areas (under the Nature Conservation Act), nature conservation areas (under the Nature Conservation Act), prefectural nature conservation areas (under the Nature Conservation Act), and natural habitat conservation areas (under the Act on Conservation of Endangered Species of Wild Fauna and Flora), protected areas in total expanded into around 1 billion km^2 in 2010, about 3.1 times larger than the previous area, corresponding to 20.3% of land and inland water areas.

The remnant ecosystems retaining genuine natural environments are important to know natural characteristics of Japan, and are regarded as core areas for biodiversity conservation. Such areas are classified as "areas that harbor natural ecosystems featuring the national lands" according to an exercise work by Ministry of the Environment of Japan (2012b). About half or more of the areas are within protected areas (Figure 8.3), mostly designated by the national government and sometimes by prefectural governments at the same time. When looking at species number of birds, gaps between highly prioritized areas for conservation and protected areas are now detected (Naoe et al., 2015).

In Kagoshima prefecture, which published prefectural biodiversity strategy in 2014, the proportion of natural parks to total prefectural area is currently 9.4%, but it is scheduled to increase up to 14.4% by fiscal year 2023. Each local government is also addressing the issue of protected area to be increased, based on local biodiversity strategies and action plans.

In contrast, only 8.3% of marine areas are covered by coastal and marine protected areas. In the past, most of marine protected areas, being as a part of national and quasi-national parks, were designated as buffer zones of protected areas. Currently, the marine protected areas are reconsidered as a means of conservation and sustainable use of biodiversity.

8.6 Challenges and Major Threats to Biodiversity in Japan

In comparison with high economic growth period and the bubble economy until the early 1990s, pressure of infrastructure development and land use has been decreasing in recent 20 years. However, development and land

	Drivers of Biodiversity Loss										
	First Crisis			Second Crisis			Third Crisis			Fourth Crisis	
	Development, alternation of ecosystems	Eutrophication	Loss of endangered species	Reduced use and management of Satochi-Satoyama	Reduced direct use of wildlife	Loss of endangered species	Invasion and establishment of alien species	Chemical substances	Loss of endangered species	Climate change	Loss of endangered species
Between 50 and 20 years ago	↑	↗	↑	↗	↗	↗	↗	↗	↗	→	?
From 20 years ago to the present	↘	↘	↗	↗	→	↗	↗	↘	↗	↗	?
Degree of impact and current trend	↘	→	↘	↗	→	↗	↗	↘	↗	↗	?

(Long-term trend of impact for first two rows; Degree of impact and current trend for bottom row)

Note: Descriptions of the terms used in the table are as follows:
- First Crisis is the impact on biodiversity caused by development, exploitation, and other human activities, including habitat alternation, direct use, and water pollution.
- Second Crisis is the impact caused by decline in human intervention in nature, including reduced use/management of Satochi-Satoyama.
- Third Crisis is the crisis brought by alien species, chemical substances, and other consequences of modern lifestyles and human activities.
- Fourth Crisis is the impact due to climate and other environmental changes, including global warming, increased occurrence of strong typhoons, change in precipitation patterns, decreased fisheries catch, and ocean acidification.

Legend	Drivers			
	Degree of impact during assessment period		Long-term and current trend of impact	
	Weak	○	Decreasing	↘
	Medium	○	Same	→
	Strong	◔	Increasing	↗
	Very strong	●	Increasing rapidly	↑

Note: Graphic symbols may not represent all of the multiple factors related to the indicators in question.
Note: Arrows circled by dotted lines indicate that information is insufficient to make accurate assessments.

Figure 8.3 Summary of the assessment result about biodiversity by JBO2 (Ministry of the Environment of Japan, 2016b).

change are still continuing in small scales, and also some wild animal and plant species are collected and harvested in unsustainable manner. Such artificial impacts could be more significant on species whose habitats were already shrunk. Many natural forests, farmlands, wetlands, and tidal flats have been displaced, and intact vegetation is remained in less than 20% of the national lands. Around 40% of tidal flats disappeared during high economic growth period.

In contrast, impacts due to decrease of human interventions are becoming more important than before. Current area of abandoned farmlands is about three times larger than that of 1975. Besides, impacts of alien species are stronger than before. At present, 429 species were selected as alien species that threaten ecosystems and biodiversity in Japan. In particular, alien species are responsible for 70% of decrease of endangered reptile species. Climate change is also affecting biodiversity more clearly than before. For instance, phenological synchronism between alpine plants and symbiotic bumblebees is disturbed. Coverage by coral reefs to a sea bottom around Okinawa Main Island was reduced to 7.5% in 2009. Shoot growth speed of stone pines (*Pinus pumila*) has been increasing over the last two decades from 30–35 mm/yr to 50–55 mm/yr (i.e., increase by 60%).

8.7 Supplemental Information

Ecosystems, fauna and flora, together with implemented measures of nature conservation in this country are visually described with maps and photos by Biodiversity Center of Japan (2010) with many case studies on affluent biodiversity. Genetic diversity of Japanese species has been rarely reviewed and summarized, though Ohsawa and Ide (2011) compiled and reviewed 63 case studies on genetic diversity of Japanese plant species, suggesting the genetic hotspots (past refugia sites), where many regionally unique species occurred. As well, Tsumura and Suyama (2015) showed maps with genetic structures of main tree species so that tree breeders can know to which extent each tree seed and seedling can be moved without genetic disturbances. As such, information on biodiversity in Japan is gradually accumulated at even the genetic level, and some of the information is now available for practical nature conservation.

Keywords

- JBO2
- Satoyama
- four crises
- Watase Line

References

Biodiversity Center of Japan, (2010). *Biodiversity of Japan*. A harmonious coexistence between nature and humankind.

Fisheries Agency of Japan, (1998). *Data Book for Japanese Rare Wild Aquatics*. Japan Fisheries. Resources Conservation Association, Tokyo. (in Japanese).

Gansert, D., (2004). Treelines of the Japanese Alps–altitudinal distribution and species composition under contrasting winter climates. *Flora*, *199*, 143–156.

Goka, K., Yokoyama, J. U. N., Une, Y., Kuroki, T., Suzuki, K., Nakahara, M., Kobayashi, A., Inaba, S., Mizutani, T., & Hyatt, A. D., (2010). Amphibian chytridiomycosis in

Japan: Distribution, haplotypes and possible route of entry into Japan. *Molecular Ecology, 18*(23), 4757–4774.

Hotta, M., (1974). *Evolutionary Biology in Plants (III) History and Geography of Plants.* Sansei- do, Tokyo, pp. 400.

Japanese Association of Benthology, (2012). *Threatened Animals of Japanese Tidal Flats: Red. Data Book of Seashore Benthos.* Tokai University Press.

Japanese Ecological Society, (2011). *A Series of Modern Ecology (11): Ecology of microbes.* Kyoritsu Publisher.

Kajihara, M., Matsuno, K., Simulundu, E., Muramatsu, M., Noyori, O., Manzoor, R., Nakayama, E., Igarashi, M., Tomabechi, D., Yoshida, R., Okamatsu, M., Sakoda, Y., Ito, K., Kida, H., & Takada, A., (2011). An H5N1 highly pathogenic avian influenza virus that invaded Japan through waterfowl migration. *Japanese Journal of Veterinary Research, 59*(2&3), 89–100.

Kaneko, S., & Sahashi, N., (1998). *Fungi Nurturing Beech Forests.* Bun-ichi Co. Ltd. (in Japanese).

Minato, M., (1966). The final stage of land bridges in the Japanese Islands. *Earth Science. 85,86,* 2–11.

Ministry of the Environment of Japan, (2002). *The New National Biodiversity Strategy of Japan.* The basic plan for nature conservation and rehabilitation. GYOSEI Corporation, Tokyo.

Ministry of the Environment of Japan, (2010). *Japan Biodiversity Outlook.* https://www.biodic.go.jp/biodiversity/activity/policy/jbo/jbo/files/Japan_Biodiversity_Outlook_EN.pdf. (Date of access: May 2017).

Ministry of the Environment of Japan, (2012a). *The National Biodiversity Strategy of Japan 2012–2020.* Roadmap towards the establishment of an enriching society in harmony with nature. http://www. env. go. jp/press/files/en/528. pdf. (Date of access: February 2017).

Ministry of the Environment of Japan, (2012b). Report of mapping of biodiversity assessment in fiscal year 2011. http://www.biodic.go.jp/biodiversity/activity/policy/map/files/h23report_all.pdf (Date of access: February 2017).

Ministry of the Environment of Japan, (2016a). *Japan Biodiversity Outlook 2.* http://www.biodic.go.jp/biodiversity/activity/policy/index.html (Date of access: February 2017). (in Japanese).

Ministry of the Environment of Japan, (2016b). *Pamphlet: Report of Comprehensive Assessment of Biodiversity and Ecosystem Services in Japan* (JBO2: Japan Biodiversity Outlook 2). http://www.env.go.jp/nature/biodic/jbo2/pamph04.pdf (Date of access: February 2017).

Naoe, S., Katayama, N., Amano, T., Akasaka, M., Yamakita, T., Ueta, M., Matsuba, M., & Miyashita, T., (2015). Identifying priority areas for national-level conservation to achieve Aichi Target 11: A case study of using terrestrial birds breeding in Japan. *J. Nature Conservation, 24,* 101–108.

Ohsawa, T., & Ide, Y., (2011). Phylogeographic patterns of highland and lowland plant species in Japan. *Alpine Botany, 121,* 49–61.

Osono, T., & Takeda, H., (2001). Organic chemical and nutrient dynamics in decomposing beech leaf litter in relation to fungal ingrowth and succession during 3-year decomposition processes in a cool temperate deciduous forest in Japan. *Ecological Research, 16*(4), 649–670.

Sakamoto, J., (2008). *Microbiology, for the Blue Planet and Human Life.* Shokabo Co., Ltd. (in Japanese).

Tsumura, Y., & Suyama, Y., (2015). *Map-Based Guideline for Seedling Transfer of Tree Species.* Bunichi-Sogo Publishers, pp. 176.

Uemura, S., (1989). Snowcover as a factor controlling the distribution and speciation of forest plants. *Vegetation, 82,* 127–137.

Plate 8.1 Ecosystem diversity in Japan; from upper-left to lower-right - (a) wetland in Oze, (b) mangrove in Iriomote, (c) rice paddy field in Sado island, (d) autumn foliage in Yokotsu-dake (Photo credits: Kazuhiko Hashimoto).

Plate 8.2   Diversity of animals in Japan; from upper-left to lower-right - (a) white
stork in Toyooka, (b) Japanese monkey in Izu, (c) Japanese freshwater
crab in Izu, (d) whooper swan in Onuma, (e) red fox in Hakodate, (f)
Shika deer in Sarobetsu (Photo credits: Kazuhiko Hashimoto).

Plate 8.3 Diversity of northern flora in Japan; from to left - (a) *Dicentra peregrina*, (b) *Comastoma pulmonarium* subsp. *sectum*, (c) *Dryas octopetala*, (d) *Prunus nipponica*, (e) *Fagus crenata*, (f) *Cryptomeria japonica* (Yaku-sugi) (Photo credits: Ichiro Hama).

Plate 8.4 Diversity of southern (tropical) flora in Japan; from to left – (a) *Hibiscus hamabo*, (b) *Euchresta japonica*, (c) *Dryopteris sieboldii*, (d) *Claoxylon centenarium*, (e) *Bruguiera gymnorhiza*, (f) *Ophioglossum pendulum* (Photo credits: Ichiro Hama).

CHAPTER 9

Biodiversity in Lebanon

SALMA N. TALHOUK,[1,2] **M. ITANI,**[2] **and M. AL-ZEIN**[2]

[1]Department of Landscape Design and Ecosystem Management, Faculty of Agricultural and Food Sciences, Bliss, Beirut, Lebanon, E-mail: ntsalma@aub.edu.lb
[2]Nature conservation center, American University of Beirut, Bliss, Beirut, Lebanon

9.1 Biogeography of Lebanon

Located along the Eastern Mediterranean littoral at crossroads between Europe, Asia, and Africa, Lebanon is predominantly mountainous consisting of five geomorphological regions. These include a narrow coast, two mountain ranges, the Mount Lebanon and Anti-Lebanon, that run parallel to the sea and are separated by a 8–12 km wide fertile plain, the Beqaa. In the southern part of the country an elevated plateau extends from the coast to the anti-Lebanon Mountain chain. Despite its small size (10,450 km^2) Lebanon contains botanical elements from temperate, arid, and subtropical biomes presenting a climatic and ecological diversity that is unique to the eastern Mediterranean region. The whole country has been continuously occupied by humans since ancient history, and practices such as fire setting, clear-cutting, ploughing, heavy browsing and grazing through the ages have led to a richness of annual and ephemeral plants, and to high plant diversities in frequently and moderately disturbed sites (Myers et al., 2000; WWF and IUCN, 1994; Post and Dinsmore, 1933; Mouterde, 1970; Blondel and Aronson, 1999). Lebanon has bioclimatic regimes and vegetation associations that fall within two major climate zones, the Mediterranean and the pre-steppe (Chouchani, 1972) (Table 9.1). A 50 km transect from the coast inland can include both Mediterranean areas with nearly 1,000 mm of annual rainfall and pre-steppe areas with < 250 mm of rainfall (METAP, 1995). The rainy season falls between November and March.

9.2 Biodiversity Overview

Lebanon is part of the Mediterranean basin, a region that harbors a large percentage of endemic plant species (50%), and is recognized as one of the 34 world hotspots for plant diversity in need of conservation support (Myers et al., 2000; Medail and Quezel, 1997). With its climatic and ecological diversity which is unique to the eastern Mediterranean the whole country

Table 9.1 Bio-Climatic Zones, Forest Types and Main Species in Lebanon

Bio-climate	Substrate/Forest habitat type
Thermo-Med. (<500 m)	Limestone: Carb-Lentisk Scrub, Pine woodlands, Evergreen oak woodlands, Mixed oak-pine woodlands
	Marl and marly-limestone: Pine forest, Mixed conifer forest, Cypress forest
	Sandstone: Pine forests
Eu-MEd. (500–1,000 m)	Limestone: Evergreen oak forests, Mixed oak-pine forests, Deciduous oak forests
	Marl & marly-limestone: Mixed conifer forests, Pine forests, Cypress forests
	Sandstone: Pine forests
Supra-Med. (1,000–1,500 m)	Limestone: Evergreen oak forests, Mixed oak and juniper forests, Deciduous oak forests, Hophornbeam mixed forests
	Sandstone: Stone pine forests, Deciduous oak forests
Mountain-Med (1,600–1,900 m)	Mixed conifer, Mixed conifer/oak forests, Oak forests, Juniper woodlands
Oro-Med (>1,900 m)	Juniper woodlands
Steppe non-forest	Hammada scrub
Steppe-Med. (900–1,500 m)	Evergreen oak forests
Steppe-Supra-Med. (1,500–1,800 m)	Mixed oak forests
Steppe. Mountain-Med. (1,800–2,400 m)	Juniper forests
Steppe-Oro-Med (>2,400 m)	Juniper woodlands
Riparian forests	Lowland Plane tree forests
	Plane tree and alder forests
	Sandstone: Alder forests

Source: Asmar (2011). http://www.fao.org/ag/agp/agpc/doc/counprof/lebanon/lebanon.html#5.

of Lebanon is recognized as a center of plant diversity (WWF and IUCN, 1994). Despite its small size (0.007% of the world's land surface area), Lebanon hosts about 1.1% of the world's plant species (Tohmé and Tohmé, 2007) and 2.63 % of reptile, bird and mammal species and its sea harbors 2.7% of the world's marine species (MOE/UNDP, 2010).This high diversity is better reflected in terms of relative number of species per area where Lebanon's vegetation presents a high species-area ratio of 0.25 species/km^2

compared to biodiversity rich countries such as Brazil (species-area ratio is 0.0044 per km²) and South Africa (species-area ratio is 0.0081 per km² (MOE/UNDP, 2010). The total number of species in Lebanon is estimated at 9,116, with 4,630 fauna species and 4,486 floral species (MOA/UNEP/GEF, 1996) (Table 9.2).

9.3 Terrestrial Floral Diversity

9.3.1 Vascular Plants

The Lebanese natural landscape is unique in its immediate context in that it is mostly mountainous (3090 m highest peak), and is considered part of a global hotspot with an estimated floristic richness of 2,600 vascular plant species of which 311 are either nationally or regionally endemic (12%) (Davis et al., 1994; Myers et al., 2000; Wolz, 1998). A recently published illustrated flora of Lebanon by Tohme and Tohme (2014), reports more than 2500 species. However, some of these species, for example, *Malus trilobata*, were found in other countries (Kurtto, 2009). Furthermore, there is a need for updating taxonomic nomenclature of all the flora especially the endemics. For example, *Colchicum libanoticum* a reported endemic has been recently reported to be a synonym of the widespread species *Colchicum szovitsii* (World Checklist of Selected Plant Families, 2010).

Although there is no recent flora for Lebanon, several studies have been conducted in various parts of the country and in protected areas and nature reserves documenting species and recording their locations. Floristic richness in Lebanon was reported as 38 species per 600 m² in coastal vegetation communities, 54 species per 600 m² in low altitude inland areas, and 34 species per 200 m² site in riparian habitats (Talhouk et al., 2005; Abboud et al., 2012). Examples of species diversity studies include work on legumes diversity in a riparian habitat reporting high level of heterogeneity across 5 km distance of the river and recording thirty species of legumes belonging to 15 genera (Atallah et al., 2008). A recent proposed check list for orchids listed a total of 51 species and subspecies for Lebanon including two new chorological records for Lebanon (*Epipactis helleborine* subsp. *praecox*, *Ophrys asiatica*) (Vela and Viglione, 2015). The authors reported that at least 14 taxa are regionally endemic, seven of which are limited to two border countries.

Native Mediterranean flora has a relatively low extinction rate of 0.11% but high percentages of threatened plant taxa, between 25 and 125 times as high as extinction rates (Greuter, 1994). Such findings are exemplified by a study on endemic *Oncocyclus irises* (Iridaceae), *Iris cedretii* and *Iris*

Table 9.2   Recorded Number of Species in Lebanon for Each Major Group

Marine and Coastal Fauna and Flora	
Phytoplankton and Macrophytic Marine Algae	586
Zooplankton	621
Crustaceans	312
Molluscs	294
Reptiles	5
Adult fish	354
Mammals	13
Vascular plants	15
Freshwater fauna and flora	
Thallophytes	264
Vascular Plants	7
Plants terrestrial and freshwater ecosystems	13
Invertebrates	656
Fish	60
Terrestrial flora	
Mushrooms	213
Lichens	48
Bryophytes	222
Pteridophytes	35
Gymnosperms	11
Angiosperms	2863
Plant parasitic bacteria	19
Plant parasitic fungi	350
Terrestrial fauna	
Annelids and nematodes	12
Molluscs	239
Crustaceans	38
Arachnids	249
Insects	839
Butterflies	377
Amphibians and Reptiles	48
Birds	338
Mammals	59

Source: MOA/UNEP/GEF, 1996.

sofarana subsp. *sofarana*, which found that the species consisted of small remnants of once larger populations or highly fragmented populations (Saad et al., 2009).

Molecular studies are also revealing high genetic diversity in various species in Lebanon. Examples of such studies include the genetic diversity assessment of 30 presumed *Astragalus* species which revealed large variation of genome sizes, an important inter-specific chromosome polymorphism and the existence of a high phylogenetic diversity (Samad et al., 2014). Genetic analysis of *Abies cilicica* populations growing in West Taurus, East Taurus, and Lebanon mountains found significant genetic differentiation between areas of species occurrence (Sękiewicz et al., 2015). Molecular analysis of the genetic diversity of the coastal species *Pancratium maritimum* revealed higher variation within than between populations (Zahreddine et al., 2004). Another molecular study presented C-values for 225 taxa belonging to 55 families and 141 genera, including novel C-values for 193 taxa of which 126 plants are endemic to the Eastern Mediterranean region and the first values for 50 genera (Bou Dagher-Kharrat et al., 2013).

With respect to species conservation status it is worth noting that redlisting has been recently initiated and 200 species have been redlisted (IUCN Red Listing Lebanese plants: a comprehensive workshop at USJ, 2014).

9.3.2 Angiosperms

Top 10 Angiosperms families in Flora of Lebanon (number of species) are Fabaceae (358 species), Asteraceae (316), Poaceae (230), Brassicaceae (161), Lamiaceae (148), Caryophyllaceae (146), Apiaceae (124), Ranunculaceae (70), Boraginaceae (66), and Rubiaceae (64) (Tohme and Tohme, 2014).

9.3.3 Gymnoperms

Gymnosperms in Lebanon include 5 species of Cupressaceae, 5 species of Pinaceae, and 1 species of Ephedraceae (Tohme and Tohme, 2014).

9.3.4 Pteridophytes

Pteridophytes in Flora of Lebanon are represented by 28 species. These include Aspleniaceae (8 species), Pteridaceae (5 species), Equisetaceae (3 speices), Dryopteridaceae (2), while other families Lycopodiaceae, Isoetaceae, Selaginellaceae, Ophioglossacae, Osmundaceae, Dennstaedtiaceae,

Cystopteridaceae, Blechnaceae, Athyriaceae, and Polypodiaceae are represented by one species each.

9.3.5 Agricultural Biodiversity

Lebanon is part of a region in the Middle East often referred to as the Fertile Crescent which is considered as one of the 8 main centers of diversity identified and where many food plants were first domesticated (Vavilov, 1949–50). Examples of plants that are grown in Lebanon and that have wild ancestors of global importance include cereals such as oat, barley and wheat, pulses such as chickpeas and lentils, and forages such as medics and vetch. Among the cereal crops, wheat ranks first and farmers use local landraces such as Hourani, Bekaii, Salamouni, Douchani, and Nabeljamal. Research centers have collected and evaluated the genetic makeup of important species such as wheat, barley and lentils and they have established seedbanks to house specimens ("ICARDA Backup Seeds from Arctic 'Doomsday Vault' are Housed at AREC," n.d.). However, with urban and agriculture expansion wild crop relatives have been restricted to marginal lands with shallow soil, abandoned agricultural lands, and field borders (Jaradat, 1998). There are ongoing efforts at enriching grazing lands with species such as *Medicago, Trifolium,* and *Vicia* and at *in situ* conservation of wild crop relatives namely species of *Aegilops, Hordeum, Lathyrus, Lens, Medicago, Trifolium, Triticum,* and *Vicia* (Chalak et al., 2011). The assessment of these plant genetic resources in Lebanon is primarily based on morphological descriptors and agronomic traits and it does not benefit from molecular studies which are limited to few crops while breeding efforts are conducted only on wheat, barley, chickpea and lentil (Chalak et al., 2011) (Table 9.3).

The conservation of the tree genetic resources of the country is equally important as the genetic variation in wild tree species is an important resource in breeding for improved traits such as drought, heat and disease resistance (Hodgkin, 1993) (Table 9.4).

Of the above listed wild tree species, almond is widely cultivated in the country and fruits are mostly used for fresh consumption (Chalak et al., 2011). However commercial varieties consist of one or two varieties, while the reported high variability of *Amygdalus communis* populations constitutes a valuable source in almond improvement Programs (Talhouk et al., 2000). Almond is among the earliest cultivated fruit trees in the Near East. The spread of almond cultivation to the Mediterranean basin, southwest Asia and middle Asia terminated the geographic isolation between *Amygdalus communis* and its wild relatives and resulted in spontaneous hybrids

Table 9.3   Crop Wild Relatives native to Lebanon

Wild wheat

Trtiticum thaoudar Reut	*T. urartu* Tum.
T. dicoccoides (Koern. ex Asch. & Graebn.) Schweinf.	*T. boeoticum* Boiss.

Wild oat

Hordeum spontaneum K. Koch	*H. leporinum* Link
H. hystrix Roth	*H. bulbosum* L.

Chickpeas

Cicer arietinum	*C. pinnatifidum*
C. incisum	*C. judaicum*

Wild lentils

Lens culinaris	*L. orientalis*
L. ervoides	

Related to wheat

Aegilops ovata L.	*A. comosa* Sibth & Sm
A. triaristata Willd.	*A. squarrosa* L.
A. columnaris Zhukovsky	*A. crassa* Boiss.
A. biuncialis Vis	*A. ligustica* (Sav.) Coss.
A. triuncialis L.	*A. speltoides* Teusch
A. kotschyi sub. *eu-variabilis* Eig.	*A. longissima* Schweinf & Muschler
A. peregrina (Hackel) Maire & Wieller	*A. searsii* Feldman & Kislev
A. cylindrica Host	*A. vavilovii* (Zhuk.) Chennav.
A. caudata L.	

Medics

Medicago granatensis Willd.	*M. polymorpha* L.
M. itertexta (L.) Miller	*M. minima* Lam.
M. murex Willd.	*M. laciriata* All.
M. turbinata (L.) Willd.	*M. praecox* D.C.
M. aculeata Gaertner	*M. rotata* Boiss.
M. constricta Durieu	*M. blancheana* Boiss.
M. rigidula (L.) Desr.	*M. rugosa* Desr.
M. truncatula Gaertn.	*M. scutellata* Mill.
M. litoralis Rohde	

Table 9.3 (Continued)

Clover

Trifolium subterraneum L.	*T. tomentosum* L.
T. cherleri L.	*T. alexandrinum* L.
T. fragiferum L.	*T. campestre* Schreb.
T. hirtum All.	*T. scabrum* L.
T. pilulare Boiss.	*T. purpereum* L.
T. resupinatum L.	

Vetch

Vicia ervilia (L.) Willd.	*V. sativa amphicarpa* Dorthes
V. monantha Retz	*V. sativa angustifolia* Roth
V. narbonensis L.	*V. sativa cordata* Wulf
V. pannonica Jacqu.	*V. villosa* Roth
V. peregrina L.	

Source: MOA/UNEP/GEF, 1996.

Table 9.4 Wild Species of Fruit Trees

Amygdalus agrestis	*Pistacia mutica*
Amygdalus communis	*Pistacia palaestina*
Amygdalus korschinskii	*Prunus cerasus*
Amygdalus orientalis	*Prunus mahaleb*
Ceratonia siliqua	*Prunus microcarpa*
Crataegus azarolus	*Prunus prostata*
Crataegus monogyna	*Prunus tortuosa*
Ficus carica	*Prunus ursine*
Ficus sycomorus	*Pyrus syriaca*
Malus trilobata	*Rubus collinus*
Myrtus communis	*Rubus hedycarpus*
Persica vulgaris	*Rubus sanctus*
Pistacia lentiscus	*Rubus tomentosus*

Source: MOA/UNEP/GEF, 1996.

especially between closely related species within the *A. communis* group (Browicz and Zohary, 1996; Kester and Gradziel, 1996; Kester et al., 1991). Six wild *Amygdalus* species have been reported in Lebanon including *A.*

communis L. However, seedling stands of hybrid trees and feral forms of *A. communis* subsp. *spontanea* found in neglected orchards are threatened by grafting or replacement with other fruit trees.

The carob tree, *Ceratonia siliqua*, a small sclerophyllous evergreen tree which is thought to have originated either in the eastern Mediterranean or the Arabian Peninsula (Battle and Tous, 1997; Zohary, 2002) is traditionally interplanted with olives, grapes, almonds and barley in low intensity farming systems in most carob producing countries (Battle and Tous, 1997; Blondel and Aronson, 1999; Ramon-Laca and Mabberley, 2004). In Lebanon, earlier reports indicate that the carob used to grow abundantly on the lower coastal hills however, agriculture and urban expansion caused the uprooting of the species and today, remaining carob populations thrive in semi-natural habitats (Talhouk et al., 2005). The national interest in carob conservation is based on the multi-purpose nature of this tree species (MOA/UNEP/GEF, 1996; Correal et al., 1987). In addition, its adaptability to all types of soils at lower and middle altitudes and its resistance to drought, make it suitable for reforestation because the species is considered beneficial in associations with low altitude conifer trees because of its resistance to fires (MOA/UNEP/GEF, 1996; Battle and Tous, 1997; Dereix et al., 1999).

In addition to wild tree resources, and according to government statistics (Chalak et al., 2011) the olive, a native tree, is the main cultivated tree species that contributes to food security. In Lebanon, the species has retained over hundreds of years its importance contributing a range of roles in local economies from highly lucrative financial opportunities on the one hand to the livelihood of marginalized communities on the other (Chamoun et al., 2009). A genetic diversity study segregated old planted tree specimens into agriculture provenances suggesting that the germplasm currently existing in old Lebanese olive groves may be the result of historic trade of plant material with neighboring regions (Chamoun et al., 2009).

Other important local fruit varieties are grape and apricot. A total of 28 local grape varieties are found in Lebanon (Riachy, 1998). Chemical analyses of these varieties revealed that the majority are suitable for fresh consumption the most famous of which are Beytamoni, Tfayfihy and Merwah. Only 2 varieties. 'Mirweih' and 'Bikaai,' are suitable for wine production and might have a potential as a Lebanese varietal wine grape. As for apricot, it is cultivated in the North of Lebanon and northern Bekaa and the fruits used for fresh consumption, as well as the production of juices and jams. Local varieties include Ajami, Byadi and Um Hussein (Chalak et al., 2011).

9.3.6   Forest and Woodland Biodiversity

The Lebanese Forests have supplied timber to the region since Phoenician times, and the wood from Lebanese forests was prized by Ancient Egyptians (Wilson, 1955), Mesopotamians (Speiser, 1955), and Romans (Meiggs, 1982). In the 7th century AD deforestation expanded to higher altitudes when religious minorities established villages in the highlands (1000–1500 m a.s.l.) and cleared forests to build agricultural terraces (Mikesell, 1969). Further forest destruction occurred under Ottoman rule (from the 16th century until after World War I as logging became an alternative form of tax payments and later more forests were destroyed to fuel the railway connecting Damascus with Aleppo and Beirut. During World War II forests were again cut for the construction of the railway connecting Tripoli and Haifa (Mikesell, 1969). The impact of this continuous and systematic extraction of Lebanese timber resources has transformed the original extensive forests to relict patches and scrub vegetation. Towards the end of the 20th century the agro-pastural practices were in decline, the youth left villages, abandoned terraces started degrading and spontaneous woodlands returned covering lands with a thorny garrigue.

Today Lebanon has 49 forest tree species and its forests cover is estimated at 13% (137,000ha) with woodlands contributing an additional 10% (FAO, 2010). About 65% of forested areas are considered dense and they are primarily located in North Lebanon (30%) and Mount Lebanon (37%) (Verheye, 1998; MOA, 2003). Oak forests occupy the largest areas (52.42%), conifers, mainly Pine forests (14.91%) and Juniper forests (8.74%) extend over 31% of the lands and mixed forests represent 17.98% of the forested lands. Despite the fact that Lebanon is known for its cedar trees, Cedar forests today cover less than 1% of the country (MoE, 2012; MOA, 2003; FAO, 2005). http://www.moe.gov.lb/Sectors/Biodiversity-Forests.aspx?lang=en-us).

Lebanon's low altitude coastal woodlands consist of Mediterranean sclerophyllous evergreen vegetation occupying semi-natural habitats and presenting various levels of degradation (Makhzoumi and Pungetti, 1999). The predominant species are pistachio (*Pistacia* spp.), oak (*Quercus* spp.), and carob (*Ceratonia siliqua*) all of which have evolved to withstand drought and high summer temperatures (Christodoulakis, 1992). The carob tree (*Ceratonia siliqua*) has been included in a national list of priority forest genetic resources as a target for the conservation and management in Lebanon. The geographic distribution of the semi-natural carob populations seems to indicate that the species has receded to zones within its climatic tolerance, to soils with poor agricultural capability, and topographies that are either inaccessible or unsuitable for agricultural and urban expansion (Talhouk et al., 2005). Given the long history of habitat disturbance and the cultivation of carob by humans, it is most likely that these populations are remnants of past forests (absence of a planting structure within a population) that were traditionally managed by agropastoral communities which grafted wild trees with more desirable cultivars. Where carob species were cultivated or managed, wild forms of the same species must have established in the surrounding areas particularly in sites inaccessible to people.

A study by Al-Qaddi et al. (2016) described the distribution of oak species in Lebanon as primarily driven by temperature and precipitation. The authors explained the distribution of all oak species in Lebanon and indicated that the country is considered the western and northern limit of *Q. look* Kotschy, and the southern limit of *Q. cedrorum* Kotschy, *Q. cerris* L., and *Q. pubescens* Willd., which are represented through endemic subspecies or varieties adapted to the local environment in isolated stands.

In the Mediterranean basin, genetic diversity is higher than other regions of the world for conifers (Fady-Welterlen, 2005). The majority of natural conifer forest associations in Lebanon occur over 1200–2000 m, and consist mainly of pure and mixed forest of cedars of Lebanon, Cilician fir and junipers. The distribution range of the former two species extends from Turkey to Lebanon, where they both reach their lower latitudinal limits (Mouterde, 1966; Chouchani, 1972; Farjon, 1998). A snapshot of each species is described in altitudinal order, from those characteristic of lower zones to those that occur above 2000 m.

There are 94 pure stands of Calabrian pine (*P. brutia*) while mixed stands in which Calabrian pine is the dominant species varies with altitude and is usually found associated with oak *Q. calliprinos* at lower altitudes, and *Q. infectoria* and *Q. cerris* at higher altitudes. At its upper altitudinal limit, Calabrian pine is often found in cedar of Lebanon and Cilician fir forests. Calabrian pine is a coloniser of hillside clearings (Chouchani, 1972). There are 91 pure stands of stone pine, on sandstone, stands are dense, by contrast they are sparse on calcareous soils. The abundance of this species is a consequence of its cultivation; the kernels are an important item in Middle Eastern cooking, and the timber is used for furniture and fuel. Aleppo pine (*P. halepensis*) is less cold tolerant than other pines (Dereix et al., 1999), and exists in sparse populations restricted to the southern part of the country. The Mediterranean cypress (*Cupressus sempervirens*), along with Calabrian pine and oak (*Q. calliprinos*), are increasingly colonizing higher altitudes of the western slopes of the Lebanon mountain range in the north, replacing *Q. infectoria*

(Chouchani, 1972). Pure Cedar of Lebanon (*Cedrus libani*) forests probably once formed a continuous band on the western slopes of the Lebanon chain, but only 15 natural stands remain, more than half of which are less than 100 ha in area and all in some state of degradation (Chouchani, 1972). A small pure cedar of Lebanon stand in the south in Niha marks the southern limit of the species. The Cilician fir (*Abies cilicia*) is restricted to the northern part of the western mountain range, where it reaches its southernmost geographic limit of its range (Zohary, 1973) and where it is geographically isolated from the rest of its range with more than a 150 km distance from *Abies cilicica* in Syria (Awad et al., 2014). The remaining populations have low genetic variation but a high recent effective population size (Awad et al., 2014). The Grecian juniper *Juniperus excelsa* grows in pure stands or as part of cedar of Lebanon and Cilician fir forests but it also occurs in xeric areas above 2000 m and, and unlike other Lebanese conifers, it extends to the eastern slopes of Mount Lebanon and the higher slopes of the Anti-Lebanon forming scattered stands with little undergrowth. On the other hand, the Prickly juniper is commonly found associated with oak, pine, and cedar of Lebanon forests, and grows mainly in forest clearings (Chouchani, 1972).

A list of priority tree species was produced to safeguard national forest genetic resources. The selected species include *C. siliqua, Quercus calliprinos, Pinus pinea, Juniperus excelsa, Cedrus libani,* and *Salix alba* all of which were characterized as having potential socio-economic and/or ecological value, and were facing threat of genetic erosion (van Breugel and Bazuin, 2001).

9.3.7 Wild Edible, Medicinal, and Aromatic Plants

The traditional harvesting of wild edible plants among Lebanese rural communities has contributed to the continued use of such plants in local recipes (Maarouf et al., 2015). Thanks to this deep rooted and intimate relationship, many indigenous people and rural populations traditionally have a wealth of knowledge of wild plants. Wild edible plants play a role in cultural and religious traditions. For instance, people abstaining from the consumption of animal products during Lent seek to diversify their diet by reverting to traditional recipes based on wild edible greens (Maarouf et al., 2015). Another promising opportunity for renewal interest in wild edible plants is Lebanon's culinary heritage. Lebanese food is part of Mediterranean diets that are globally recognized for their health-promoting properties (Trichopoulou and Vasilopoulou, 2000) and a large number of Mediterranean recipes are still based on wild edible plants. On a commercial level the Lebanese domestic market of herbal, medicinal and aromatic plants is estimated at US$35 million per year (UNDP, 2008). Around 365 Medicinal and Aromatic Plants (MAPs) are found and exploited, 47 of which are endemic to the region (UNDP/LARI/GEF, 2014). However these plants are threatened by destructive harvesting and overharvesting with 98% of the market supplemented by wild stocks. Research efforts have focused on incorporation of conservation dimensions into the gathering, processing and marketing of globally significant species namely *Salvia fruticosa, Origanum syriacum, Cyclotricium origanifolium, Micromeria libanotica, Viola libanotica, Alcea damascene,* and *Origanum ehrenbergii* (UNDP, 2011). Wild plants were also found in a national study to contribute significantly to the health status of the Lebanese as two out of five people use complementary medicine most of which is in the form of "FolkHerbs" (Naja et al., 2015).

9.3.8 Coastal and Urban Floral Diversity

Among the most threatened areas in Lebanon is the coastline that has become a chaotic matrix of tourist resorts, private beaches, agricultural lands, industrial and urban development. This problem is not unique to Lebanon. The loss of native coastal vegetation due to unregulated urban and tourist development is well known in the Mediterranean Basin. An estimated 64% of Lebanon's population is clustered along the coast (UNH, 2009). Furthermore, in less than 25 years, 3.2 Million square meters along the coastline were lost to unlicensed developments (SOE, 2010). Among these vegetation communities, the most threatened ones are those occurring in prime tourist areas, such as sandy beaches and coastal sand dunes, which are targeted by developers for the establishment of tourist resorts. These habitats have already been classified as threatened by extensive tourism throughout the Mediterranean (IUCN, 1994). In Lebanon sandy beaches and coastal sand dunes are typically 'cleaned' of all garbage, debris, and vegetation every spring in anticipation of the tourist season.

Studies on the biodiversity of the Lebanese coast have shown that native species are becoming scarcer in coastal landscapes (Zahreddine et al., 2004; Talhouk et al., 2004). More alarmingly, a survey of the Beirut waterfront showed the number of taxa occupying the coastal zone has, in comparison to the 1930's, significantly decreased (Chmaitelly, 2007). As a result, the distribution of some native plant species with limited natural range has become restricted to urban areas. Examples include three species of plants endemic to the Lebanese coast namely, *Limonium mouterdei, Limonium postii,* and *Matthiola crassifolia* (Itani, 2015). The latter species was observed to utilize both semi-natural and anthropogenic habitats and was recognized as an umbrella species (Itani, 2015). Urban encroachment of the coast has contributed in some instances to the extinction of several species some of which are regional endemics such as *Rumex occultans, Campanula sulphurea, Gagea dayana*, and *Trifolium billardieri* (Tohme and Tohme, 2007, Author's observations).

9.3.9 Non-Vascular Plants

While the vascular flora of Lebanon is relatively well-studied, there are no new revisions of the non-vascular flora. In a recent paper, Kurschner (2010) reported four liverwort and 12 moss species presumed new to the flora of Lebanon. However, these species were based on a collection dating back to the 1960s.

As for the Lycophytes, most of the recent work has focused on quillworts (Isoetaceae). A new species of *Isoetes* (Bolin et al., 2011) and two new records (Bolin et al., 2008; Musselman and Al-Zein, 2009) were recently published. *Isoetes olympica*, extinct in Turkey and endangered in Syria, was found in the Akkar area, Northern Lebanon, adjacent to the Syrian border.

Recent study on the bryophyte flora of Lebanon reports the presence of 252 species including two hornworts, 43 liverworts and 207 mosses (Kürschner, 2010). The author sheds light on outstanding new records such as *Saccogyna viticulosa* and *Tritomaria polita*, both new to the bryoflora of South-West Asia, and the range extension of *Philonotis laxitexta*, previously known only from Iran.

9.3.9.1 Fungi

In Lebanon there are no red-listed fungi and no fungi are protected by law. In 1996 Lebanese national experts compiled information about fungi (Table 9.5) and till today this information remains the only official source for the country (MOA/UNEP/GEF, 1996). The situation of the mycota of Lebanon is the same in the region which according to Helfer (2008) is only known at 10–30 % of their estimated final diversity (Helfer, 2008). Fungi of Lebanon are not included in the country's national biodiversity study reports nor in the Lebanon state of the environment reports.

With respect to mushrooms, the latest comprehensive study in Lebanon is by Modad (2006) who conducted field samplings throughout Lebanon and included more than 200 described mushrooms belonging to 52 families.

Table 9.5 Fungi of Lebanon

Family (Number of species)	
Agaricaceae (97)	Caloceraceae (1)
Paxillaceae (10)	Auriculariaceae (1)
Hygrophoraceae (13)	Phallaceae (2)
Russulaceae (10)	Secotiaceae (3)
Boletaceae (18)	Hymenogastraceae (1)
Polyporaceae (14)	Nidulariaceae (1)
Corticiaceae (2)	Lycoperdaceae (10)
Hydnaceae (12)	Morchellaceae (2)
Cantharellaceae (1)	Helvellaceae (1)
Clavariaceae (3)	Pezizaceae (7)
	Tuberaceae (4)

Source: MOA/UNEP/GEF, 1996.

With respect to economic value of lower organisms, a recent study exploring the biodiversity of oleaginous microorganisms, including bacteria, yeasts and filamentous fungi which are capable of accumulating intracellular lipids under certain culture conditions, revealed that most of the filamentous fungi isolates can use xylose and carboxymethyl cellulose (CMC) as a sole carbon source for maximum lipid production and are considered as promising single cell oils producers on using agro-industrial waste materials (El-haj et al., 2015).

9.4 Terrestrial Fauna Diversity

9.4.1 Mammals

A review of recorded species of mammals in Lebanon reported the presence of 51 species belonging to 37 genera and 7 groups (MOA/UNEP/GEF, 1996). Updated information on prevalence of mammals is available from studies in protected areas using non destructive methods. For example, in the nature reserve of Ehden, a 253,120 m² area surveyed in 20 days over a period of nine months (one quadrat per day) led to the recording of 12 medium to large size mammals (Nader et al., 2011). In the reserve of Jabal Moussa, Abi Said and Amr (2012) using camera traps reported thirteen species of mammals belonging to five orders and 12 families. In this study,

Erinaceus concolor, was the only extant species of the order Insectivora that was encountered in Jabal Moussa. The carnivores were represented by seven species, accounting for more than half of the recorded species.

With respect to bats, the latest published study in Lebanon is by Lewis and Harrison (1962), which included more than 300 bats sampled throughout the country revealed the presence of fourteen species and subspecies.

9.4.2 Birds

Lebanon is considered as an important migratory route for birds between Europe, Asia and Africa. The total number of birds species in Lebanon was reported to be 337 species belonging to 101 genera and 20 groups (Table 9.6) (MOA/UNEP/GEF, 1996). An updated checklist revealed the presence of 399 bird species in Lebanon 35% of which reportedly use the country as breeding grounds and of these 65 species are exclusively residents (Serhal and Khatib, 2014). Spring migration route through Lebanon occurs along the Eastern flanks of the Mount Lebanon Range and the western half of the Bekaa (Serhal and Khatib, 2014). During spring most soaring birds pass down the eastern flanks of the Lebanon Mountains while some travel higher up on both sides of the ridge. Depending on species different habitats are targeted and the integrity of these habitats in turn affects the bird species. Table 9.6 shows examples of bird species that target preferentially different natural habitats including marine and coastal areas, wetlands, forest habitats, subalpine habitats and semi-arid habitats (Table 9.6).

The checklist of the birds of Lebanon encompasses 10 threatened species and 15 near-threatened species according to IUCN criteria. Important Bird Areas (IBAs) are defined nationally following local data collection based on global criteria. The process of IBA engages local support and builds on consensus. Sites are recognized as IBA if they hold significant numbers of one or more globally threatened species, if they are part of a set of sites that together hold a suite of restricted-range species, or if they have exceptionally large numbers of migratory or congregatory species. In Lebanon between 2005 and 2008, a total of 15 IBAs were declared of which 5 are nature reserves, 8 are conserved by NGOs and local communities, and 4 have no current protection (Serhal and Khatib, 2014).

9.4.3 Herpetofauna

With respect to Amphibians, field work by Hraoui-Bloquet et al. (2001) revealed the presence of six species on Lebanese territory. These are

Table 9.6 Bird Species Targeting Specific Natural Habitats in Lebanon

Marine habitats	Coastal habitats	Wetland habitats
Puffinus yelkouan	*Falco eleonorae*	*Anas platyrhynchos*
Pelecanus onocrotalus	*Himantopus himantopus*	*Tachybaptus ruficollis*
Morus bassanus	*Vanellus spinosus*	*Nycticorax nycticorax*
Phalacrocorax carbo	*Charadrius hiaticula*	*Ardea cinerea*
Larus audouinii	*Tringa nebularia*	*Gallinula chloropus*
Larus michahellis	*Tringa ochropus*	*Fulica atra*
Larus fuscus	*Actitis hypoleucos*	*Alcedo atthis*
Larus ridibundus	*Calidris minuta*	*Cettia cetti*
Sterna hirundo	*Philomachus pugnax*	*Acrocephalus scirpaceus*
Stercorarius parasiticus	*Sylvia conspicillata*	*Cinclus cinclusrufiventris*
Forest habitats	**Subalpine habitats**	**Semi-arid habitats**
Scolopax rusticola	*Pyrrhocorax graculus*	*Alectoris chukar*
Strix aluco	*Eremophila alpestris*	*Cursorius cursor*
Dendrocopos syriacus	*Turdus viscivorus*	*Athene noctua*
Lanius nubicus	*Phoenicurus ochrurossemirufa*	*Upupa epops*
Garrulus glandarius	*Oenanthe isabellina*	*Lanius meridionalisaucheri*
Periparus ater	*Carpospiza brachydactyla*	*Poecile lugubris*
Lullula arborea	*Petronia petronia*	*Calandrella rufescens*
Turdus merula	*Anthus similis*	*Scotocerca inquieta*
Turdus philomelos	*Anthus pratensis*	*Sitta neumayer*
Fringilla coelebs	*Serinus syriacus*	*Turdus pilaris*
	Pyrrhocorax graculus	*Oenanthe hispanica*
	Eremophila alpestris	*Emberiza cia*

Source: Serhal and Khatib, 2014.

Salamandra infraimmaculata, Triturus vittatus, Hyla savignyi, Rana bedriagae, and *Bufo viridis arabicus*. The authors indicated that *Bufo bufo* found in the Nahr-Ibrahim valley is new to both Lebanon and the Middle East and that a previously reported species, namely *Pelobates syriacus syriacus* was not found. Recent data on herpetofauna of Lebanon by Hraoui-Bloquet et al. (2002) reported the presence of 5 amphibians and 44 reptiles derived from museum collections and recent field observations.

9.4.4 Hexapoda

Recent work on insects has focused on pests of agricultural crops, vectors of human diseases, in addition to surveys of specific groups (mosquitoes, non-frugivorous fruit flies, etc.).

Considering the abundance of amber in Lebanon there has been ongoing research to explore historical species prevalence. Azar (1997) developed a new method for the extraction of plant and insect fossils from Lebanese amber. Since then, numerous species of fossil insects belonging to different extinct and extant families have been identified and described. The author indicates that study of Lebanese amber inclusions often constitute 'missing links' between the old fauna and the modern one.

In the context of studying the transmission of leishmaniasis in Lebanon, 19 species of sandflies, belonging to two genera were identified (Haddad et al., 2003).

A revision by Schneider (2004) of dragonflies of the Levant reports six species from Lebanon, of which one is endangered, and one is probably vulnerable; however, the distribution and the taxonomic status of the later species is unclear, rendering its red listing difficult.

A survey of the mosquitoes of Lebanon, conducted in 1999–2001, revealed the existence of 12 species, of which *Culex pipiens* was the pre-dominant species (Knio et al., 2005). Research on insecticide resistance in this species yielded low resistance level to organophosphates perhaps due to switching to a new class of pesticides by municipalities (Osta et al., 2012). Ongoing work on insecticide resistance in *C. pipiens* to pyrethroids showed that mosquitoes from several Lebanese cities showed high resistance, perhaps as a result of using these pesticides extensively and at high doses (Masri, 2016).

A survey of non-frugivorous fruit flies of Lebanon, conducted in 1995–1998, revealed the existence of 17 species, of which 15 were new records for Lebanon (Knio et al., 2002). Since then, research on non-frugivorous flies has focused on understanding host race formation in selected species (Knio et al., 2007; Smith et al., 2009; Sayar et al., 2009; Haddad et al., 2017).

A recent study reported that Lebanon harbors only 7% of the reported Dictyoptera diversity in the region that includes Lebanon, Syria, Turkey, Iraq and Jordan (Caesar et al., 2015). The authors noted that besides the small size of the country, other reasons for this low diversity may be explained by difficult access to natural sites, and the lack of scientists interested in these groups (Caesar et al., 2015).

Interestingly the Lebanese amber has also contributed to knowledge on spiders with the oldest known linyphiidae reported from Lebanon (Penney and Selden, 2002). Amber research in Lebanon has also led to the description of the oldest known records of Penicillata that belong to the family Polyxenidae (Duy-Jaquemin and Dany, 2004).

An inventory of selected insect groups in different plant communities along the Lebanese coast in which more than 2000 insects were collected revealed that 77% of the collected insect species belonged to the Hymenoptera (28%), Lepidoptera (11%), Diptera (19%), and Coleoptera 19%) orders. The study revealed a change in the status of 25 butterflies of which 10 became rare while 7 became abundant (El Hachem, 2000) as compared to earlier references such as Larsen's Butterflies of Lebanon (1974).

9.4.5 Molluscs

Information on terrestrial land snails produced by Tohme and Tohme (1988) was recently updated by Bößneck (2011) who collected specimens from various parts of the country.

9.4.6 Crustaceans

38 species of Crustacea belonging to 17 genera have been reported from Lebanon (MOA/UNEP/GEF, 1996).

9.4.7 Arachnids

With respect to Arachnids, knowledge is still limited in Lebanon even though the first spider reported from the Middle East since the beginning of zoological nomenclature is *Araneus galilaea* collected in May 1751 from Lebanon (Levy, 1997). In 1872, a field investigation was conducted in Palestine and Lebanon reporting 278 species (Zonstein and Marusik, 2013).

9.4.8 Nematods and Annelids

There is little information on Nematodes and Annelids. With respect to nematodes, studies have focused on lizards (*Agama stellio*) to collect and describe species including *Thelandros* (*Parapharyngodon*) *tyche* (Nematoda: Oxyuroidea) and *Abbreviata adonisi* (Nematoda: Physalopteroidea) (Sulahian and Schacher, 1968b), *Foleyella philistinae* (Nematoda: Filarioidea) (Schacher and Khalil, 1967), *Bryogoofilaria agamae* gen. et sp. n. (Filarioidea: Onchoceridae) (Sulahian and Schacher, 1968a), and *Saurositus baal* (Filarioidea: Onchocercidae) (Sulahian and Schacher, 1969).

9.4.9 Fresh Water Biodiversity

Estimates in annual precipitation in Lebanon range between 8.6 billion m³ and 9.7 billion m³ (MOE/ECODIT, 2002; CDR, 2004; MOEW, 2010). Lebanon houses 38 permanent and seasonal flow rivers and about 2,000 springs with a total annual exploitation volume of 637 Mm³ (MOE/UNDP/ECODIT, 2011). Some of these springs occur in the sea off the coast of Tyre, Saida, Jbeil, and Chekka (El-Hajj, 2008). Water resource in Lebanon is decreasing as evidenced by a study that recorded a decrease in snow days and in area covered by snow (Shaban, 2009; Shaban and Kayrouz-Khalf, 2013). The reported fresh water biodiversity of Lebanon includes an estimated 987 species (MOA/UNEP/GEF, 1996).This baseline study conducted in 1996 lists 64 fish species, 61 fresh water worms, 41 fresh water mollusks, 60 freshwater crustaceans, 639 freshwater insects, and 284 freshwater flora.

Recent studies include investigations on Mediterranean freshwater endemic fish diversity which is considered amongst the most threatened biota in the world given substantial reductions in precipitation coupled with pollution and the increased presence of invasive species (Hermoso and Clavero, 2011).

Assessment of fish fauna of Lebanon was included in a recent study by Geiger et al. (2014) who presented an analysis of the first DNA barcode library for the freshwater fish fauna of the Mediterranean BH (526 spp.), with virtually complete species coverage (498 spp., 98% extant species) (Geiger et al., 2014). On the other hand, recent field investigations reduced fish diversity in Lebanon to 25 species belonging to the families Cyprinidae, Cyprinodontidae, Cobitidae, Salmonidae, Anguillidae, Cichlidae, Mugilidae, Puciliidae, Blenniidae, and Lutjanidea (El Zein, 2002; El Zein and Khalaf, 2012).

With respect to other freshwater fauna, updates on species diversity are reported following localized field investigations. With respect to aquatic invertebrates, a survey (2002–2003) conducted in one of the few wetlands in Lebanon, the Aammiq Marsh (Bekaa Valley), recorded 78 invertebrate taxa with species-level identifications completed for most of the Coleoptera (beetles) and Odonata (dragonflies and damselflies). The Coleoptera comprised 18 genera and 22 identified species, of which 9 were new records for Lebanon (Storey, 2007).

Ozcan et al. (2012) provided an annotated list of decapod crustacean fauna of Orontes River, primarily based on previous studies. The list included 6 species (3 Natantia, 3 Brachyura) belonging 3 families however, only *Potamon setigerum* Rathbun was reported from Lebanon.

With respect to molluscs, Bößneck (2011) reported a total of 17 freshwater gastropod species as well as 6 small clams based on field investigations conducted in central and north Lebanon. The author reported the following new records for Lebanon: *Stagnicola* cf. *berlani, Planorbis carinatus, Potamopyrgus antipodarum, Musculium lacustre, Pisidium amnicum, Pisidium tenuilineatum stelfox, Orculella mesopotamica riedeli, Buliminus damascensis, Turanena benjamitica, Sphincterochila fimbriata*, and *Monacha* cf. *compingtae*.

Leeches are amongst the best studied invertebrates in Lebanon with a reported 13 species (Dia, 1983; Ruckert, 1985; Moubayed, 1986; Bromley, 1994).

9.5 Marine Biodiversity

Lebanese waters represent less than 1% of the world's ocean surface. However, almost 6% of all global marine species are found in those waters (MOE/UNDP, 2010).

The Mediterranean sea is almost closed; it communicates with the Atlantic Ocean through a 14 km wide and 320 m deep channel, with the Black sea through the Bosphorus, and with the Red Sea through the relatively recent digging of the Suez Canal which ended a division that existed for 12 million years between the two seas (Boudouresque, 2004). The opening of the Suez Canal, aquaculture and ship transport, have resulted in the introduction of alien species into the Mediterranean Sea and particularly the eastern basin (Crocetta and Russo, 2013). The number of animal and macrophyte species (including cyanobacteria) in the Mediterranean are estimated at 8000 and 1500 respectively and are mostly confined to the 0–50 m bathymetric zone (Boudouresque, 2004). This diversity is significant considering that the Mediterranean represents less than 0.8% of the world ocean area and less than 0.3% of its volume while its fauna and flora represent 7% of recorded species. Biodiversity of the Lebanese waters is considered low as species diversity is higher in the Western part of the Mediterranean, and in the Eastern part, it is greater in the Aegean Sea than the Levatine basin (Boudouresque, 2004).

The Lebanese coastline includes beaches (sandy, rocky and pebbly), three bays, prominent headlands, several river deltas, rocky off-shore islands (i.e., Palm Island Nature Reserve – El Mina, El Zire – Saida, etc.), sandstone and limestone cliffs (i.e., Jbeil, Ras Al Shakaa and Ras El Bayada, etc.), coastal sand dunes (i.e., unstabilized: Tyre, stabilized: Airport, etc.), and coastal caves (i.e., Beirut, Amchit, Ras Al Shakaa, etc.) (MOE/GEF, 2016). Marine coastal habitats stretch from the shoreline to the edge of the 100 meter deep continental shelf which is the widest in North Lebanon, becomes very narrow or non-existent at the center of the country, then widens again towards the southern part of the country (Lakkis, 2011a). Unique features of the Lebanse coast includes vermetid platforms that are unique to warm Mediterranean coasts and that harbor species classified as endangered or threatened by the Barcelona Convention and strictly protected fauna species by the Bern Convention (UNEP-MAP-RAC/SPA, 2010). The coast also includes deep canyons (Elias, 2007; Würtz, 2012), hydrothermal vents (Shaban, 2013) and submarine freshwater springs (Bakalowicz, 2014).

Recent assessment of Opisthobranchs, a diverse group of specialized gastropod molluscs, from Lebanon reported 35 taxa including 22 native (~63%), 12 alien (~34%) and one cryptogenic (~3%) taxa (Crocetta et al., 2013a).

An updated checklist of bivalves of Lebanon revealed a poor fauna relative to the Mediterranean yielding a total of 114 species (96 native and18 alien taxa) (Crocetta et al., 2013b).

An updated check-list of polyplacophorans, scaphopods, and cephalopods from Lebanon was also recently reported by Crocetta et al. (2013).

Earlier studies of fish biodiversity in Lebanon reported the presence of 44 cartilaginous fish species and 313 bony fish species (George et al., 1964, 1971; George and Athanassiou, 1965, 1966, 1967; Boulos, 1968). The number of cartilaginous species was updated to 22 during the 70s and 80s (Mouneimne, 1977, 1978, 1979, 2002). More recent inventories recorded the presence of 25 cartilaginous fish species (Lteif, 2015).

The first description of phytoplankton along the Lebanese coast was described based on monthly plankton hauls between 1973 and 1975 (Lakkis and Novel-Lakkis, 1981). A total of 263 taxa were identified and of these 107 (45 genera) were diatom species and 157 dinoflagellates (25 genera); many of them being of Indo-Pacific origin as a result of known migration pattern through the Suez Canal (Lakkis and Novel-Lakkis, 1981). The spatial and temporal variability in plankton communities along the Lebanese coast is mainly correlated to seasonal hydro-climatic factors prevailing in the area. A long-term survey between 1965 and 2003 revealed the presence of about 400 phytoplankton species, including 160 diatoms and 230 dinoflagellates (Lakkis, 2007).

One thousand zooplankton taxa have been identified in the Lebanese seawaters, including all the planktonic groups from protozoans up to prochordates (Larvacea and Fish larvae). Two hundred and fifty microplankton species were found of which 141 *Tintinnids*, 25 *Foraminifera*, 10 *Acantharia*, 25 *Radiolaria Spumellaria*, 30 *Nasselaria*, 6 *Phaeodaria* and 1 *Heliozoa* (Lakkis, 2007) (Table 9.7). Macrozooplankton includes all species from *Hydromedusae* up to Tunicates and fish larvae (Lakkis, 2011b).

A recent national report documenting previous literature indicated that there is an estimated 230 species of marine macroalgae including native and invasive species (MOE/GEF, 2016). The document also indicates that many marine invertebrates are unknown with the exception of some information on Sponges, Cnidaria, Nemerta, Polychaetes, Sipunculiens, Mollusks, Brachiopoda, Crustacea, Echinoderma, and Asidies. With respect to marine mammals, the report indicates that dolphin species are the most represented mammals in Lebanese marine waters including seven species namely Short-beaked common dolphin (*Delphinus delphis*), Common bottlenose dolphin (*Tursiops truncates*), fin whale (*Balaenoptera physalus*), Cuvier's beaked whale (*Ziphius cavirostris*), Risso's dolphin (*Grampus griseus*), Striped dolphin (*Stenella coeruleoalba*), and the Mediterranean monk seal (*Monachus monachus*). As for marine herpetofauna reported species include the Loggerhead turtle (*Caretta caretta*), the Green sea turtle (*Chelonia mydas*) the Leatherback turtle (*Dermochelys coriacea*, the Hawksbill Sea turtle (*Eretmochelys imbricata*) which is rarely recorded in the Eastern Mediterranean, and the Nile softshell turtle (*Trionyx triunguis*) which is a large fresh water turtle.

Table 9.7 Taxonomical Composition of Zooplankton Community in the Lebanese Seawaters During 1965–2003

Group	Number of species	Group	Number of species
Euphausiacea	5	Mysidaceae	4
Amphiopda	5	Cirripedia (larvae)	4
Ostracoda	6	Decapoda (larvae)	110
Cladocera	6	Chaetognatha	10
Copepoda	173	Pteropoda	9
Siphonophora	28	Heteropoda	4
Scyphozoa	5	Polycheata (larvae)	8
Hydromedusae	68	Polycheata (adults)	4
Tintiniidae	141	Appendicularia	15
Actinopoda	66	Thaliacea	6
Foraminifera	12		

Source: Lakkis, 2007.

A recent study on marine benthos focusing on bryozoans along the coastal zone of Lebanon (3–35 m) revealed the presence of seven species of Smittinidae belonging to two genera, *Parasmittina* and *Smittina* (Harmelin et al., 2009).

9.6   Conservation Actions

9.6.1   *Protected Areas and Nature Reserves*

Following the end of the civil war in 1992 Lebanon has experienced political instability, political assassinations, and wars. Furthermore, since 2011, the country has been struggling with a sudden increase in human pressure estimated at 1.5 M Syrian refugees constituting 25 to 30% of the Lebanese population. More than half the Lebanese population resides in cities and towns along the coast while towns and villages in the mountains serve primarily as permanent residences for farming communities, and weekend and summer homes for city dwellers originating from these villages. Despite hardship and instability the Lebanese civil society and the government of Lebanon are very active pursuing legal venues to achieve conservation of sites (Tables 9.8–9.13). Lebanon has engaged in signing international

Table 9.8   Nature Reserves in Lebanon

Nature Reserve	Legal Instrument	Surface Area (ha)	Nature Reserve	Legal Instrument	Surface Area (ha)
Horsh Ehden	Law 121, 1992	1,740	Tannourine Cedar Forest	Law 9, 1999	195
Palm Islands	Law 121, 1992	417	Wadi Al Houjeir	Law 121, 2010	3,595
Karm Chbat	Decision 14/1, 1995	520	Mashaa Chnaniir	Law 122, 2010	27
Al Shouf Cedars	Law 532, 1996	15,647	Kafra	Law 198, 2011	40
Tyre Coast	Law 708, 1998	3,889*	Ramya	Law 199, 2011	20
Bentael	Law 11, 1999	75	Debl	Law 200, 2011	25
Yammouni	Law 10, 1999	2,100	Beit Leef	Law 201, 2011	20
			Jaj Cedars	Law 257, 2014	20

*Land: 176, Sand: 6, Water: 3,706.
Source: MoE/UNEP/GEF, 2016.

Table 9.9 Natural Sites in Lebanon Protected by Decision or Decree of the Minister of Environment

Forests and Landscape	
Kadisha Valley site (Kanobeen & Kizhaya)	Decision 151/1997
Coastal Front Rocks of Wata Silm (Tabarja)	Decision 200/1997
Forests between Ain El Hour- Daraya- Debiyé- Bérjin; Sheikh Osman Forest; Deir al Mokhalis surrounding; Ain w Zein Hospital surrounding; Dalboun forest; Al Mal valley; Kafra wells; Ainbal valley sites	Decision 132/1998
Al Makmel Mountain (Black summit)	Decision 187/ 1998
Al Kammoua Area (Akkar)	Decision 19/ 2002
Dalhoun Forest (Al Shouf)	Decision 22/ 2002
Al Qaraqeer Valley (Zgharta)	Decision 21/ 2002
Baatara Sinkhole site (Tannourine)	Decision 8/ 2004
Jabal Moussa Biosphere Reserve (Kesrouan)	Decree7494/ 2012
Kassarat Grotto (Metn)	Decree 11949/ 2014
River Streams	
Ibrahim River till outlet	Decision 34/1997
Al Jawz River till outlet	Decision 22/1998
Al Kalb River till outlet	Decision 97/1998
Al Damour River till outlet	Decision 29/1998
Beirut River till outlet	Decision 130/1998
Al Awali River till outlet	Decision 131/1998
Arka River till outlet	Decision 188/1998
Al Assi River till outlet	Decision 189/1998

Source: "Protected Areas in Lebanon," n.d.

agreements related to biodiversity and has taken step to increase the number of protected natural sites that include the country's most diverse areas. In 2002, the Government of Lebanon expanded protection criteria to include, in addition to nature reserves, natural parks, defined as a vast rural territory, partially inhabited, with exceptional natural and cultural heritage, recognized nationally and deserving protection on the long-term, natural Sites and Monuments, areas containing one or more natural features of exceptional importance which deserve protection because of their rarity, representativeness or beauty, and Hima, defined as Community Based Natural Resources Management (CBNRM) System that promotes Sustainable Livelihood, Resources Conservation, and Environmental Protection for

Table 9.10   Hima and Forests Declared by Minister of Agriculture Decision

National hima from Maaser Al Shouf to Dahr El Baydar	Decision 127/1991
National Marine hima at the Marine Sciences Center (Batroun)	Decision 129/1991
National hima in Al Kammoua mountain (Akkar)	Decision 165/1991
National hima in Kfar Zabad village (Zahle)	Decision 71/1992
National hima in Hbaleen	Decision 152/1992
Tannourine, Hadath El Jebbe, Jaj and Al Arz forests	Decision 499/1996
Cedar forest in Swaysi area (Hermel)	Decision 587/1996
Cedar, Shouh, juniper forest in A Kammoua (Akkar)	Decision 588/1996
Cedar, Shouh and juniper forest in Karm Chbat (Akkar)	Decision 589/1996
Cedar, Shouh, juniper, oak, ofis, and malloul forest in Bezbina (Akkar)	Decision 591/1996
Knat forest	Decision 592/1996
Bkassine forest	Decision 3/1997
Cedar, Shouh, juniper, oak, ofis, and malloul forest in Ain AlHokaylat and AlKeif Kirnet and shalout (Al Diniyé)	Decision 8/1997
Cedar and juniper forest in Jurd Al Njass – Al Arbaen mountain (Al Diniyé)	Decision 9/1997
Cedar and ofus forest in Sfiné village (Akkar)	Decision 10/1997
Cedar, Shouh, juniper forest in Marbine – Jhanam Valley (Akkar)	Decision 11/1997
Chebaa valley	Decision 174/1997

Source: "Protected Areas in Lebanon," n.d.

Table 9.11   Natural Sites and Monuments Declared by Decree

Cedars in Bcharre
Deir El Qalaa site
Bolonia Forest
Al Mourouj Oak site
Horsh Beirut Site
Al Yammouneh Lake
Natural Bridge on Al-Laban river
Baalbeck historical monuments

Source: "Protected Areas in Lebanon," n.d.

the Human Wellbeing (UNU-INWEH). The draft Protected Areas Framework Law (decree No. 8045 dated 25/4/2012) awaits final endorsement by the parliament (reference). In addition the country has developed national

Table 9.12 Sites of Natural and/or Ecological Importance in Need for Protection

Wetlands	Holes
Ammiq Wetland	Kateen Azar Hole (Tarshish)
Grottos	Fawar Dara Hole (Tarshish)
Al Rwaess grotto	Meshemshiyit Hole (Tarshish)
Afqa grotto	AlBadwiyi Hole
Ain Labne grotto	AlAbed Hole (Tannourine)
Salem grotto	AlKadaha Hole (Tannourine)
Al Tarash grotto	Osman AlRamhi Hole
Kfarhim grotto	Smokhaya Sinkhole (Rmeich)
Dahr El Ain grotto	Katmeen Sinkhole (Rmeich)
Al Rahwa Spring grotto	Natural Bridges and Rocks
Zoud grotto	Faqra Natural Bridge and Rocks
Al Motran grotto	Al Laqlouq Natural Bridge
Al Hawa grotto	Calcerous- Karstic rocks ex. Faytroun and Rayfoun in Kesrouan; Douma in Batroun
AlShatawi spring grotto	AlQellé Forest (Akkar)
Al Rihan grotto	Jaj Cedar Forest(Jbeil)
Deir Amess grotto	Al Ozor (Fneidek)
Haris grotto	Bshaalé Olives
Debl grotto	Sir Doniyeh Valley
Jeita grotto	Sheikh Zinad area
Al Kasarat grotto	Gold river – Moussa Mountain

Source: "Protected Areas in Lebanon," n.d.

strategies (NS) for effective nature conservation measures. Examples of studies include National Strategy for forest Fires Management (Mitri, 2009), National Strategy for Marine Protected Areas (MOE/IUCN, 2012), National Strategy for Rural Tourism (MOT, 2014), and a national master plan for the protection of mountain peaks, natural areas, coastal zones, green spaces, and agricultural lands.

9.6.2 Microreserves

PMrs are small land plots (up to 20 ha) of peak value in terms plant species richness endemism or rarity, given over to long-term monitoring and conservation of plant species and vegetation types. The legal frame confers

Table 9.13   International Protection of Natural Sites

Biosphere Reserves UNESCO-MAB
The Shouf Biosphere Reserve (2005), which includes Al-Shouf Cedar Nature Reserve and Ammiq Wetland as well as 22 surrounding villages
The Jabal Moussa Biosphere Reserve (2008)
The Jabal Al Rihane Biosphere Reserve (2007).
Cultural landscapes UNESCO World Heritage
The Valley of Qannoubine
The Arz El Rab Cedar Forests
Wetlands of international importance Ramsar sites
Ammiq Wetland
Raas Al-Chaqaa
Tyre Coast Nature Reserve
Palm Islands Nature Reserve
Important Bird Areas (IBA)
Horsh Ehden Nature Reserve
Palm Islands Nature Reserve
Aamiq Wetland
A-Shouf Cedar Nature Reserve
HimaAnjar/ KfarZabad
Lake Qaraoun
Riim/ Sannine Mountain
Bentael Nature Reserve
Tannourine Cedars Nature Reserve
Hima Ebel es-Saqi
Semi Deserts of Ras Baalbek
Beirut River Valley
Upper Mountains of Akkar-Donnieh
Jabal Moussa Mountain
Ramlieh Valley
Regional biodiversity sites Specially Protected Areas of Mediterranean Importance
Tyre Coast Nature Reserve
Palm Islands Nature Reserve

PMrs a permanent status and provides strong protection to plants and substrates while allowing traditional activities compatible with plant conservation. According to IUCN criteria, PMrs would fall into type Ib and

IV categories, that is, designations where the administrations and/or the landowners play a major role in conservation by means of active management. There are ongoing efforts to establish three Plant Micro-Reserves (PMR) in Baskinta (12 ha; Mount Lebanon; privately owned land; *Drosera rotundifolia* L.) (USJ/CEPF, n.d.a), Ehmej (Al Dichar Natural Site; 55 ha; mount Lebanon; publicly owned land; *Iris sofarana keserwana*) (USJ/CEPF, n.d.b), and Sarada (10 ha; south of Lebanon; Religious Waqf; *Iris bismarckiana*) (USJ/CEPF, n.d.c).

9.6.3 Seed Banks

Established in 2004, the American University of Beirut established the first seed bank in Lebanon. The AUB seed bank along with the cold store facilities has been used by ICARDA (International Center for Agricultural Research in the Dry Areas) for the last three years to store part of ICARDA valuable seed stock. Recently, and as a consequence of the war in Syria, ICARDA has made a significant seed withdrawal from the Svalbard Global Seed Vault, known as the "Doomsday Vault." This seed bank contains seeds from every known crop in the world as a backup in the event of a catastrophe that devastates crops. Eight Thousand and Six Hundred accessions have been relocated and stored at AREC's seed bank and a total of 30,000 accessions are now housed at AREC. Part of the seeds are planted at AREC the AUB research station to replenish the stock and make it available for researchers and other seed vaults ("ICARDA Backup Seeds from Arctic 'Doomsday Vault' are Housed at AREC," n.d.). https://www.aub.edu.lb/fafs/arec/Pages/2015NewsandEvents.aspx.

In 2013, a Seed Bank at the Lebanese Agricultural Research Institute (LARI) was officially launched and assigned as the National Seed Bank of Lebanon. The facility hold 1,380 seed collections representing 881 different Lebanese wild species which are stored under long-term conditions, with duplications held at Kew's Millennium Seed Bank of the Royal Botanic Gardens. A great number of wheat and barley landraces, improved varieties of wheat, barley, lentil, chickpea, and vetch are also conserved as *ex situ* collections at LARI and are regularly regenerated every five years ("Official opening of the Seed Bank in Lebanese Agricultural Research Institute," n.d.).

Recently, Saint-Joseph University in collaboration with a national NGO (Jouzour Loubnan) established the Laboratory for Seed Germination and Conservation (LSGC) with the aim of supporting the regeneration and management of woodlands in the Lebanese mountains ("Welcome to Jouzour Loubnan SEED Bank," n.d.).

9.6.4 Botanic Gardens

Countries of the Arab League, of which Lebanon is a member, have the lowest number of botanic gardens, the lowest number of gardens per total area and the lowest number of gardens per number of Individuals (Talhouk et al., 2014). Botanic gardens are integral to the process of plant conservation and development, but the establishment of a botanic garden is challenging in countries such as Lebanon, where land is limited and real-estate value is high. Recent efforts are ongoing to address this shortcoming through a new concept which calls for the establishment of a complementary category of botanic gardens, ancillary botanic gardens (ABGs). ABGs are informal, deregulated gardens for the conservation of plant diversity and cultural plant knowledge; they are established by local communities in open sites which have existing levels of land protection owing to their primary purpose as archaeological sites, educational institutions, religious landholdings, private institutions and touristic sites (Talhouk et al., 2015). In 2016, the campus of the American University of Beirut was announced an ABG (Jafari Safa, 2016).

University campuses are sites where young people in both rural and urban areas spend the majority of their young formative lives. Educational institutions provide an excellent opportunity to sensitize young people to ecological processes through ABGs.

9.6.5 Important Plant Areas (IPAs)

Important plant areas (IPAs) are internationally significant sites for plant diversity, identified at national level, using internationally standardized criteria, they provide a vital tool for conserving wild plants and their habitats *in situ*.

IPAs of Lebanon are distributed throughout the country and are representative of its major ecosystems and habitats. Although as expected, the majority of the IPA sites lie on the western slopes of the Mount Lebanon range and include Qornet Es-Sawda (the highest peak in the country at 3088 m) (Yazbek et al., 2010). IPAs are also found on the Eastern Mediterranean shore, the Anti-Lebanon mountain range, the semi-arid areas of the Bekaa valley and the marshes of West Bekaa. Endemic and/or threatened species are found in almost every designated IPA, most of them contain more than 10 nationally endemic species and some sites are exceptionally rich in endemics: Bcharreh-Ehden (50 species), Chouf (32), Makmel (25) and Keserwan (25). They include threatened endemics represented in a single IPA: *Vicia canescens* Labill (in Mount Makmel IPA), *Chaerophyllum aurantiacum* Post (Tannourine IPA), *Centaurea mouterdii* Wagenitz. (Rihane) and *Tulipa lownei* Baker

(Chouf) or within more than one IPA for example: *Matthiola crassifolia* Boiss. et Gaill., *Melissa inodora* Boiss., *Viola libanotica* Bornm. and *Iris sofarana* Foster. In addition to endemics, the designated IPAs include some species that are at the edge of their distribution range such as *Abies cilicica* (Antoine & Kotschy) Carr found in Bcharreh-Ehden IPA and *Ostry carpinifolia* Scop. in Jabal Moussa (Wadi Jannah IPA). Although all identified IPAs contain species with trans-boundary distributions, many contain (very local) steno-endemics (see section IV on restricted range species).

Qammouaa-Dinnyeh-Jurd Hermel, Palm Islands, Bcharreh-Ehden, Tannourine and Tyre-Naqoura IPAs include entire nature reserves or parts thereof. Aamiq, Palm Islands, Ras Chekka and Tyre Naquora include Ramsar sites. Qadisha valley is a natural heritage site, Wadi Jannah, Chouf and Rihane are bioreserves and Qammouaa-Dinnyeh- Jurd Hermel is under consideration for some kind of protection.

Three Lebanese IPAs that are priorities for the conservation action are:

1. *Makmel:* This IPA comprises a chain of high mountain peaks in the North of Lebanon covered with snow for long periods, sometimes more than six months. Al-Qournet es-Sawda, the highest peak along the Eastern Mediterranean coast is within this site. The vegetation is predominantly oro-mediterranean and the area is very rich in plant species (ca. 200 recorded). It is home to 47 species restricted to the Eastern Mediterranean, 6 endemic to Lebanon, Syria and Turkey, 36 to Lebanon and Syria and 25 to Lebanon.

2. *Qammouaa-Dinnyeh:* Qammouaa-Dinnyeh IPA is situated mostly in the Akkar district in North Lebanon and contains the largest continuous stands of natural forests in Lebanon. A huge diversity of forest types occur at this site: Calabrian pines, mixed cedar, fir and juniper, mixed fir and cedar, pure fir, evergreen oak and relic turkey oak stands. The area covers four vegetation series: the Eu-, Supra-, Mountainous and Oro-Mediterranean and it is characterized by a wide variety of landscapes: valleys, forests, rivers, gorges, rocky cliffs and mountains. 320 species plant species have been recorded: 82 species are restricted to the Eastern Mediterranean, 6 are endemic to Lebanon, Syria and Palestine, 17 to Lebanon and Syria, 9 to Lebanon, Syria and Turkey, 10 to Lebanon and 2 threatened species according to experts' opinion.

3. *Jabal Moussa-Nahr Ibrahim:* Situated on the western slopes of Mount Lebanon, in the central part of the country, this IPA extends along the southern banks of Nahr Ibrahim (Ibrahim River). The site has been continuously inhabited for more than a thousand years and contains Roman

inscriptions, deserted houses and wells. The importance of this IPA resides not only in its species and ecosystem diversity (deciduous oak-pine woodlands, mixed evergreen deciduous woodlands and garrique) but it is also the southernmost limit of *Ostrya carpinifolia* in the Eastern Mediterranean. The core area of the Jabal Moussa Biosphere, one of Lebanon's globally recognized Important Bird Areas (IBA) is included and it is one of Lebanon's sites for old growth trees. The IPA is very rich in plant species (216 species) and is home to 61 species restricted to the Eastern Mediterranean, 8 endemic to Lebanon, Syria and Palestine, 2 endemic to Lebanon and Syria, 8 to Lebanon, Syria and Turkey, 13 to Lebanon and 10 threatened species according to experts' opinion.

9.6.6 IBAs Important Bird Areas

Lebanon lies on a migration route, between sub Saharan Africa in the south and Western Palearctic breeding grounds to the north. It is known as a "bottle neck site" playing host to large concentrations of migrating soaring birds each spring and autumn and harboring 15 important bird areas (Serhal and Khatib, 2014; BirdLife International, 2017).

Now, Lebanon has an inventory of 15 IBAs in total. These IBA sites are: Ehden Forest Nature Reserve, Palm Islands Nature Reserve, Aammiq Wetland, Shouf Cedar Nature Reserve, Hima Aanjar/ Kfar Zabad, Lake Qara-aoun, Riim/ Sannine Mountain, Tannourine Cedars Nature Reserve, Hima Ebel es-Saqi, Semi Deserts of Ras Baalbek, Beirut River Valley, Upper Mountains of Akkar-Donnieh, Jabal Moussa Mountain, Bentael Forest Nature Reserve, and Ramlieh Valley. Of the 15 sites declared, 5 are Government declared nature reserves, 2 are conserved by SPNL in collaboration with local communities through the Hima approach, 4 have active conservation NGOs and 4 have no current protection.

9.6.7 Hima

In Arabic, Hima means protected area. It is a traditional system of resource tenure that has been practiced for more than 1400 years in the Arabian peninsula (Kilani et al., 2007). This concept was revived in 2004 in Lebaon and since then Himas were established in the villages of Ebeles Saqi in South Lebanon, Anjar, Kfar Zabad wetlands, Qolieleh and Mansouri marine, Upper Akkar, and El Fekha in the Bekaa Valley, Ain Zebdeh Kherbet Anafar ("Hima," n.d.). This and similar community based approaches to natural resource management exist in West Asia and North Africa and they have received global endorsement through a resolution that recognizes the various forms and names of Community Conservation, such as Al Hima, Mahjar, Agdal, Qoroq, Adat, etc.(Kilani et al., 2007)

9.7 Threats to Biodiversity

Lebanon occupies a relatively small area that offers high diversity, however recent anthropogenic activities are threatening this natural heritage and the cost of this environmental degradation that is linked to land and wildlife resources is estimated at $100 million per year, or 0.6% of Lebanon's GDP (World Bank, 2004). There are global threats to biodiversity the information below illustrates examples of threats to Lebanon's biodiversity.

9.7.1 Urban Expansion

One of the main reasons of forest and woodland loss is urban expansion which not only appropriates wooded lands for built areas but also puts them at risk of practices associated with urbanization such as agriculture, quarries, and the construction of urban infrastructure and water reservoirs. Most significantly, urbanization is causing a high level of forest fragmentation as evidenced by a recent study that showed an increase in the number of small forest patches (0–100 ha) (Jomaa et al., 2008). Urban expansion also increases incidences of forest fires especially in pine forest ecosystems; roads and infrastructure established in proximity of forests leads to a proliferation of houses that create a glut of solid waste that is dumped and burned near forests. Urbanization also contributes to the decline of trees that may be partially attributed to the complex and cumulative effects of air pollution, to which Mediterranean conifers are thought to be particularly sensitive (Bussotti and Ferretti, 1998). Lack of compliance to laws and regulations, and the sudden flood of refugees into the country have further exacerbated the negative impact of urbanization on the natural heritage of the country (UNHCR, 2014; MOE/UNDP, 2010).

9.7.2 Over-Grazing

It is estimated that 800,000 sheep and goats are using woodlands and degraded highlands for at least 2 months per year (FAO, 2011). Both sheep and goats are managed under nomadic and semi-nomadic systems, feeding on native pastures, woodland species and crop residues. Overgrazing has prevented regeneration of forests, especially the slow growing conifers, and compounded the effects of deforestation.

9.7.3 Climate Change and Biodiversity

Lebanon has many unique formations that are especially vulnerable to climate change risk, such as the cedar forests (AFED, 2009). Furthermore, Lebanon's high altitudes, which provide refuges for many specialized species and niche ecosystems, will undoubtedly witness distribution shifts and in some cases disappearance of species. Two coniferous tree species, the cedar of Lebanon and the Silician fir reach their southernmost distribution limit in Lebanon and their distribution range will recede with increasing temperature to higher latitudes and altitudes in the region (AFED, 2009). Ecosystems that comprise drought resistant species will adapt more easily to climate change compared to other ecosystems. Warmer climates are expected to cause an increase in rodents (field mice, house mice, rats, etc.) all through the Lebanese territories. This phenomenon will eventually lead to an increase in rodent predators such as jackals, foxes, stone martins, etc. On the other hand, marginal mammals will become extinct due to the loss of habitat and food. This is the case for otters (*Lutra lutra*) (such as those in the Aammiq wetlands) and other mammals that depend on water bodies whose habitat will be harshly reduced due to the decrease in water resources (MoE, 2009). Increased temperatures will also cause the spread and proliferation of insect pests and disease vector populations. The shift of bioclimatic zones to higher altitudes due to climate change will affect various reptiles and amphibians (Farajalla, 2008). Other climate change impacts include modifications in: (i) population physiology, (ii) ecosystem phenology, and (iii) geographical distribution of species. Bird populations whose distribution is restricted by cold temperatures will be forced to expand beyond their natural number with warmer temperatures. Inspections of the Lebanese avifauna suggest that few bird species from hot desert climate have started to colonize the vulnerable zone of the semi-arid Qaa by competing with native avifaunal species (MOE/UNDP, 2010). The arrival of numerous new semi-desertic bird species in Lebanon is expected to occur (MOE/UNDP, 2010). In addition, increased temperatures will allow some plant species to become resistant to herbicides and pesticides.

9.7.4 Invasive Alien Species

According to the Global Invasive Species Database there are 55 alien and invasive species that negatively impact biodiversity in Lebanon. Of these, 10 are invasive in freshwater and marine habitats.

A recent study shed light on the impact of IAS on marine bioidiversity due in large part to tropicalization and to the capacity of these species to outcompete the native species (Bitar, 2010). The mold *Mytilus galloprovincialis* is replaced by the exotic species *Brachidontes pharaonis*. The *Posidonia oceanica* has not been seen in Lebanon since 1977 and is replaced by *Cymodocea nodosa* and *Halophila stipulacea*. For herbivorous fishes, the two Licepsian species *Siganus rivulatus* and *S. luridus* are in the process of replacing the local fish *Sarpasalpa*. The crab *Charybdis helleri* is becoming more and more abundant at the expense of *Eriphia verrucosa*, *Pilumnus hirtellus,* and *Pachygrapsus marmoratus*.

9.7.5 Quarries

The information below is from a recent publication by Darwish et al. (2011) which indicates that the number of quarries in 2005 was 1278 with a total land area 5267 ha, that is, one quarry every eight square kilometer. These quarries destroyed 738 ha of grasslands, 676 ha of arable lands, and 137 ha of forest area and many occur on fissured karst bedrock, which increases the risk of adverse effects on groundwater quality. The predominance of degraded ecosystems throughout the rugged landscape, have rendered spontaneous ecosystem recovery very slow and in some cases very complex (Khater et al., 2003; Khater and Martin, 2007). Quarrying activities accelerate the erosion processes and subsequent destruction of existing arable lands, modify preexisting ecosystems, change landscape patterns and integrity, destroy natural habitat and interrupt natural succession (Khater et al., 2003; Khater, 2004), as well as modify genetic resources (El-Fadel et al., 2000; ESCWA, 2001).

Keywords

- biodiversity threats
- conservation actions
- marine biodiversity
- terrestrial fauna diversity

References

Abboud, M., Makhzoumi, J., Clubbe, C., Zurayk, R., Jury, S., & Talhouk, S. N., (2012). Riparian habitat assessment tool for Lebanese rivers (RiHAT): Case study Ibrahim River. *BioRisk, 7*, 99.

Abi-Said, M., & Amr, Z. S., (2012). Camera trapping in assessing diversity of mammals in Jabal Moussa Biosphere Reserve, Lebanon. *Vert Zool, 62*, 145–152.

Al-Qaddi, N., Vessella, F., Stephan, J., Al-Eisawi, D., & Schirone, B., (2016). Current and future suitability areas of kermes oak (*Quercus coccifera* L.) in the Levant under climate change. *Regional Environmental Change*, 1–14.

Asmar, F. R., (2011). *Country Pasture/Forage Resource Profiles*. Retrieved from http://www. fao.org/ag/agp/agpc/doc/counprof/lebanon/lebanon.html#5.

Atallah, T., Hajj, S., Mehanna, M., Aoun, N., Darwish, T., & Rizk, H., (2008). Legumes diversity in the South bank of Nahr-Ibrahim river in Lebanon. *Lebanese Sci. Journ., 9*(2), 17.

Awad, L, Fady, B., Khater, C., Roig, A., & Cheddadi, R., (2014). Genetic structure and diversity of the endangered Fir tree of Lebanon (*Abies cilicica* Carr.): Implications for Conservation. *PLoS ONE, 9*(2), (e90086. doi: 10.1371/journal.pone.0090086).

Azar, D., (1997). A new method for extracting plant and insect fossils from Lebanese amber. *Palaeontology, 40*(4), 1027–1060.

Bakalowicz, M., (2014). Karst at depth below the sea level around the Mediterranean due to the Messinian crisis of salinity – Hydrogeological consequences and issues. *Geologica Belgica, 17*(1), 96–101.

Battle, I., & Tous, J., (1997). Carob tree: *Ceratonia siliqua* L. International Plant Genetic Resources Institute. Gatersleben.

BirdLife International (2017) *Country Profile: Lebanon*. Available from http://www.birdlife. org/datazone/countrylebanon.

Bitar, G., (2010). Impact des changements climatiques et des espèces exotiques sur la biodiversité et les habitats marins au Liban. *Rapp. Comm. int. Mer Médit, 39*, 452.

Blondel, J., & Aronson, J., (1999). *Biology and Wildlife of the Mediterranean Region*. Oxford University Press, USA.

Bolin, J. F., Bray, R. D., & Musselman, L. J., (2011). A new species of diploid quillwort (Isoëtes, Isoëtaceae, Lycophyta) from Lebanon. *Novon, 21*(3), 295–298.

Bolin, J. F., Bray, R. D., Keskin, M., & Musselman, L. J., (2008). The genus *Isoetes* L. (Isoetaceae, Lycophyta) in South-Western Asia. *Turkish J. Bot., 32*(6), 447–457.

Bößneck, U., (2011). New records of freshwater and land molluscs from Lebanon: (Mollusca: Gastropoda & Bivalvia). *Zoology in the Middle East, 54*(1), 35–52.

Bou Dagher-Kharrat, M., Abdel-Samad, N., Douaihy, B., Bourge, M., Fridlender, A., Siljak-Yakovlev, S., & Brown, S. C., (2013). Nuclear DNA C-values for biodiversity screening: Case of the Lebanese flora. *Plant Biosystems, 147*(4), 1228–1237.

Boudouresque, C. F., (2004). Marine biodiversity in the Mediterranean: Status of species, populations and communities. *Travaux Scientifiques Du Parc National De Port-Cros, 20*, 97–146.

Boulos, I., (1968). Contribution à l'océanographie des pêches au Liban. Thèse, Fac*ulté des Sciences, Université de Paris.*

Bromley, H. J., (1994). The freshwater leeches (Annelida, Hirudinea) of Israel and adjacent areas. *Israel J. Zool., 40*(1), 1–24.

Browicz, K., & Zohary, D., (1996). The genus *Amygdalus* L. (Rosaceae): Species relationships, distribution and evolution under domestication. *Genet. Resour. Crop Evol., 43*, 229–247.

Bussotti, F., & Ferretti, M., (1998). Air pollution, forest condition and forest decline in Southern Europe: An overview. *Environmental Pollution, 101*(1), 49–65.

Caesar, M., Roy, R., Legendre, F., Grandcolas, P., & Pellens, R., (2015). Catalogue of Dictyoptera from Syria and neighboring countries (Lebanon, Turkey, Iraq and Jordan). *Zootaxa, 3948*(1), 71–92.

CDR, (2004). *National Physical Master Plan of the Lebanese Territory*, CDR, Dar Al Handasah.

Chalak, L., Noun, J., El Haj, S., Rizk, H., Assi, R., Attieh, J., & Sabra, N., (2011). Current status of agro-biodiversity in Lebanon and future challenges. *Gene Conservation,10*, 23–41.

Chamoun, R., Baalbaki, R., Kalaitzis, P., & Talhouk, S. N., (2009). Molecular characterization of Lebanese olive germplasm. *Tree Genetics & Genomes, 5*(1), 109–115.

Chmaitelly, H. A. W., (2007). Urban floral diversity in the eastern Mediterranean Beirut coastal landscape (PhD Thesis).

Chouchani, B. (1972) *Le Liban: Contribution a Son Etude Climatique et Phytogeographique.* Memoire du doctorat de 3eme cycle, Universite de Toulouse, Toulouse.

Christodoulakis, N. S., (1992). Structural diversity and adaptations in some Mediterranean evergreen sclerophyllous species. *Environ. Exp. Bot., 32* (3), 295–305.

Correal, E., Sanchez, P., & Alcareaz, F., (1987). *Woody Species (Trees and Shrubs) of Multiple Value for the Arid and Semi-Arid Zones of North Mediterranean EEC countries.* Seminar on 'Les Especes Ligneuses A Usages Multiples Des Zones Arides Mediterraneennes,' Instituto Agronomico Mediterraneo de Zaragoza.

Crocetta, F., & Russo, P., (2013). The alien spreading of *Chama pacifica* Broderip, (1835). (Mollusca: Bivalvia: Chamidae) in the Mediterranean Sea. *Turkish Journal of Zoology, 37*(1), 92–96.

Crocetta, F., Bitar, G., Zibrowius, H., & Oliverio, M., (2013b). Biogeographical homogeneity in the eastern Mediterranean Sea. II. Temporal variation in Lebanese bivalve biota. *Aquatic Biology, 19*(1), 75–84.

Crocetta, F., Zibrowius, H., Bitar, G., Templado, J., & Oliverio, M., (2013a). Biogeographical homogeneity in the eastern Mediterranean Sea-I: The opisthobranchs (Mollusca: Gastropoda) from Lebanon. *Mediterranean Marine Science, 14*(2), 403–408.

Darwish, T., Khater, C., Jomaa, I., Stehouwer, R., Shaban, A., & Hamzé, M., (2011). Environmental impact of quarries on natural resources in Lebanon. *Land Degradation & Development, 22*(3), 345–358.

Davis, S. D., Heywood, V. H., & Hamilton, A. C., (1994). Centers of plant diversity. *Natural History, 111*(1), 1.

Dereix, C., Ohannessian-Charpin, A., Khouzami, M., Safi, S., Zreik, R., El-Hanna, C., Hanna, K., Assaf, N., El-Riachy, R., Habr, A., Munzer, M., & Fortunat, L., (1999). Les principaux arbres du Liban. Project d'assistance a la protectio de la couverture vegetale au Liban. Union Europeenne, *Ministere de L'agriculture.*

Dia, A., (1983). Recherches sur l' ecologie et la biogeographie des cours d' eau du Liban meridional. PhD Thesis, Univ. Marseille.

Duy-Jaquemin, M. N., & Dany, A. Z. A. R., (2004). The oldest records of Polyxenida (Myriapoda, Diplopoda): New discoveries from the Cretaceous ambers of Lebanon and France. *Geodiversitas, 26*(4), 631–641.

El Hachem, R. C., (2000). *Insect Diversity Along the Lebanese Coast in Selected Plant Communities.* MSc dissertation, American University of Beirut. Interfaculty Graduate Environmental Sciences Program (Ecosystem Management)).

El Hajj, A., (2008). L'aquifère carbonaté karstique de Chekka (Liban) et ses exutoires sous-marins. Caractéristiques hydrogéologiques et fonctionnement. Doctorat Université Montpellier (PhD Thesis).

El Zein, G., & Khalaf, G., (2012). La composition et la distribution du peuplement des poissons marins migrateurs dans quelques rivières libanaises et l'impact des aménagements sur la migration de ces poisons.

El Zein, G., (2002). L'Ichtyofaune du Lac de Karaoun dans le bassin supérieur du Litani au Liban. *XXVèmes Journées de la Société Française d'Ichtyologie.*

El-Fadel, M., Zeinati, M., & Jam, D., (2000). Framework for environmental impact assessment in Lebanon. *Environmental Impact Assessment Review, 20,* 579–604.

El-haj, M., Olama, Z., & Holail, H., (2015). Biodiversity of oleaginous microorganisms in the Lebanese environment. *Int. J. Curr. Microbiol. App. Sci., 4*(5), 950–961).

Elias, A., Tapponnier, P., Singh, S. C., King, G. C., Briais, A., Daëron, M., & Klinger, Y., (2007). Active thrusting offshore Mount Lebanon: Source of the tsunamigenic AD 551 Beirut-Tripoli earthquake. *Geology, 35*(8), 755–758.

ESCWA, (2001). *Development of Guidelines for Harmonized Environmental Impact Assessment Suitable for the Escwa Region.* United Nations Economic and Social Commission for West Asia: New York.

Fady-Welterlen, B., (2005). Is there really more biodiversity in Mediterranean forest ecosystems? *Taxon, 54,* 905–910.

FAO, (2005). *Global Forest Resources Assessment Country Report–Lebanon.* FRA2005/059. Rome. FAO.

FAO, (2010). *Global Forest Resources Assessment*, Main Report.

FAO, (2011). *FAOSTAT.* Food and Agriculture Organization of the United Nations.

Farajalla, N., (2008). *Effect of Climate Change on Biodiversity in Lebanon.* Report IndyACT/SPNL.

Farjon, A., (1998). *World Checklist and Bibliography of Conifers.* Royal Botanic Gardens, Kew.

Geiger, M. F., Herder, F., Monaghan, M. T., Almada, V., Barbieri, R., Bariche, M., & Denys, G. P., (2014). Spatial heterogeneity in the Mediterranean Biodiversity Hotspot affects barcoding accuracy of its freshwater fishes. *Molecular Ecology Resources, 14*(6), 1210–1221.

George, C., & Athanassiou, J. V., (1965). On the occurrence *Scomberomorus commerson* (Lacépède) in St George Bay, Lebanon. *Doriana, 4*(157), 1–4.

George, C., & Athanassiou, J. V., (1966). Additions to the check list of the fishes of the coastal waters of Lebanon. Misc. pap. *Natural Science, 5,* 6–8.

George, C., & Athanassiou, J. V., (1967). A two year study of the fishes appearing in the seine fishery of St George Bay, Lebanon. *Annali del Museo Civico di Storia Naturale, Genova, 76,* 237–294.

George, C., Athanassiou, J. V., & Boulos, I., (1964). The fishes of the coastal waters of Lebanon. Misc. pap. *Natural Science, 4,* 27.

George, C., Athanassiou, J. V., & Tortonese, E., (1971). The presence of a third species of the genus *Sphyraena* (Pisces) in the marine water waters of Lebanon. *Annali del Museo Civico di Storia Naturale, 78,* 256–263.

Global Invasive Species Database, (2005). *Rattus rattus.* Available from: http://www.issg.org/database/species/ecology.asp?si=19&fr=1&sts=sss [Accessed 1st September 2011].

Greuter, W., (1994). Extinctions in Mediterranean areas. *Philosophical Transactions of the Royal Society of London B: Biological Sciences, 344*(1307), 41–46.

Haddad, N., Léger, N., & Sadek, R., (2003). Sandflies of Lebanon: Faunistic inventory. *Parasite(Paris, France), 10*(2), 99–110.

Haddad, S. G., Smith, C. A., Al-Zein, M. S., & Knio, K. M., (2017). Genetic and morphometric variations in the Lebanese populations of the flower-head-infesting fruit fly, *Terellia serratulae* (Diptera: Tephritidae). *The Canadian Entomologist, 1–16*.

Harmelin, J. G., Bitar, G., & Zibrowius, H., (2009). Smittinidae (Bryozoa, Cheilostomata) from coastal habitats of Lebanon (Mediterranean sea), including new and non-indigenous species. *Zoosystema, 31*(1), 163–187).

Helfer, S., (2008). Mycota of southwest Asia. *Turkish, J. Bot., 32*(6), 481–484.

Hermoso, V., & Clavero, M., (2011). Threatening processes and conservation management of endemic freshwater fish in the Mediterranean basin: A review. *Marine and Freshwater Res., 62*, 244–254.

Hodgkin, T., (1993). Wild relatives. *Naturopa, 73*, 18.

Hraoui-Bloquet, S., Sadek, R. A., Sindaco, R., & Venchi, A., (2002). The herpetofauna of Lebanon: New data on distribution. *Zoology in the Middle East, 27*(1), 35–46.

Hraoui-Bloquet, S., Sadek, R., & Geze, R., (2001). Les Amphibiens du Liban: inventaire, répartition géographique et altitudinale. *Bulletin de la Société Herpétologique de France, 99*, 19–28.

Jafari, S. S., (2016). *AUB Campus Announced a Botanic Garden*. Retrieved from https://www.aub.edu.lb/news/2016/Pages/botanic-garden.aspx.

Jaradat, A. A., (1998). Biodiversity and sustainable agriculture in the Fertile Crescent. Transformations of Middle Eastern natural environments: *Legacies and Lessons*, 31–57.

Jomaa, I., Auda, Y., Saleh, B. A., Hamzé, M., & Safi, S., (2008). Landscape spatial dynamics over 38 years under natural and anthropogenic pressures in Mount Lebanon. *Landscape and Urban Planning, 87*(1), 67–75.

Kester, D. E., & Gradziel, T. M., (1996). Almonds. In: *Fruit Breeding*, Janick, J., & Moore, J. N., (eds.). John Wiley and Sons, New York, *vol. III*.

Kester, D. E., Gradziel, T. M., & Grasselly, C., (1991). Almonds (*Prunus*). In: *Genetic Resources of Temperate Fruit and Nut Crops*, Moore, J. M., & Ballington, J. R., (eds.). The International Society for Horticultural Science, Wageningen, The Netherlands, pp. 701–758.

Khater, C., & Martin, A., (2007). Application of restoration ecology principles to the practice of limestone quarry rehabilitation in Lebanon. *Lebanese Science Journ., 8*, 19–28.

Khater, C., (2004). Dynamiques végétales post perturbation sur les carriers calcaires au Liban. Stratégies pour l'écologie de la restauration en régions méditerranéennes. Thèse de doctorat. Académie de Montpellier, *Université Montpellier II*, (PhD Thesis).

Khater, C., Martin, A., & Maillet, J., (2003). Spontaneous vegetation dynamics and restoration prospects for limestone quarries in Lebanon. *Applied Vegetation Science, 2*, 199–204.

Kilani, H., Serhal, A., & Llewlyn, O., (2007). *Al-Hima: A Way of Life*. IUCN West Asia regional.

Knio, K. M., Kalash, S. H., & White, I. M., (2002). Flower head-infesting fruit flies (Diptera: Tephritidae) on thistles (Asteraceae), in Lebanon. *J. Nat. Hist., 36*(5), 617–629.

Knio, K. M., Markarian, N., Kassis, A., & Nuwayri-Salti, N., (2005). A two-year survey on mosquitoes of Lebanon. *Parasite, 12*(3), 229–235.

Knio, K. M., White, I. M., & Al-Zein, M. S., (2007). Host-race formation in *Chaetostomella cylindrica* (Diptera: Tephritidae): Morphological and morphometric evidence. *J. Nat. Hist., 41*(25–28), 1697–1715.

Kürschner, H., (2010). Amendments to the bryophyte flora of Lebanon, based on collections of CB Arzeni 1962–1964. *The Bryologist, 113*(4), 717–720.

Kurtto, A., (2009). Rosaceae (pro parte majore). In: *Euro+Med Plantbase – The Information Resource for Euro-Mediterranean Plant Diversity.*

Lakkis, S., & Novel-Lakkis, V., (1981). Composition, annual cycle and species diversity of the phytoplankton in Lebanese coastal water. *J. Plankton Res., 3* (1), 123–136. DOI: https://doi.org/10.1093/plankt/3.1.123.

Lakkis, S., (2007). Dataset and database biodiversity of plankton community in Lebanese seawater. *Proceedings of Ocean Biodiversity Informatics, an International Conference on Marine Biodiversity Data Management.* Hambourg, Germany 29 Nov.–1 Dec. Paris. UNESCO/IOC, VLIZ, BSH, 2007. IOC Workshop report, 202), Vanden Bergh et al. (eds.). VLIZ Special publication, *37*, pp. 99–113.

Lakkis, S., (2011a). Le Phytoplancton Marin du Liban (Méditerranée Orientale), *Biologie, Biodiversité, Biogéographie.*

Lakkis, S., (2011b). Le Zooplancton Marin du Liban (Méditerranée Orientale), *Biologie, Biodiversité, Biogéographie.*

Larsen, T. B., (1974). *Butterflies of Lebanon.* NCSR.

Levy, G., (1997). 12 genera of orb-weaver spiders (Araneae, Araneidae) from Israel. *Israel J. Zool., 43*(4), 311–365.

Lewis, R. E., & Harrison, D. L., (1962). Notes on bats from the Republic of Lebanon. *Proc. Zool. Soc. London,138*(3), 473–486.

Lteif, M., (2015). *Biology,* Distribution and diversity of cartilaginous fish species along the Lebanese coast, eastern Mediterranean. Doctoral dissertation, Université de Perpignan.

Makhzoumi, J., & Pungetti, G., (1999). *Ecological Landscape Design and Planning: The Mediterranean Context.* Spon-Routledge, London.

Marouf, M., Batal, M., Moledor, S., & Talhouk, S. N., (2015). Exploring the practice of traditional wild plant collection in Lebanon. *Food, Culture & Society, 18*(3), 355–378.

Medail, F., & Quezel, P., (1997). Hot-spots analysis for conservation of plant biodiversity in the Mediterranean Basin. *Annals of the Missouri Botanical Garden, 84*(1), 112–127.

Meiggs, R., (1982). *Trees and Timber in the Ancient Mediterranean World.* Clarendon Press, Oxford.

METAP (Mediterranean Environmental Technical Assistance Program), (1995). *Lebanon: Assessment of the State of the Environment, for the Ministry of Environment.* Final Report, EU/EIB/WB/UNDP, Prepared by Environmental Resources Management Resources.

Mikesell, M. W., (1969). The deforestation of Mount Lebanon. *Geographical Review, 19,* 1–28.

Mitri, G., (2009). *Lebanon's National Strategy for Forest Fire Management.* Beirut, Lebanon, AFDC.

MOA, (1996). *Biological Diversity of Lebanon*–Country Study Report. UNEP, Project GF/6105–92–72.

MOA, (2003). *National Action Program to Combat Desertification,* Beirut, Lebanon.

Modad, N. M., (2006). *Survey and Identification of Wild Mushrooms in Lebanon.* American University of Beirut.

MOE, (2009). *Fourth National Report of Lebanon to the Convention on Biological Diversity.*

MOE, (2012). *Lebanon's National Report to the United Nations Conference on Sustainable Development – RIO+20*. Lebanon: Wide Expertise Group.

MOE/ECODIT, (2002). The State of the Environment Report.

MOE/GEF, (2016). Updating the SAP-BIO National Report for the Country of Lebanon. Prepared by M. Nader and S. Talhouk.

MOE/IUCN, (2012). Lebanon's marine protected area strategy: Supporting the management of important marine habitats and species in Lebanon.

MOE/UNDP, (2010). State and Trends of the Lebanese Environment.

MOE/UNEP/GEF, (2016). National Biodiversity Strategy and Action Plan–NBSAP.

MOEW, (2010). National Water Sector Strategy: Supply/Demand Forecasts, Draft.

MOT, (2014). Rural Tourism Strategy for Lebanon.

Moubayed, Z., (1986). Recherches sur la faunistique, l' ecologie et la zoogeographie de trois reseaux hydrographiques du Liban: l' Assi, le Litani et le Beyrouth. *PhD. Thesis*, Univ. Toulouse.

Mouneimné, N., (1977). Liste des poissons de la côte du Liban (Méditerranée orientale). *Cybium 3e Série 1*, 37–66.

Mouneimné, N., (1978). *Poissons des Côtes du Liban (Méditerranée Orientale).* Biologie et pêche. Thèse Doct. ès-Sci., Université Pierre et Marie Curie (Paris VI), France.

Mouneimné, N., (1979). *Poissons Nouveaux Pour les Côtes Libanaises., 6*, 105–110.

Mouneimné, N., (2002). *Poissons Marins du Liban et De la Méditerranée Orientale.* Beyrouth.

Mouterde, P., (1966). *Nouvelle Flore de Liban et de La Syrie. Dar el–Machreque*, Beirut, Tome I.

Mouterde, P., (1970). *Nouvelle Flore du Liban et de la Syrie, El Machreq, éditeurs Beyrouth*, Distribution Librairie Orientale Beyrouth, Liban. Tome II.

Musselman, L. J., & Al-Zein, M. S., (2009). *Isoetes duriei* new to Lebanon. *American Fern Journ., 99*(4), 333–334.

Myers, N., Mittermeier, R. A., Mittermeier, C. G., Da Fonseca, G. A., & Kent, J., (2000). Biodiversity hotspots for conservation priorities. *Nature, 403*(6772), 853–858.

Nader, M. R., El Indary, S., Salloum, B. A., & Dagher, M. A., (2011). Combining non-invasive methods for the rapid assessment of mammalian richness in a transect-quadrat survey scheme–Case Study of the Horsh Ehden Nature Reserve, North Lebanon. *Zoo Keys, 119*, 63.

Naja, F., Alameddine, M., Itani, L., Shoaib, H., Hariri, D., & Talhouk, S. N., (2015). The use of complementary and alternative medicine among Lebanese adults: Results from a national survey. *Evidence-Based Complementary and Alternative Medicine*, Article ID 682397, http://dx.doi.org/10.1155/2015/682397, pp. 9.

Osta, M. A., Rizk, Z. J., Labbé, P., Weill, M., & Knio, K., (2012). Insecticide resistance to organophosphates in *Culex pipiens* complex from Lebanon. *Parasites & Vectors, 5*(1), 132.

Ozcan, T., Ozcan, G., & Erdogan, H., (2012). Checklist of the freshwater decapod crustaceans from the Orontes River. *Arthropods, 1*(3), 118.

Penney, D., & Selden, P. A., (2002). The oldest linyphiid spider, in lower Cretaceous Lebanese amber (Araneae, Linyphiidae, Linyphiinae). *J. Arachnology, 30*(3), 487–493).

Post, G. E., & Dinsmore, J. E., (1933). *Flora of Syria Palestine and Sinai,* Second Edition, vol. I–II. American University Press. Beirut, Lebanon.

Ramon-Laca, L., & Mabberley, D. J., (2004). The ecological status of the carob-tree (*Ceratonia siliqua,* Leguminosae) in the Mediterranean. *Bot. J. Linn. Soc., 144*(4), 431–436.

Riachy, I. E., (1998). *Germplasm Characterization and Climatic Zoning for Viticulture in Lebanon.* MSc dissertation, American University of Beirut. Department of Crop Production and Protection.

Ruckert, F., (1985). Egel aus den Levante-Landern (Clitellata: Hirudinea). *Senckenbergiana Bioi., 66,* 135–152.

Saad, L., Talhouk, S. N., & Mahy, G., (2009). Decline of endemic *Oncocyclus irises* (Iridaceae) of Lebanon: survey and conservation needs. *Oryx, 43*(1), 91–96.

Samad, F. A., Baumel, A., Juin, M., Pavon, D., Siljak-Yakovlev, S., Médail, F., & Kharrat, M. B. D., (2014). Phylogenetic diversity and genome sizes of *Astragalus* (Fabaceae) in the Lebanon biogeographical crossroad. *Plant Systematics and Evolution, 300*(5), 819–830.

Sayar, N. P., Smith, C. A., White, I. M., & Knio, K. M., (2009). *Terellia fuscicornis* (Diptera: Tephritidae): Biological and morphological adaptation on artichoke and milk thistle. *J. Nat. Hist., 43*(19–20), 1159–1181.

Schacher, J. F., & Khalil, G. M., (1967). *Foleyella philistinae* sp. n.(Nematoda: Filarioidea) from the Lizard, *Agama stellio,* in Lebanon, with Notes on *Foleyella agamae* (Rodhain, 1906). *J. Parasitol., 53*(4), 763–767.

Schneider, W., (2004). Critical species of Odonata in the Levant. *Intern. J. Odonatol., 7*(2), 399–407.

Sękiewicz, K., Dering, M., Sękiewicz, M., Boratyńska, K., Iszkuło, G., Litkowiec, M., et al. (2015). Effect of geographic range discontinuity on species differentiation—East-Mediterranean *Abies cilicica*: a case study. *Tree Genetics & Genomes, 11*(1), 1–10.

Serhal, A. A., & Khatib, B. C., (2014). *State of Lebanon's Birds and IBAs.* Ministry of Environment.

Shaban, A., & Kayrouz-Khalf, L., (2013). The Geological controls of the geothermal groundwater sources in Lebanon. *Intern. J. Energy Environ. (IJEE), 4,* 787–796.

Shaban, A., (2009). Indicators and aspects of hydrological drought in Lebanon. *Water Resources Management,23*(10), 1875–1891.

Shaban, A., Darwich, T., & El Hage, M., (2013). Studying Snowpack-related Characteristics on Lebanon Mountains. *Intern. J. Water Sci., 2.*

Smith, C. A., Al-Zein, M. S., Sayar, N. P., & Knio, K. M., (2009). Host races in *Chaetostomella cylindrica* (Diptera: Tephritidae): Genetic and behavioral evidence. *Bull. Entomol. Res., 99*(4), 425–432.

Speiser, E. A., (1955). Akkadian myths and epics. In: An*cient Near Eastern Texts Relating to the Old Testament* (Pritchard, J. B., ed.), 2nd edn., Princeton Press, Princeton, pp. 72–99.

Storey, R., (2007). *Aquatic Invertebrate Diversity and Distribution at Aammiq Marsh,* Lebanon. Arocha Lebanon.

Sulahian, A., & Schacher, J. F., (1968a). *Brygoofilaria agamae* gen. et sp. n.(Nematoda: Filarioidea) from the lizard *Agama stellio* in Lebanon. *J. Parasitol., 831–833.*

Sulahian, A., & Schacher, J. F., (1968b). *Thelandros (Parapharyngodon) tyche* sp. n. (Nematoda: Oxyuroidea) and *Abbreviata adonisi* sp. n.(Nematoda: Physalopteroidea) from the lizard *Agama stellio* in Lebanon. *J. Helminthol., 42*(3–4), 373–382.

Sulahian, A., & Schacher, J. F., (1969). *Saurositus baal* sp. n., a Filarial Worm from the Lizard *Agama stellio* in Lebanon, with Notes on *Saurositus macfiei* Fitzsimmons. J. Parasitol., 55(1), 104–107.

Talhouk, S. N., Dardas, M., Dagher, M., Clubbe, C., Jury, S., Zurayk, R., & Maunder, M., (2004). Patterns of floristic diversity in semi-natural coastal vegetation of Lebanon and implications for conservation. *Biodiversity and Conservation, 14*(4), 903–915.

Talhouk, S. N., Lubani, R. T., Baalbaki, R., Zurayk, R., AlKhatib, A., Parmaksizian, L., & Jaradat, A. A., (2000). Phenotypic diversity and morphological characterization of *Amygdalus* L. species in Lebanon. *Genetic Resources and Crop Evolution, 47*(1), 93–104.

Talhouk, S. N., Van Breugel, P., Zurayk, R., Al-Khatib, A., Estephan, J., Ghalayini, A., Debian, N., & Lychaa, D., (2005). Status and prospects for the conservation of remnant semi-natural carob *Ceratonia siliqua* L. populations in Lebanon. *Forest Ecology and Management, 206*, 49–59.

Talhouk, S., Abunnasr, Y., Hall, M., Miller, T., & Seif, A., (2014). Ancillary botanic gardens in Lebanon–Empowering local contributions to plant conservation. *Sibbaldia: The Journal of Botanic Garden Horticulture, 12*, 111–128.

Tohme, G., & Tohme, H., (1988). Les coquillages terrestres du Liban. Publications de l'Université Libanaise, *Section des Sciences Naturelles, 20*, 1–113.

Tohme, G., & Tohme, H., (2007). *Illustrated Flora of Lebanon*. National Council for Scientific Research.

Tohme, G., & Tohme, H., (2014). *Illustrated Flora of Lebanon*. National Council for Scientific Research.

Tohmé, G., & Tohmé, H., (2015). Les Plantes De La Cote Libanaise. *Lebanese ScienceJourn., 16*(2), 119.

Trichopoulou, A., & Vasilopoulou, E., (2000). Mediterranean diet and longevity. *British J. Nutrition, 84*(S2), S205–S209.

UNEP-MAP-RAC/SPA, (2010). *Fisheries Conservation and Vulnerable Ecosystems in the Mediterranean Open Seas, Including the Deep Seas*. By de Juan, S., & Lleonart, J., (eds.) RAC/SPA, Tunis.

UNHCR, (2014). *Syria Regional Refugee Response*.

USJ/CEPF (n. d. a). Process frame work – Baskinta plant micro-reserve – District of Metn. Retrieved from http://www.cepf.net/SiteCollectionDocuments/mediterranean/63257-Safeguard-ProcessFramework-Baskinta.pdf.

USJ/CEPF (n.d.b). Process framework – Ehmej plant micro-reserve – District of Jbeil. Retrieved from http://www.cepf.net/SiteCollectionDocuments/mediterranean/63257-Safeguard-ProcessFramework-ehmej.pdf.

USJ/CEPF (n.d.c). Process framework – Sarada plant micro-reserve – District of marjayoun. Retrieved from http://www.cepf.net/SiteCollectionDocuments/mediterranean/63257-Safeguard-ProcessFramework-Sarada.pdf.

Van Breugel, P., & Bazuin, T. O. M., (2001). *Development of Research Activities on the Conservation and Use of Forest Genetic Resources in Lebanon*. Workshop Report. International Plant Genetic Resources Institute, Rome, Italy.

Vavilov, N. I., (1949–1950). The phytogeographic basis of plant breeding. In: *The Origin, Variation, Immunity, and Breeding of Cultivated Plants*, translated by Chester, K. S., Waltham, Chronica Botanica, pp. 13–54.

Vela, E., & Viglione, J., (2015). Recent inputs to the Lebanese orchid flora and proposal of a national checklist for Orchidaceae family. *J. Acta Botanica Gallica., 162*, 4.

Verheye, W. H., (1998). Land use, *Land Cover and Soil Sciences, vol. I.*

Wilson, J. A., (1955). Egyptian historical texts. In: An*cient Near Eastern Texts Relating to the Old Testament,* (Pritchard, J. B., ed.), 2nd edn., Princeton Press, Princeton, pp. 227–264.

Wolz, K. D., (1998). *Protection and Remediation of Rivers and River Landscapes in the Republic of Lebanon Volume One* (report). Ministry of Environment, Lebanon.

World Bank, (2004). *World Development Indicators.* Washington, DC: World Bank. Retrieved from http://documents.worldbank.org/curated/en/517231468762935046/World-development-indicators-2004.

World Checklist of Selected Plant Families, (2010). The board of trustees of the Royal Botanic Gardens, Kew.

Würtz, M., (2012). *Mediterranean Submarine Canyons: Ecology and Governance.* IUCN.

WWF & IUCN, (1994). *Centers of Plant Diversity.* Gland, Switzerland.

Yazbek, M., Machaka-Houri, N., Al-Zein, M. S., Safi, S. Sinno, N., & Talhouk, S. N. T., (2010). Lebanon In: *Important Plant Areas of the South and East Mediterranean Region: Priority Sites for Conservation,* Radford, E. A., Catullo, G., & De Montmollin, B., (eds.). pp. 53–57.

Zahreddine, H., Clubbe, C., Baalbaki, R., Ghalayini, A., & Talhouk, S. N., (2004). Status of native species in threatened Mediterranean habitats: The case of *Pancratium maritimum* L. (sea daffodil) in Lebanon. *Biological Conservation, 120*(1), 11–18.

Zohary, D., (2002). Domestication of the carob (*Ceratonia siliqua* L.). *Israel J. Plant Sci., 50*(sup1), 141–145.

Zohary, M., (1973). *Geobotanical Foundations of the Middle East.* Gustav Fischer Verlag, Stuttgart, Germany.

Zonstein, S., & Marusik, Y. M., (2013). Checklist of the spiders (Araneae) of Israel. *Zootaxa, 3671*(1), 1–127.

Websites

"Hima" (n.d.). Retrieved from http://www.spnl.org/hima/.

"ICARDA Backup Seeds from Arctic 'Doomsday Vault' are Housed at AREC" (n.d.). Retrieved from https://www.aub.edu.lb/fafs/arec/Pages/2015NewsandEvents.aspx.

"IUCN Red Listing Lebanese plants: a comprehensive workshop at USJ" (July 8, 2014). Retrieved from http://www.cepf.net/news/top_stories/Pages/To-grow-hope-for-endangered-plants.aspx.

"Official opening of the Seed Bank in Lebanese Agricultural Research Institute" (n.d.). Retrieved from http://lebanon.plantgenetic.com/Pages/Events.aspx?I=0&DId=0&CId=0&lang=EN&CMSId=5002863&id=1185.

"Protected Areas in Lebanon" (n.d.) Retrieved from http://www.moe.gov.lb/protectedareas/categories.htm.

"Welcome to Jouzour Loubnan SEED Bank" (n.d.). Retrieved from http://www.lebanon-flora.org/seeds.htm.

Figure 9.1 *Centaurea mouterdei* Wagenitz (Photo: Hicham El Zein).

Figure 9.2 *Alkanna prasinophylla* Rech. f. (Photo: Hicham El Zein).

Figure 9.3 *Matthiola crassifolia* Boiss. & Gaill (Photo: Hicham El Zein).

Figure 9.4 *Silene reuteriana* Boiss. (Photo: Hicham El Zein).

Figure 9.5 *Cytisus syriacus* Boiss. & Bl. (Photo: Hicham El Zein).

Figure 9.6 *Origanum ehrenbergii* Boiss. (Photo: Hicham El Zein).

Figure 9.8 *Ptyodactylus puiseuxi* (Photo: Rami Khashab).

Figure 9.9 *Natrix tessellata* (Photo: Rami Khashab).

Figure 9.7 *Stachys hydrophila* Boiss. (Photo: Hicham El Zein).

Figure 9.10 *Pseudopus apodus* (Photo: Rami Khashab).

Figure 9.11 *Circaetus gallicus* (Photo: Fouad Itani)

Figure 9.12 *Turdus viscivorus* (Photo: Fouad Itani)

Figure 9.13 *Cinnyris osea* (Photo: Fouad Itani)

Figure 9.14 *Alcedo atthis* (Photo: Fouad Itani)

Figure 9.15 Cedars of Lebaon.

Figure 9.16 Arid highlands, northern Bekaa (Photo: Lama-Yasmin Tawk)

Figure 9.17 Lebanon, a mountainous country (Photo: Lama-Yasmin Tawk)

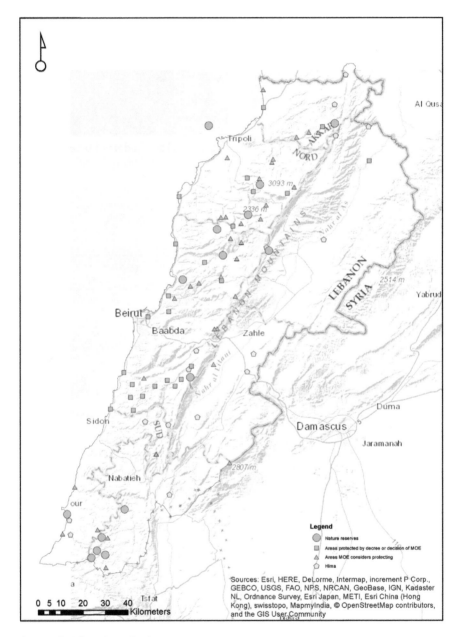

Figure 9.18 Map of Lebanon.

CHAPTER 10

Biodiversity in Malaysia

A. LATIFF

School of Environmental and Natural Resource Science, Universiti Kebangsaan Malaysia, 43600 Bangi, Selangor, Malaysia. E-mail: pakteh48@yahoo.com

10.1 Introduction

Malaysia's 329,613 km^2 land area, about 40% constitutes Peninsular Malaysia and 60% Sabah and Sarawak and nearly 4,800 km of coastal line exhibit a notable diversity of habitats and ecosystems due to significant variations in altitude and topography that cause minor seasonal changes in vegetation. They include a wide range of ecozones like mountains, highlands, tropical forests, wetlands, riverine areas as well as island archipelago. It hosts a few biodiversity hotspots: the Mount Kinabalu, Danum Valley, Maliau Basin, Imbak Canyon, in Sabah, Mount Mulu, Mount Murud, Lanjak-Entimau Wildlife Sanctuary, Bako National Park, in Sarawak, the Langkawi Archipelago, Royal Belum State Park, the hilly ranges that straddle the National Park in Peninsular Malaysia. These hotspots have numerous threatened, rare and endemic species of plants and animals. Though comparatively small in area, Malaysia is home to a great diversity of forests, ranging from mangrove swamp forests, limestone hill forest, the rich lowland and hill dipterocarp forests and montane forests, quite typical of Southeast Asia. In Peninsular Malaysia, the highest points are Mt. Korbu (2,183 m) and Mt. Tahan (2,187 m), but Mt. Kinabalu in Sabah reaches 4,095 m above sea level. Malaysia is regarded as one of 17 megabiodiversity country in the world for their exceptional number of unique and endemic species, endowed with an immense variety of natural resources by way of its rich animal diversity, plant diversity and microbial diversity.

Forest cover in Peninsular Malaysia extends over 5.89 million ha constituting about 44.7% of total land areas, with 4.92 million ha or 83.5% of this forested area gazetted as Permanent Reserve Forests (PRFs). Within the PRFs, 2.83 and 2.09 million ha of the forests area are managed as a Production and Protection Forests, respectively. Production Forests are forested lands that are set aside by the Federal government for the purpose of supplying wood and non-wood production sustainably, at reasonable rates of all forms of forest produce which can be economically produced within the country and are required for agricultural, domestic, industrial, and for export purposes. The Production Forests have been producing the hard and heavy tropical timbers for foreign exchanges. Whereas, the Protection Forests afford services, such for ensuring favorable climatic and physical conditions of the country, safeguarding of water resources, soil fertility, environmental quality, preservation of biological diversity and minimization of damage by floods and erosion to rivers and agricultural lands.

The PRFs are further classified into 12 forest functional classes. These functional classes reflect the multiple values and uses of forest products and services that the forest can offer to communities under the implementation of Sustainable Forest Management practices since Independence. Production Forests are managed sustainably for timber production as well as other non-forest products such as rattans, bamboos, fruits, and medicines. Hence, the Protection Forests are managed in accordance with the 11 functional classes as shown in Table 10.1.

From Table 10.2, it was obvious that Peninsular Malaysia has two major administrations with regards to the forest resources. The state or alienated forests are under the jurisdiction of states in Malaysia and the rest are under the Federal departments. The permanent Reserved Forests are administrated by the Forest Department and the National Parks, Wildlife and Birds sanctuary forests come under the Department of Wildlife and National Park. These

Table 10.1   Extent of Permanent Reserved Forest by Functional Classes in Peninsular Malaysia*

Purpose of Management	Functional classes	Extent (ha)
Production Forest	i. Timber Production Forest	283,6016
Protection Forest	ii. Soil Protection Forest	587,414
	iii. Soil Reclamation Forest	6,804
	iv. Flood Control Forest	11,119
	v. Water Catchment Forest	812,337
	vi. Forest Sanctuary For Wild Life	90,986
	vii. Virgin Jungle Reserve Forest	30,258
	viii. Amenity Forest	61,296
	ix. Education Forest	31,079
	x. Research Forest	40,886
	xi. Forest For Federal Purposes	15,231
	xii. State Park Forest	
Total		393,444

*The forests in Sabah and Sarawak are not classified as above, hence not included.

Table 10.2 Major Forest Types and Their Distribution in Peninsular Malaysia

State or Alienated Lands	Permanent Reserved Forests	National Parks, Wildlife and Birds Sanctuary Forests
5.3%	84.6%	10.1%
0.309 Million ha	Production—Protection	0.581 Million ha
	2.90 Million ha—1.99 Million ha	
For production and development		Totally protected

I--I   I--I

55.5%	44.5%
3.21 Million ha	2.57 Million ha

exclude those in Sabah and Sarawak, which come under the respective states' administration and departments (Figure 10.1 and Table 10.3).

Malaysia's coastline is about 4,800 km stretching along the Peninsular Malaysian states of Perlis in the northwest to Johor in the south and Kelantan

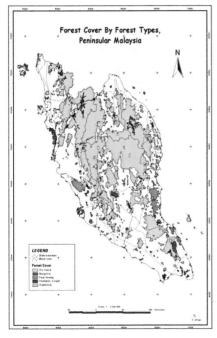

Figure 10.1 Map showing the distribution of forest cover in Malaysia (in green, as of 2008).

Table 10.3 The Distribution of the Major Forest Types

No.	Forest types	Area (Million ha)	Distribution
Tropical Forests			
1	Wet evergreen inland forests	5.53	Lowland to montane of the central part of Peninsular Malaysia, central and southern parts of Sabah and along the Kalimantan border of Sarawak
2	Peat swamp forests	0.21	Behind the beach both west and east coasts of Peninsular Malaysia; west, east and northern coasts of Sabah and west coast of Sarawak
3	Mangrove forests	0.1	Along the sheltered coast, lagoons, estuaries, somewhat inland of rivers

in the east, to Sabah and Sarawak bounds much of the southern part of Sulu Sea, South China Sea and the Straits of Malacca is one of the world's most fascinating and productive seas. The entire area was above sea level during the last glacial period when waters were 150 m below their present level, some 12,000 years ago. The areas including the whole of Borneo, Sumatera, Java and all the smaller islands is known as Sundaland. As such the Malaysian seas are relatively young and they are very shallow. However, the seas and the coastline support thousands of species of plants and animals, most of them the marine biodiversity is quite unknown. These range from the smallest planktonic lives, the foods of fishes to large mammals such as the whales and the endangered dugongs.

10.2   Types of Forest in Malaysia

With respect to the type of forest basically, Malaysia has two types, namely the climatic forests and the edaphic forests (Table 10.4).

10.2.1   Climatic Forests

10.2.1.1   Lowland Dipterocarp Forests

This is the most luxuriant of all the forests in Malaysia as the plant communities are dense, the trees are tall and characterized by clustering and gregarious consociations of the canopy and emergent trees. Basically, there are five strata of forests, the emergent, the main canopy, the main storey, the substorey, and the ground flora. The species that demonstrated this

Table 10.4 Vegetation Types in Peninsular Malaysia

Climatic forests	Edaphic forests
Lowland evergreen rainforest	i) Heath forest
a) Lowland dipterocarp forest 0–300 m	j) Limestone hill forest
b) Hill dipterocarp forest 300–800 m	k) Vegetation over quartzite
c) Upper hill dipterocarp forest 800–1200 m	l) Forest on Ultramafic hills
d) Submontane forest 1200–2500 m	m) Coastal forest
e) Montane oak forests 1200–1500 m	n) Mangrove swamp forest
f) Montane ericaceous forest 1500–2188 m	o) Gelam forest or Forest on BRIS soil
g) Subalpine forest >3000 m	p) Peat swamp forest
h) Semi-deciduous forest	q) Freshwater swamp forest

phenomenon of emergent layer reaching more than 40 m tall are *Shorea curtisii, S. leprosula, Shorea faguetiana, Dipterocarpus baudii, Dryobalanops aromatica* (all Dipterocarpaceae) among many others. The forests consist of at least five strata dominated by members of the Dipterocarpaceae in the emergent and canopy layers. Several genera of the dipterocarps especially *Anisoptera, Dipterocarpus, Parashorea, Shorea, Hopea, Vatica, Dryobalanops* predominate. There are 9 genera and more than 160 species of the dipterocarps and Sabah and Sarawak have 10 genera and more than 271 species of these gigantic trees. No other forest in the world shows such abundance and diversity of a single family of large trees. The other families of plants include Leguminosae, Ebenaceae, Euphorbiaceae, Sapindaceae, Lauraceae, Sapotaceae, and many others. This is where much of the large, medium and small mammals, including the primates are found. However, much of these forests were logged since the colonial days to make ways for rubber and oil-palm plantations and other socio-economic development such as townships, highways, and industrial areas.

10.2.1.2 Hill Dipterocarp Forests

As the elevation increases above 750 m above sea level, the dipterocarps get lesser and the trees are also getting smaller. The members of dipterocarps such as species of *Anisoptera, Dipterocarpus, Parashorea, Shorea, Hopea, Vatica, Dryobalanops* are still dominant and this forest also become the source of timbers. Other families such as the Verbenaceae, Leguminosae, Ebenaceae, Euphorbiaceae, Sapindaceae, Lauraceae, Sapotaceae occur densely. As Malaysia is losing much of the lowland dipterocarp forests, the socio-economic developers are now looking at this forest for timber source.

10.2.1.3   Submontane Dipterocarp Forests

In this high altitude forest, the ground is covered with bryophytes and lichens and ferns and the trees are smaller in stature. All forests over 1,250 m above sea levels have smaller and stunted vegetation. It is only in this type, the Vegetation Mountain resorts were developed for recreation and holidays. In addition, much of the highlands were also developed for fruits and vegetable farmings as well as floriculture.

10.2.1.4   Oak Forests

In this forest, the ground is covered with bryophytes and lichens and ferns, the trees are small and stunted. The cloud line is about 1,200 m and the lower montane forest develops on well-drained soils mainly at elevation 1,000–1,500 m. The flora is rich as the lowland forests. Kiew (1998), for example, listed over 900 species of seed plants alone from the lower montane forests of Fraser's Hill, Pahang. Members of the oak family, the Fagaceae (*Castanopsis, Quercus, Lithocarpus*) and the laurel family, Lauraceae (*Litsea, Beilschmiedia*) are common in this type of forests. Other genera that frequently occur in this forest formation include *Adinandra* (Theaceae), *Engelhardtia* (Juglandaceae), *Garcinia* (Guttiferae), *Toona* (Meliaceae), *Syzygium* (Myrtaceae), *Santiria* (Burseraceae) and species of *Dipterocarpus, Vatica, Shorea* (Dipterocarpaceae). Gingers (*Alpinia, Amomum, Camptandra, Etlingera, Geostachys, Globba, Hornstaedtia, Zingiber*) are also well represented among the ground flora. Among the epiphytes, the orchids (*Calanthe, Liparis, Hylophila, Malaxis*) and ferns are many.

10.2.1.5   Ericaceous Forests

This forest formation develops over 1,500 m above the cloud line. In this forest too, the ground is covered with bryophytes (especially *Sphagnum*) and lichens and ferns. The members of the Ericaceae such as *Rhododendron, Vaccinium,* and *Gaultheria* are common in this type of forest. The other common families in this vegetation include Araliaceae (*Schefflera*), Ericaceae (*Vaccinium, Rhododendron*), Theaceae (*Eurya, Gordonia*), Piperaceae (*Piper*), Cunoniaceae (*Weinmannia*), Symplocaceae (*Symplocos*), Lauraceae (*Lindera*), Myrtaceae (*Syzygium, Leptospermum, Tristaniopsis*), Rutaceae

(*Melicope*), Podocarpaceae (*Podocarpus*, *Dacrydium*), Sapindaceae, Moraceae (*Ficus*) and Rosaceae (*Rubus*, *Prunus*). The tree ferns (*Cyathea*) are very conspicuous. Other notable plants are species of Nepenthaceae (*Nepenthes sanguinea, N. macfarlanei*), and *Pandanus klosii,* among others.

10.2.1.6 Subalpine Forests

The ground is also covered with bryophytes and lichens and ferns and the trees are small with crooked stems. Herbaceous species of *Ranunculus* (Ranunculaceae), *Potentilla* (Rosaceae), *Euphrasia* (Scrophulariaceae), *Gentiana* (Gentianaceae), *Trigonotis* (Boraginaceae), *Vernonia* (Compositae), *Viola* (Violaceae), *Astilbe* (Saxifragaceae), *Argostemma, Galium*, (Rubiaceae), *Haloragis, Gunnera* (Haloragaceae), *Impatiens* (Balsaminaceae), *Drosera* (Droseraceae), *Sonerila* (Melastomataceae). In this forest-type, many species of *Nepenthes* (Nepenthaceae) also occur and they have spectacular pitchers.

10.2.1.7 Semi-Deciduous Forests

This forest formation is also known as the White meranti-gerutu forest and only occurs on the Langkawi Archipelago and the northwestern states of Perlis and Kedah that experience about 2–3 months of monsoon climate where the total annual rainfall is between 200–3,000 mm. This forest occurs on well-drained soil and not subjected to waterlogging. In this forest formation, the Burmese-Tai elements are very conspicuous. The dominant species are the White meranti group, *Shorea assamica, S. henryana, S. hypochroa,* and *S. roxburghii*. Other common species include *Anisoptera costata, Hopea ferrea, H. latifolia, Dipterocarpus baudii, D. dyeri, D. kerrii, D. grandifolia, Parashorea stellata,* and *Vatica cinerea*. Other non-dipterocarps include *Schima wallichii* (Theaceae), *Gigantochloa latifolia* (Gramineae), *Dillenia obovata* (Dilleniaceae), *Parkia timoriana, Koompassia malaccensis, Intsia palembanica* (Leguminosae).

10.2.2 Edaphic Forests

10.2.2.1 Heath Forests

These forests are found on poor soil especially the sandstones and the thin layer of peat may also form above the podzolic soils (Anderson, 1963; Brunig, 1974). The heath forest has a very simple structure as the main storey comprises of saplings and small trees. The ground commonly has rich mosses and liverworts, large woody climbers and climbing palms are more common and large trees are very rare. Myrmecophytes are common too. Other notable plants are species of *Nepenthes* such as *N. ampullaria, N. x hookeriana,* orchids and also palms. In the heath forest members of the dipterocarps are also present, *Shorea glauca, S. materialis, Hopea semicuneata, H. griffithii* among others. Other notable plants are *Gluta* spp. (Anacardiaceae), *Irvingia malayana* (Irvingiaceae), *Madhuca utilis* (Sapotaceae), *Mesua ferrea, Garcinia hombroniana, G. nigrolineata* (Guttiferae), *Syzygium* spp., *Tristaniopsis obovata* (Myrtaceae), *Eurycoma longifolia* (Simaroubaceae).

10.2.2.2   Limestone Hill Forests

Craggy limestone hills form a striking part of the landscape in many parts of Peninsular Malaysia, Sabah, and Sarawak. The limestone tower karsts have a diversity of habitats and soils. The structure and physiognomy of the vegetation of the limestone hill forests depend on the microhabitats and develops from a combination of the soils, substrate, and climate. Many species of plants occur in this type of forests, and in Peninsular Malaysia alone over 1,200 species have been recorded. For example, *Mammea calciphila, Gonystylus nervosus, Cycas clivicoloa, Caryota mitis, Maxburretia gracilis, Arenga westerhoutii, Casuarina nobilis, Dacrydium beccari* are among others. Some species of dipterocarps also occur on the limestone hills such as *Shorea siamense, S. guiso, S. isopteran, Vatica kanthanensis, V. cinerea, Dipterocarpus caudiferus, Hopea helferi, H. latifolia,* and others. Many species of gesneriads such as species of *Boea, Chirita, Monophyllea,* and *Paraboea* occur in these forests. Common shrubs and trees such as *Bridelia, Cleistanthus* (Euphorbiaceae), *Callicarpa angustifolia* (Verbenaceae), *Dehassia pauciflora* (Lauraceae), *Pentaspadon curtisii* (Anacardiaceae), and among the herbs include *Impatiens mirabilis* (Balsaminaceae) (Chin, 1977).

10.2.2.3   Forest on Ultramafic Hills

This forest is quite restricted to some localities such as in northeast Ranau, Sabah and near Raub, Pahang. Some species of dipterocarps also occur, *Shorea andulensis, S. kunstleri, S. lasa, Dipterocarpus geniculatus, D. lowii.* Others include *Leptospermum recurvum, Rhododendron ericoides, Dacrydium gibbsiae,* and others.

10.2.2.4   Coastal Forests

Basically, there are two kinds of beach vegetation, that formed along the accreting coasts where new sands are being deposited and the other is where the herbaceous plants cover a broad sand strands. Among the common plant species occur in this forests are *Ipomoea pes-caprae, Scaevola taccada, Casuarina equisetifolia, Cyperus pedunculatus, Canavalia microcarpa, Fimbristylis sericea, Ischaemum muticum, Spinifex littoralis, Lepturus repens, Calophyllum inophyllum, Cycas edentata, Pandanus tectorius, Syzygium grande, Pouteria obovata, Terminalia catappa, Vigna marina, Barringtonia asiatica, Crinum asiaticum, Sophora tomentosa, Tacca leontopetaloides, Pongamia pinnata, Canavalia cathartica,* and others.

10.2.2.5   Mangrove Swamp Forests

The mangrove forests in Malaysia are very extensive and in Malaysia they are the sources of charcoal, firewood, and poles. The most notable plants are species of *Rhizophora, Ceriops, Aegiceras, Bruguiera, Avicennia, Sonneratia, Xylocarpus, Intsia bijuga, Brownlowia argentata* that form the 104 species of exclusive mangroves, non-exclusive species and also those associated with the mangrove swamp forests (Latiff and Faridah-Hanum, 2013). Two common species of ferns are *Acrostichum aureum* and *A. speciosum* (Parkeriaceae).

10.2.2.6   Gelam Forests or Forests on BRIS Soil

In this type of forest, only *Melaleuca cajuputi* (Myrtaceae) predominates as pure stands. However, of late much of this forest formation especially in the states of Kelantan, Terengganu and Pahang have been cleared to make way for shrimp cultures, oil palm plantations, and other socio-economic development.

10.2.2.7   Peat Swamp Forests

The peat swamp forest is a special type with a rather restricted flora because of its acidic, waterlogged area with peat or loosely textured soil with organic matter. For Peninsular Malaysia, a total of 238 plant species have been enumerated and 221 species were recorded from the Southeast Pahang alone (Faridah-Hanum et al., 2009). Some of the noteworthy species are *Vatica pauciflora, Anisoptera marginata, Shorea platycarpa* (Dipterocarpaceae),

Durio carinata, Neesia malayana (Bombacaceae), *Dialium indum, Koompassia malaccensis, Archidendron clypearia* (Leguminosae), *Santiria rubiginosa, Dacryodes macrocarpa* (Burseraceae), *Litsea grandis, Cryptocarya impressa, Cinnamomun altissimum* (Lauraceae), *Alstonia angustiloba* (Apocynaceae), *Garcinia nigrolineata, G. parvifolia, G. urophylla, Calophyllum ferrugineum, C. sclerophyllum* (Guttiferae), *Diospyros argentea, D. lanceifolia, D. maingayi* (Ebenaceae), among others.

10.2.2.8   Freshwater Swamp Forests

The soil surface covered by this type of forest is regularly inundated with mineral-rich fresh water of fairly high pH and the water level fluctuates. Among the dominant species are *Metroxylon sagu, Campnosperma coriaceum, Mallotus leucodermis, M. muticus* and the secondary elements consist of *Melaleuca cajuputi, Ilex cymosa, Macaranga pruinosa, Ploirarium alternifolium, Randia dasycarpa, Alstonia spatulata, Cratoxylon cochinensis,* among others (Corner, 1978).

10.3   Forest Biodiversity Conservation Programmes

Like in many other countries in southeast Asia, in Malaysia there are two approaches of conservation programmes that had been implemented in order to sustain forest biodiversity from disappearing from the lands and seas. The two programmes are *in situ* conservation and *ex situ* conservation where the *in situ* conservation means maintaining plants, animals, and microorganisms in their original habitats in nature.

10.3.1   *In Situ Conservation*

A total of 87 Virgin Jungle Reserves (VJRs), which comprise 23,002 ha, was established in Peninsular Malaysia to conserve the richness of forest biodiversity. These VJRs which range in size from as small as 3 ha to 1600 ha representing various types of forest were established to serve as permanent nature reserves and natural arboreta. There are also similar VJRS in Sabah and Sarawak. There are also the National Parks, the State Park Forests which are also gazetted to conserve forest biodiversity such as the Mount Kinabalu National Park, Crocker Range National Park (Sabah), Gunung Mulu National Park, Gunung Gading National Park (Sarawak), Perlis State Park (Perlis), Gunung Stong State Park (Kelantan), Endau-Rompin State Park, (Pahang), Bukit Panchor State Park (Penang), Selangor State Park

(Selangor) to name some. In Peninsular Malaysia, about 747,310 ha were reserved as Water Catchment Forests, Genetic Resource Area (GRA) in Ulu Sedili Forest Reserve (4806 ha) and Seed Production Area (SPA) about 159 ha of total forest cover in Peninsular Malaysia which are also examples of *in situ* conservation areas.

10.3.2 *Ex Situ Conservation*

In Peninsular Malaysia, a few places are reserved as *ex situ* conservation areas where some of the endemic, new findings, rare and critical endangered species are maintained outside their original habitat to preserve as a heritage from losing. As of 2008, the Forestry Department of Peninsular Malaysia had established some *ex situ* conservation areas such as Rimba Herba Perlis with 700 species of medicinal herbs, Taman Herba Papan, Perak stored 450 species of medicinal herbs, Arboretum at Ulu Sat Forest Reserve, Kelantan with 1000 species of higher and lower plants, Arboretum Sultan Haji Ahmad Shah, Raub, Pahang contains 40 species of forest fruit trees, Cameron Highlands Montane Park, Pahang contains 387 montane species and also Arboretum Ayer Keroh Botanical Park, Melaka covering an area of 37.0 ha of forest, planted with a variety of timber trees to locate not only higher plants, but also herbs and also flowering plants that are said to be rare and endangered.

All the plants and trees that planted at *ex situ* conservation area are monitored by the expert group on the growth, phenology, and fertility. Each and single plant that is placed in the *ex situ* conservation area are recorded and whenever new species included in this area, the list is updated. The record shows that the largest groups of forest plant species under *ex situ* conservation are orchids (1,639 species), rare fruit trees (434 species), timber species (364 species), and medicinal plants (115 species). Orchids become the highest number of species conserved at *ex situ* conservation because of their beauties, aesthetic values, and also they have high demand in the market.

10.4 Floral Diversity in Malaysia

In terms of plant diversity, Malaysia ranks tenth in the world and fourth in Asia, after China, India, and Indonesia. With over 15,000 plant species, Malaysia ranks high among the small countries in term of the world's known floral diversity (Bidin and Latiff, 1995; Latiff, 1997; 2009, 2010). As elsewhere in the world, many organisms especially in lower groups such as bacteria, fungi, algae, lichens, and bryophytes are yet to be described and

Table 10.5 The Flora of Peninsular Malaysia Comparison

Enumeration	Plant group	Families	Genera	Species
Ridley (1922–1925)	Ferns and Fern-allies	16	86	417
	Gymnosperms	3	5	23
	Dicotyledons	132	1,048	5,009
	Monocotyledons	31	354	1,734
	Total	**182**	**1,493**	**7,183**
Turner (1995)	Ferns and Fern-allies	34	133	632
	Gymnosperms	4	8	27
	Dicotyledons	165	1,092	5,529
	Monocotyledons	45	418	2,010
	Total	**248**	**1,651**	**8,198**

remote geographical areas are to be comprehensively explored and bota-
nized. The richness of Malaysian plant species as compared to the world is
shown in the Table 10.5.

It must be noted after the monumental Flora of the Malay Peninsula
(Ridley, 1922–1925), Malaysia had embarked on the Revised Flora (Gilliland, 1971; Holttum, 1950). Subsequently, after 1995 Malaysia embarked
on the Flora of Peninsular Malaysia project (Kiew et al., 2010; Kiew et al.,
2011) as to update the Tree Flora of Malaya, which was completed in 1989
(Whitmore, 1972, 1973; Ng, 1978, 1989). At the same time the Tree Flora of
Sabah and Sarawak project is still in progress (Soepadmo and Wong, 1995;
Soepadmo et al., 1995, 2007, 2014). Many new species to science from Peninsular Malaysia and new records to Peninsular Malaysia were described
(Saw and Chung, 2007; de Kok et al., 2009; Jaman, 2014).

Amongst the various families of flowering plants in Malaysia the dominant dicotyledons are Dipterocarpaceae, Acanthaceae, Phyllanthaceae, Labiatae, Euphorbiaceae, Leguminosae, Rubiaceae and the monocotyledons are

Table 10.6 Ferns and Fern-Allies Checklist Enumerated in 1997

Region	Number of species
Peninsular Malaysia	647
Sabah	750
Sarawak	615
Total	1,165

Table 10.7   Comparing Revision of Similar Families of the Tree Flora of Sabah and Sarawak (TFSS) with the Tree Flora of Malaya (TFM)

TFSS			TFM			
Families	Genera	Species	Families	Genera	Species	Common species to both
31	99	312	31	91	227	152
23	75	247	21	63	186	116
4	29	358	4	27	246	139
6	21	292	6	21	202	106
4	25	361	4	27	225	132

Orchidaceae (Holttum, 1953), Cyperaceae, Marantaceae (Holttum 1951), Palmae, and Zingiberaceae (Holttum, 1950). Needless to state that the Euphorbiaceae and the Phyllanthaceae were once one family with more than 365 species. A few families such as Sterculiaceae, Bombacaceae, and Tiliaceae have been merged under Malvaceae and the Verbenaceae has been merged with the Labiatae but not adopted here in the analysis. At the same time many new but smaller families have been splitted from the traditional families, for example, *Acorus* has been transferred from the Araceae to Acoroaceae, *Cleome* from Capparaceae to Cleomaceae, *Maesa* from Myrsinaceae to Maesaceae, *Memeycylon* from Melastomataceae to Memycylaceae, *Muntingia* from Tiliaceae to Muntingiaceae, *Ruppia* from Potamogetonaceae to Ruppiaceae, and several others. The diversity of dicotyledons and monocotyledons in Peninsular Malaysia are given in Table 10.8. A similar analysis is yet to be carried out for the families in Sabah and Sarawak.

Some genera are also very species and are given in Table 10.9.

10.4.1   Endemicity

When the Tree Flora of Malaya was completed in 1989, a total of more than 2,830 species of trees were listed and out of these 746 are endemics to Peninsular Malaysia, mostly the dipterocarps (Ng, 1989; Ng et al., 1990). The same is waiting for the completion of the Tree Flora of Sabah and Sarawak. However, the endemicity of seed plants in Malaysia is very high, suffice to demonstrate with a few examples. In Sabah and Sarawak, 59 species of the dipterocarps are endemic out of 257 species. In the genus *Durio* (Bombacaceae) of the 30 species known, 16 species or about 53% occurring in Sabah and Sarawak only. In another genus, *Mangifera* (Anarcadiaceae) out of 50

Table 10.8 Large families of Angiosperms in Peninsular Malaysia

Family	Number of genera	Number of species
Dipterocarpaceae	9	150
Acanthaceae	29	159
Cyperaceae	29	162
Phyllanthaceae	12	167
Zingiberceae	18	176
Labiatae	28	193
Euphorbiaceae	57	198
Gesneriaceae	23	201
Annonaceae	36	202
Lauraceae	16	214
Myrtaceae	11	215
Palmae	30	230
Apocynaceae	64	235
Gramineae	92	238
Leguminosae	81	298
Rubiaceae	79	562
Orchidaceae	149	880

species known, 18 species are known from Sabah and Sarawak. Kiew (2005) showed that of the 52 species of *Begonia* in Peninsular Malaysia, 26 of them are known from a single locality and endemic and for Sabah and Sarawak, there are 82 and 96 species, respectively and many are also endemic (Kiew et al., 2015).

10.4.2 The Dipterocarps

The dipterocarps (Dipterocarpaceae) has a total of 386 known species, 291 species or 75% are recorded in Borneo, of which 257 species or about 66% occur in Sabah and Sarawak. Of the 291 species occurring in Borneo, 156 or 54% are endemic and 59 species or about 20% are restricted to Sabah and Sarawak (Meijer and Wood, 1964). In Peninsular Malaysia, there are more than 150 species (Foxworthy, 1932; Symington, 1943; Ashton, 1982). Most of them produce timber for the local wood industries and exports. Ashton (1988) attempted to do the same for the non-dipterocarps in Sabah and Sarawak (Table 10.10).

Table 10.9 Some of the Large Genera (More Than 20 Species) in Peninsular
Malaysia

Genera	Family	Number of species
Ferns and Fern-allies		
Selaginella	Selaginellaceae	29
Asplenium	Aspleniaceae	29
Cyathea	Cyatheaceae	20
Lindsaea	Dennstaedtiaceae	23
Diplazium	Woodsiaceae	30
Dicotyledons		
Ilex	Aquifoliaceae	20
Schefflera	Araliaceae	24
Dischidia	Asclepiadaceae	24
Dipterocarpus	Dipterocarpaceae	30
Hopea	Dipterocarpaceae	32
Shorea	Dipterocarpaceae	59
Vatica	Dipterocarpaceae	24
Diospyros	Ebenaceae	63
Elaeocarpus	Elaeocarpaceae	29
Antidesma	Euphorbiaceae	23
Aporusa	Euphorbiaceae	24
Cleistanthus	Euphorbiaceae	30
Macaranga	Euphorbiaceae	27
Glochidion	Euphorbiaceae	24
Phyllanthus	Phyllanthaceae	20
Lithocarpus	Fagaccae	38
Dimocarpus	Gesneriaceae	89
Paraboea	Gesneriaceae	20
Mesua	Guttiferae	20
Calophyllum	Guttiferae	42
Garcinia	Guttiferae	46
Alseodaphne	Lauraceae	23
Cinnamomum	Lauraceae	21
Litsea	Lauraceae	56
Memecylon	Memecylonaceae	26
Sonerila	Melastomataceae	45

Table 10.9 (Continued)

Genera	Family	Number of species
Horsfieldia	Myristicaceae	22
Knema	Myristicaceae	27
Syzygium	Myrtaceae	195
Piper	Piperaceae	78
Argostemma	Rubiaceae	45
Lasianthus	Rubiaceae	54
Hedyotis	Rubiaceae	38
Ophiorrhiza	Rubiaceae	21
Tarenna	Rubiaceae	39
Madhuca	Sapotaceae	29
Symplocos	Symplocaceae	22
Monocotyledons		
Homalomena	Araceae	34
Fimbristylis	Cyperaceae	31
Dioscorea	Dioscoreaceae	20
Bulbophyllum	Orchidaceae	110
Coelogyne	Orchidaceae	28
Dendrobium	Orchidaceae	86
Eria	Orchidaceae	48
Oberonia	Orchidaceae	29
Thrixspermum	Orchidaceae	29
Calamus	Palmae	60
Daemonorops	Palmae	32
Licuala	Palmae	26
Pandanus	Pandanaceae	48
Alpinia	Zingiberaceae	23
Scaphochlamys	Zingiberaceae	20

10.4.3 The Orchids

In the World, the Orchidaceae consists of 25,000–30,000 species, though most of them have been described and named, many others will perish due to the loss of various habitats especially the highlands and the peat swamp forests where the orchids predominate as epiphytes. The orchids are

Table 10.10   The Dipterocarps in Malaysia

Genus	Number of species in Peninsular Malaysia	Number of species in Sabah and Sarawak	Number of species in Malaysia
Anisoptera	6	5	8
Cotylelobium	2	3	3
Dipterocarpus	30	39	49
Dryobalanops	2	7	7
Hopea	32	43	58
Neobalanocarpus	1	0	1
Parashorea	3	6	9
Shorea	59	130	143
Vatica	25	37	47
Upuna	0	1	1
Total	**160**	**271**	**326**

known as the Mother of the Plant Kingdom or the Pandas of the Plant World because of their beauty and aesthetic values as well as their role in horticultural industries. More than 880 species have been recorded in Peninsular Malaysia and many are known from Sarawak and Sabah, especially Mount Kinabalu and its vicinity (Holttum, 1953; Seidenfaden and Wood 1992; Ong et al., 2011). Conservatively, the Orchidaceae in Malaysia have been classified into six different subfamilies, Apostasioideae, Cypripedioideae, Neottioideae, Orchidoideae, Epidendroideae, and Vandoideae. A recent phylogenetic classification of Chase et al. (2003) had recognized only five subfamilies, Apostasioideae, Cypripedioideae, Orchidoideae, Epidendroideae, and Vanilloideae (Table 10.11, Figure 10.2).

10.4.4 The Rafflesiaceae

In southeast Asia, there are about 26 species of *Rafflesia,* the holoparasitic genus with the largest flower in the Plant Kingdom. The species are very restricted to geographical localities and regions. For example, there are only four species endemic to Peninsular Malaysia, *R. cantleyi, R. kerri, R. azlanii,* and *R. su-meiae* (Wong and Latiff, 1994), in Sabah there are *R. pricei, R. keithii,* and *R. tengku-adlinii* and in Sarawak there are *R. pricei, R. keithii, R. tuan-mudae,* and *R. haseltti.* The genus *Rhizanthes* is only represented by *R. infanticida* in Peninsular Malaysia and *R. lowii* in Sabah and Sarawak.

Table 10.11 The Subfamilies, Genera and the Number of Species of
Orchidaceae in Peninsular Malaysia

Subfamily	Genera	Number of species
Apostasioideae	*Apostasia, Neuwiedia*	3, 3
Vanilloideae	*Cyrtosia, Erythrorchis, Galeola*	1, 2, 1
	Lecanorchis, Vanilla	3, 5
Cypripedioideae	*Paphiopedilum*	6
Orchidoideae	*Anoectochilus, Cheirostylis, Corybas*	9, 3, 11
	Cryptostylis, Cystorchis, Erythrodes	2, 4, 3
	Goodyera, Habenaria, Herminium	7, 9, 1
	Hetaeria, Hylophila, Kuhlhasseltia	9, 2, 1
	Lepidogyne, Ludisia, Macodes	1, 1, 1
	Orchipedum, Pecteilis, Peristylus	1, 1, 8
	Platanthera, Pristiglottis	1, 1
	Spiranthes, Vrydagzynea, Zeuxine	1, 3, 10
Epidendroideae	*Abdominea, Acampe, Acanthephillium*	6, 4, 2
	Acriopsis, Adenoncos, Aerides	8, 3
	Agrostophyllum, Aphyllorchis	13, 3, 1
	Appendicula, Arachnis, Arundina	1, 1, 1
	Adcidieria, Ascocentrum, Ascochilopsis	2, 4, 1
	Ascochilus, Biermannia, Brachypeza	11, 133, 21
	Bromheadia, Bulbophyllum, Calanthe	2, 11
	Cephalantheraopsis, Ceratostylis	1, 1, 3
	Chelonistele, Chilochista, Chroniochilus	1, 1, 1
	Chrysoglossum, Claderia, Cleisomeria	20, 27, 1
	Cleisostoma, Coelogyne, Collabium	5, 1, 11
	Cordiglottis, Corymborkis, Cymbidium	86, 14
	Dendrobioum, Dendrochilum	1, 1, 1
	Dilochia, Diplocaulobium, Dipodium	2, 1, 5
	Doritis, Epigeneium, Epipogium, Eria	1, 4, 1, 51
	Eulophia, Flickingeria, Gastrochilus	6, 11, 3
	Gastrodia, Geodorum, Grammatophyllum	2, 2, 2
	Grosourdya, Hippeophyllum, Kingiadium	4, 1, 1
	Liparis, Luisia, Macropodanthus	23, 5, 2
	Malaxis, Malleola, Micropera	13, 6, 4
	Microsaccus, Microtatorchis, Michobulbum	5, 1, 2

Table 10.11 (Continued)

Subfamily	Genera	Number of species
	Nephelaphyllum, Nervilia, Oberonia	2, 4, 31
	Octarrhena, Ornithochilus, Pachystoma	2, 1, 1
	Panisea, Papilionanthe, Pelatantheria	1, 1, 2
	Pennilabium, Phaius, Phalaenopsis	4, 5, 9
	Pholidota, Phreatia, Placoglottis	9, 8, 4
	Poaephyllum, Podochilus, Polystachya	1, 4, 1
	Pomatocalpa, Porpax, Porphyroglottis	7, 1, 1
	Pteroceras, Renanthera, Renantherella	8, 2, 1
	Rhynchogyna, Rhynchostylis, Robiquetia	1, 2, 2
	Sarcoglyphis, Sarcostoma, Schoenorchis	1, 1, 4
	Smitinandia, Spathoglottis, Ataurochilus	1, 5, 2
	Stereosandra, Taeniophyllum, Tainia	1, 16, 5
	Thecopus, Thecostele, Thelasis	2, 1, 6
	Trixspermum, Thunia, Trichoglottis	33, 1, 7
	Trichotosia, Tropidia, Tuberolabium	12, 4, 1
	Vanda, Vandapsis, Ventricularia	1, 1, 1

The species *Mitrastema yamamotoi* occurs also on the highlands in Sabah and Sarawak (Figure 10.3).

10.4.5 Diversity of Crop Plants

Malaysia is believed to be the home of some species of crops and species of wild relatives of cultivated plants. This include wild relatives of rice *(Oryza)*, mango *(Mangifera)*, banana *(Musa)*, pepper *(Piper)*, turmeric *(Curcuma)*, ginger *(Zingiber)*, cardamom *(Elettaria, Amomum)*, Jack fruit *(Artocarpus)*, durians (*Durio*), rambutans (*Nephelium*), yams *(Dioscorea, Alocasia, Colocasia)*, vines *(Trichosanthes)*, cinnamon *(Cinnamomum)*, nutmeg *(Myristica)*, *Amaranthus*, and several others.

10.4.5.1 Gymnosperms

The gymnosperms are currently placed in five distinct and widely divergent orders namely Cycadales, Ginkgoales, Taxales, Coniferales, and Gnetales and consisting of 1026 species in 84 genera in the world in almost all continents, except Antarctica. For Malaysia, Keng (1978) reported three families,

Figure 10.2   Some of the common orchid species in Malaysia. A. *Apostasia nuda,*
flower; B. *Bulbophyllum linearifolium,* flowers; C. *Bulbophyllum
pileatum,* flower; D. *Corybas carinatus,* plant with flower; E.
Dendrobium singaporense, flower; F. *Flickingeria forcipata,* flower;
G. *Paphiopedilum barbatum,* flower; H. *Spathoglottis plicata,* flower;
I. *Trichotosia vestita,* flower.

four genera and a total of 19 species indigenous to Peninsular Malaysia,
excluding the cycads. This number has been revised by Turner (1995) who
reported a total of four families, 8 genera and 27 species. Recently, two
cycads, *Cycas clivicola* and *C. cantafolia* were added, the former was mis-
identified as *C. siamense* and the latter was discovered in 2009. Hence the
number of species has increased to 29. For Sabah and Sarawak, there is an
additional genus, *Phyllocladus* which is restricted to Borneo. The conifers

Figure 10.3 The Rafflesia species in Malaysia. Clockwise from top left: *R. keithii, R. kerri, R. azlanii, R. pricei, R. parvomaculata, R. cantleyi, R. hasselttii, R. tuan-mudae.*

flora of Malaysia is considered very poor. Many species of *Podocarpus, Dacrydium, Agathis,* and *Phyllocladus* are confined to higher altitude except the *Gnetum.* The species of *Gnetum* are climbers except for *G. gnemon,* which inhabit evergreen lowland tropical rainforests.

10.4.5.2 Pteridophytes

The precursor was established by Holttum (1954) and no documentation of Malaysia's rich pteridophyte flora has previously been prepared until Parris and Latiff (1997) recorded a total of 1165 species, 647 of which occur in Peninsular Malaysia, 750 in Sabah and 615 in Sarawak. Since then several new taxa have been described and added (Jaman, 2014; Latiff, 2015). Currently, a new Flora of Peninsular Malaysia project has started with Series 1: Ferns and Lycophytes (Parris et al., 2010, 2013). For Peninsular Malaysia a total of 40 families and 138 genera are listed for taxonomic revision. In the first and second volumes, about 9 families each were revised. A total of 43 species in 17 genera and 10 families are believed to be endemic to Peninsular Malaysia, another 17 species are endangered (Table 10.12).

10.4.5.3 Bryophytes

The most comprehensive checklist of mosses for Peninsular Malaysia was given by Yong et al. (2013) who reported 46 families, 159 genera, 522 species, 20 subspecies, 72 varieties, and 7 forms. There are seven endemic taxa also reported that include *Calymperes woodii, Fissidens benitotanii, F. subdiscolor, Schistomitrium sparei, Trichosteleum fleischeri, T. fruticola,* and *Splachnobryum temasekensis.* Earlier, Mohamed and Tan (1988) produced a checklist for Malaya and Singapore and Suleiman et al. (2006) produced a revised catalog of Bornean mosses, which include mostly from the states of Sabah and Sarawak. A checklist of Hepaticae and Anthocerotae of Malaysia was given by Chuah-Petiot (2011) who reported a total of 38 families, 122 genera and 758 taxa of Hepaticae and one family, 5 genera and 6 taxa of Anthocerotae. The full catalog of the Lejeuneaceae of Malaysia was given by Lee et al. (2011, 2013) who reported a total of 266 species and 13 infraspecific taxa. The details are given in Table 10.13.

10.4.5.4 Algae

Very little is known about the algal diversity except they are commonly found growing in a variety of habitats ranging from freshwater, brackish water,

Table 10.12 The Dominant Families and Genera of the Lower Plants

Dominant families	Dominant genera
Lycophytes	**Lycophytes**
Selaginellaceae (74)	*Selaginella* (62)
Lycopodiaceae (27)	*Huperzia* (19)
Ferns	**Ferns**
Schizaeaceae (135)	*Asplenium* (56)
Grammitidaceae (110)	*Tectaria* (51)
Polypodiaceae (108)	*Grammitis* (42)
Dryopteridaceae (103)	*Cyathea* (39)
Hymenophyllaceae (86)	*Lindsaea* (30)
Thelypteridaceae (83)	*Pteris* (30)
Dennstaedtiaceae (70)	*Sphaerostephanos* (28)
Aspleniaceae (56)	*Selliguea* (24)
Cyatheaceae (39)	*Pronephrium* (22)
	Diplazium (19)

Table 10.13 The Bryophytes of Malaysia

Mosses	Liverworts and Hornworts
Peninsular Malaysia	Malaysia
522 species; 159 genera, 46 families	764 species; 127 genera; 39 families
Dominant Families (Peninsular Malaysia):	**Dominant Families:**
Calymperaceae (74 taxa)	Lejeuneaceae (282 taxa)
Semantophyllaceae (69 taxa)	Lepidoziaceae (110 taxa)
Hypnaceae (43 taxa)	Frullaniaceae (67 taxa)
Dicranaceae (32 taxa)	Plagiochilaceae (52 taxa)
Fissidentaceae (31 taxa)	Geocalycaceae (36 taxa)
Neckeraceae (21 taxa)	Lophoziaceae (35 taxa)
Leucobryaceae (19 taxa)	Radulaceae (35 taxa)
Dominant Genera in Peninsular Malaysia:	**Dominant Genera:**
Fissidens (30)	*Cololejeunea* (84 taxa)
Syrrhopodon (27)	*Frullania* (67)
Acroporium (23)	*Bazzania* (53)
Ectropothecium (21)	*Plagiochilla* (47)
Calymperes (20)	*Radula* (35)

marine, terrestrial and also soil. According to Phang et al. (2010) freshwater algae are dominated by Chlorophyceae (green algae), Bacillariophyceae (diatoms) and Cyanophyceae (blue-green algae) represent the major portion of Malaysia algal flora accounting for ca. 390 genera and 4500 species followed by terrestrial algae (125 genera and 615 spp.); soil algae (80 genera and 1500 spp.); marine algae (169 genera and 680 spp.). The economically important algae in the freshwater ecosystem are *Chlorella, Spirulina, Synechococcus* (Single Cell Protein), *Euglena, Chlamydomonas, Scenedesmus, Ankistrodesmus, Nitzschia* (diatom-rich in vitamin A), diatomaceous earth; production of Diatomite (Fire-proof material). Species of *Sargassum* yield Sodium alginate – a raw material for textile and paper industries, *Gelidiella, Gracilaria, Hypnea* – yield agar and carrageenans used in pharmaceutical industries, which are distributed in the marine ecosystem.

10.4.5.5 Seaweeds

Malaysia has an extensive coastline and numerous islands form clusters along the coastlines. These established numerous habitats such as rocky shores and sandy bays alternate with mudflats which harbor niches for a variety of seaweed species. More than 386 species of seaweeds occur in Malaysia, comprise of Chlorphyta (13 families), Rhodophyta (27 families), Phaeophyta (8 families), Cyanophyta (8 families). Among those that exhibit a great diversity of forms are species of *Halymenia, Gracilaria, Ulva, Eucheuma, Enteromorpha, Chaetomorpha, Chlorodermis, Caulerpa, Halimeda, Cladophora,* among others (Japar Sidik and Muta Harah, 2011).

10.4.5.6 Seagrasses

Seagrasses are submerged monocotyledonous plants that form patches, beds or meadows in shallow coastal water. As Malaysia has about 4,800 km of coastlines stretching from Perlis to Kelantan in Peninsular Malaysia, Sabah, and Sarawak and including many islands, the seagrass communities also contribute to the productivity and biodiversity of coastal Malaysia. Japar Sidik and Muta Harah (2011) gave a comprehensive review of seagrass diversity comprises of 15 species in three families.

10.4.5.7 Medicinal Plants

Both the indigenous and non-indigenous Malaysians practice traditional medicine as an alternative to medicine especially in the rural areas where the latter is not readily available. Out of the total of more than 1,200 species of plants nearly 10% is available as traditional herbs (Mat-Salleh and Latiff, 2002). The only indigenous and naturalized plants that have been commercialized in a small way for domestic consumption are *Ficus deltoidea, Eurycoma longifolia, Brucea javanica, Morinda citrifolia, Citrus aurantifolia, Momordica charantia, Ageratum conyzoides, Cosmos caudatus, Centella asiatica, Labisia pumila, Orthosiphon aristatus, Albizia myriophylla, Phyllanthus niruri,* among others. However, the country still depends on the imports of medicinal plants from India, China, and even Europe.

10.4.5.8 Forests

Forests are an important reservoir of biodiversity. Ancient and frontier forests, because of their long-standing and relatively lower levels of human disturbance, are typically richer in biodiversity than other natural or semi-natural forests. An illustration of the conservation importance of forests relative to biodiversity is found in the recent analysis of biodiversity hotspots (Latiff, 1998b). Loss of forest will inevitably result in the reduction of biodiversity as a direct result of the loss of habitat. There are many anthropocentric reasons to preserve biodiversity (direct use value, option value, etc.) but one such principal reason considered here is the indirect use value in the form of ecosystem services. A loss in biodiversity affects the stability of an ecosystem resulting in a reduction of its resistance to disruption of the food web (by the loss of the weak interaction effect), resistance to species invasion and resilience to global environmental change (Latiff, 1988a; Latiff 2001; Latiff, 2007). Therefore, the argument for conserving biodiversity is, two-fold:

1. Biodiversity is essential for maintaining ecosystem services; and
2. Biodiversity increases the stability and resilience of an ecosystem to a disturbance.

10.4.6 Biosphere Reserves

As a common knowledge biosphere reserve is a specific type of a conservation area which accommodates and benefits both the natural environments and communities living around it. This is possible as the reserve consists of three different but associated zones core, buffer and the transitional zone. Core zone is ecologically sensitive but a pristine area, where nature conservation is a priority with allowed low impact activities. A buffer zone is less ecologically sensitive but a natural area where recreational and sustainable utilization of a natural product can be accommodated. Finally, the transition zone is less ecologically sensitive where a great variety of land user occur. All zones are interdependent and managed according to the definitions above. It is here, by linking conservation, sustainable utilization and utilization of natural resources occur. The details of Malaysian Biosphere reserves are given in Table 10.14.

Both the above Biosphere reserves in Malaysia have been included in the World Network of Biosphere Reserves of UNESCO. Malaysia envisages more areas will be gazetted as biosphere reserves in future.

10.4.7 Protected Area Network in Malaysia

Malaysia's first National Park was established in 1938/1939 as King George V National Park to commemorate the coronation of the English King. Its total area is about 4,343 km^2 which straddle three east-coast states of Pahang, Terengganu, and Kelantan. It is very rich in biodiversity as it has never been disturbed and managed by the Department of Wildlife and National Park. The second National Park in Malaysia is Pantai Acheh National Park in Penang which is about 1,100 ha.

10.4.8 National Parks in Sabah

The National Parks in Sabah were enacted through the State Legislature hence they are fundamentally State Parks (Table 10.15). They came under four departmental jurisdictions, namely the Department of Wildlife, the Sabah Foundation, Department of Forestry, and the Sabah Parks. What is important in this context all these are protected areas for the protection and conservation of both the terrestrial, aquatic, and marine flora and fauna. In particular, the Gunung Kinabalu National Park is the UNESCO's World

Table 10.14 The Biosphere Reserves in Malaysia

No.	Name of the Biosphere Reserve	Notes
1	Tasik Chini, Pahang, Peninsular Malaysia	A fresh-water wetland which is one of a few natural lake in Peninsular Malaysia.
2	Crocker Range, Sabah	A hilly and submontane forests which form part of the Sabah National Park in Sabah. Rich in biodiversity.

Table 10.15 The National Parks, Protected Areas and the Conservation Areas in Sabah

Name	Year established	Area (ha)
Wildlife Department		
Tabin Wildlife Reserve	1984	120,521
Kalumba Wildlife Reserve	1984	20,682
Lower Kinabatangan Wildlife Reserve		41,765
Sepilok Orang Utan Rehabilitation Centre	1931	1,235
Gomantong and Supa Forest Reserve	1984	1,816
Pulau Sipadan Bird Sanctuary		11
Kota Belud and Tempasuk Plane Bird Sanctuary		12,000
Pulau Mantanani Bird Sanctuary		–
Sabah Foundation		
Danum Valley Conservation Area	1976	43,800
Maliau Basin Conservation Area	1981	39,000
Imbak Canyon Conservation Area	2009	30,000
Tumunong Hallu Conservation Area	2003	580
Forestry Department		
Protection Forests		342,216
Amenity Forests		20,767
Mangrove Forests		316,042
Virgin Jungle Reserves		90,382
Wildlife Reserved Forests		132,653
Sabah Parks		
Gunung Kinabalu National Park	1964	75,400
Crocker Range National Park	1984	350,585
Pulau Tiga National Park	1978	15,800
Tunku Abdul Rahman National Park	1974	4,929
Tun Sakaran Marine Park	2004	305,000
Tun Mustafa Marine Park	2016	980,000
Tawau Hills Park	1979	27,972
Turtke Islans Park	1965	1,740

Heritage Site and the Crocker Range National Park is the UNESCO's Man and Biosphere Reserve. Sabah has aspired to gazette a million ha under its Protected Areas policy.

10.4.9 National Parks in Sarawak

Similarly, the concept of National Park in the state of Sarawak was accepted in the early 1950s as a result of the National Park Ordinance passed in 1956 and the first National Park, Bako National Park was established by the Forestry Department in 1974. Since then many more were established for biodiversity conservation and environmental protection (Table 10.16). In 1990, the National Parks Ordinance was amended to include a new category of land-use, namely nature reserves, though of smaller size they are equally important for habitat protection and biodiversity conservation (Latiff and Zakri, 1998; Latiff, 2011).

10.4.10 National Parks in Johor

The state of Johor also follows the governance as shown by Sabah and Sarawak by gazetting her protected areas under the Johor National Park Ordinance. Hence five areas have been gazetted as its National Park System, namely Endau Rompin National Park (80,000 ha), the Tanjung Piau National Park, the Kukup National Park, the Gunung Ledang National Park and the Islands of Mersing National Park.

10.4.11 Other State Parks

There are other state parks in Peninsular Malaysia which were gazetted under the purview of the Department of Forestry for biodiversity conservation (Table 10.17).

10.4.12 Ramsar Sites

Currently there are only six Ramsar sites established in Malaysia by the Federal government and put under the management of state government agencies (Sabah, Johor and Sarawak) and the Department of Wildlife and National Park (Tasik Bera) (Table 10.18). As Malaysia lies on the fly-ways of the migratory birds, more wetland areas should be declared as RAMSAR sites in future to support bird protection and conservation.

10.5 Faunal Resource

10.5.1 Mammals

There are 307 species belonging to 108 genera and 32 families of non-marine mammals in Malaysia, this total includes 229 species in Peninsular Malaysia

Table 10.16 The National Parks in Sarawak

Name	Year established	Area (ha)
Bako National Park	1957	2,728
Gunung Mulu National Park	1974	52,866
Niah National Park	1974	3,140
Lambir Hills National Park	1975	6,952
Similanjau National Park	1978	7,067
Gunung Gading National Park	1983	4,106
Kubah National Park	1989	2,230
Loagan Bunut National Park	1991	10,736
Batang Ai National Park	1991	24,040
Tanjung Datu National Park	1994	1,379
Total		115,244

Table 10.17 Other State Parks in Peninsular Malaysia

Name	State	Year established	Area (ha)
Perlis State Park	Perlis	2005	5,000
Bukit Panchor State Park	Penang	2012	12,000
Royal Belum State Park	Perak	2007	117,000
Pulau Sembilan State Park	Perak	2014	10
Selangor State Park	Selangor	2010	115,000
Endau-Rompin State Park	Pahang	2000	80,000
Gunung Stong State Park	Kelantan	2012	24,400
Setiu State Park (Proposed)	Terengganu	–	4,000
Total			**357,410**

Table 10.18 Ramsar Sites in Malaysia

Name of sites	Area (ha)
Tasik Bera, Pahang	38,446
Pulau Kukup, Johor	647
Sungai Pulai, Johor	9,126
Tanjung Piai, Johor	526
Kuching Wetlands, Sarawak	6,610
Lower Kinabatangan-Segama Wetlands, Sabah	78,803
Total	**134,158**

and 221 species in Sabah and Sarawak, of which 152 species are common (Tweedie and Harrison, 1956; Medway, 1977, 1983). At least 30 of them are endemic. Two genera, *Pithecheirops* and *Diplogale* are endemic to Malaysia. The most speciose orders are the bats (106) and rodents (55) and the most specious genera are *Rhinolophus* (18), *Hipposideros* (18), and *Myotis* (9).

10.5.2 Primates

The iconic *Pongo pygmaeus* occurs in Malaysian Bornean states of Sabah and Sarawak, whose population in Sabah has been estimated at 10,000–20,000 and in Sarawak fewer than 2,000. There had been an attempt to subdivide the *P. pygmaeus* populations into three subspecies, namely, *P. p. pygmaeus*, *P. p. morio* and *P. p. wurmbii* on DNA differences in three different geographical localities (Latiff and Mashhor, 2013). The proboscis monkey, *Narsalis larvatus* is only found in Sabah and Sarawak too. The gibbons are represented throughout the country by *Hylobates syndactylus, H. lar, H. agilis, H. muelleri*. The family Loridae with a lone species of the genus *Tarsius*, the western tarsier, *T. bancanus* and the slow loris, *Nycticebus coucang*. The family Cercopithecidae, subfamily Cercopithicinae, includes 3 species of the genus *Macaca, M. fascicularis, M. nemestrina* and the rare *M. arctoides*. The leaf monkeys are represented by *Presbytis cristata, P. obscura, P. melanophos, P. rubicund* and the subfamily Colobinae includes the genus *Semnopithecus* and 4 species of *Trachypithecus*. The ape family Hylobatidae includes a lone species viz., *Bunopithecus hoolock, Hyalobates hoolock* (Marsh and Wilson, 1981).

10.5.3 Large Mammals

Malaysia can also claim to have the large mammals in *Elephas maximus maximus* and *E. maximus borneense* (endemic to Borneo), the widespread *Tapirus indicus, Dicerorhinus sumatrensis,* is now confined to a small population in Sabah and the wild populations *Bos gaurus* are still many in Peninsular Malaysia but *B. javanicus* is only found in Sabah and Sarawak (Medway, 1977, 1983). The intermediate and small mammals are many too. The pigs are represented by *Sus scrofa* and *S. barbatus* and the family Cervidae, the deers are represented by *Cervus unicolor, Muntiacus muntjak, M. antherodes, Tragulus napu* and *T. javanicus*. Many species of the order Carnivora are reported from Malaysia under some families viz., Canidae (one species of *Cuon, C. alpinus)*, Felidae (includes species of *Felis silvestris, Prionailurus bengalensis, Catopuma temminckii, C. planiceps, C. badii, Neofelis nebulosa, Panthera pardus, Panthera tigris corbetti, Parado marmorata).* The family Mustelidae (*Lutra lutra, L. sumatrana, L. perspicillata, Amblonyx cinereus,* 12 species of civets, *Viverra zibetha, V. megaspila, V. malaccensis, V. tangalunga, Hemigalus derbyanus, H. hosei, Prionodon linsang, Paguma larvata, Cynogale bennetti, Arctogalidia trivirgata, Arctictis binturong, Paradoxurus hermaphroditus.* The family Herpestridae includes 7 species of mongoose, *Herpestes semitorquatus, H. urva, H. hosei, H. javanicus, H. edwardsii, H. brachyurus,* and *H. auropunctatus.*

10.5.4   Rodentia

This is the largest order of mammals in the world comprising 2277 species in 481 genera under 33 families (Wilson and Reeder, 2005), includes squirrels, rats, mice, voles, gerbils, hamsters, dormice, porcupines etc. A total of 26 species of both the diurnal and ground squirrels has been recorded, including *Ratufa bicolor, R. affinis, Rheithrosciurus macrotis, Callosciurus notatus, C. caniceps, C. prevostii* and *Sundasciurus tenuis.* The flying squirrel is represented by *Ratufa bicolor.* The flying lemur, *Cynocephalus variegatus* is also present. The family Hystricidae has 4 species, *Thecurus crassipinis, Hystrix brachyura, Trichys fasciculata* and *Atherurus macrourus.* The shrews, rats, and mice are also common, a total of 9 shrew species (e.g., *Sunchus murinus, S. etruscus, S. ater, Crocidura fulignosa),* 10 treeshrew species (e.g., *Ptilocercus lowii, Tupaia glis),* 27 rat species (e.g., *Rattus norvegicus, R. rattus, R. tiomanicus, R. argentiventer, Bandicota indica, B.bengalensis)* and 6 mouse species (e.g., *Mus musculus, M. caroli)* have been recorded (Davison and Akbar, 2007).

10.5.6   Chiroptera

The bats are the largest group of mammals in Malaysia comprising about 40% of the 280 species of mammals. They include both the Megachiroptera and the Microchiroptera. A total of 18 species of fruit bats have been recorded and these include *Pteropus vampyrus, Eonycteris spelaea, Cynapoterus brachyotis, B. horsfiledii, Panthetor lucasi* and *Balionycteris maculata.* A total of 72 species of insectivorous bats has been recorded including *Scotophilus kuhlii, Myotis muricola, Tylonycteris robustula, T. pachypus,* and *Rhinolophus trifoliatus.*

10.5.7   Sea Mammals

Almost 20 species of sea mammals have been recorded in Malaysian waters, the main are dolphins, whales and the dugongs. All of them are endangered, but the dugongs (*Dugong dugon*) has one of the highest pro-files among the Malaysian endangered species. Incidentally, the dugongs also were hunted for meat. Of late the Irrawady dolphins (*Orcaella bre-virostris*) have been observed frequently at the estuary of Sungai Matang, Perak.

10.5.8 *Aves*

Birds evolved about 150 million years ago, occupied diverse ecological niches and are distributed in all available habitats. In the World, it is esti-mated there are more than 8,600 species of birds. The Malaysia avifauna represents the Oriental, and Australasian zoogeographic elements with 785 species with 43 endemics belonging to 85 families (Jeyarajasingam, 2007). It includes about 426 residents, and the others are migrants which use the East Asian Flyway to migrate from the northern hemisphere at the begin-ning of September-October annually. Some winter in Malaysia, flying north again at the beginning of northern summer; others stop for only a short time before continuing their long journey south. Malaysia has 41 species of diurnal birds of prey (raptors) belonging to the order Falconiformes which include *Heliastus indus, Elanus caeruleus, Spilornis cheela, Haliaeetus leucogaster, Spizaetus alboniger, S. cirrhatus, Pernis ptilorhynchus, Pan-dion halaitus, Aviceda leuphotes,* among others. Among the nocturnal birds are *Otus lempiji, Ketupa ketupu, Strix leptogrammica, Bubo sumatranus,* and *Phodilus badius.*

Of the World's 290 pigeons and dove species, Malaysia has 26 spe-cies which include the green pigeons such as *Treron curvirostra, T. olax, T. gading* and *Ptilinopus jambu.* Among the common ground doves include *Columba livia, Caloenas nicobarica, Macropygia ruficeps, Streptopelia chi-nensis, Geopelia straita* and *Chalcophaps indica.* There is also the impe-rial pigeon such as *Ducula aenea, D. bicolor* and *D. badia.* Malaysia has 15 species of kingfishers including *Alcedo atthis, A. meninting, Halycyon chloris, H. smyrnensis, H.pileata* and *H. capensis* and 26 species of wood-peckers and piculets. Among the common woodpeckers are *Caleus brachy-urus, Dryocopus javensis,* and *Dinopium javanesis.* The piculets are small, solitary birds, relative to woodpeckers which forage for insects and these include *Sasia abnormis* which is about 9 cm long.

The iconic hornbills are represented by 10 species out of about 54 spe-cies in the World. These include *Buceros vigil, B. rhinoceros, B. bicor-nis, Anorrhinus galeritus, Aceros undulates, A. subruficolis, A. comatus, A. corrugatus* and *Anthracoceros albirostris.* In Sarawak, in particular, the hornbills have a special place in the customs and ceremonies of the local indigenous inhabitants, especially the Ibans (Davison, 1995). The water-birds in Malaysia consist of 156 species including 12 species of ducks (*Dendrocyhna javanica, Nettapus coromandelianus, Ardea purpurea)* and 10 species of herons (*Nycticorax nycticorax, Mycteria cinerea, Butorides striatus*), 6 species of egrets (*Egretta alba, E. garzetta, Bubulcus ibis*). Other waterbirds include waterhens (*Amauronis phoenicurus*) and sand-piper (*Actitis hypoleucos*).

10.5.9 Reptilia

An updated checklist of 567 species in 28 families of reptiles includes two species of crocodiles (*Crocodylus porosus, Tomistoma schlegelii)* out of 25 known in the World, 4 species of marine turtles and 18 species of non-marine turtles and tortoises, 143 species of lizards, 4 species of *Varanus* and 140 species of land snakes, including 17 poisonous snakes, 4 species of pythons recorded till this date from Malaysia. The world fauna of reptiles includes 9,230 known species.

10.5.10 Sea Snakes

There are a total of 22 species of sea snakes in Malaysian waters, and all of them are poisonous. Among them are *Enhydrina schistosa, Hydrophis cyanocinctus, H. melanosoma, H. ornatus, H. spiralis, H. caerulescens, Laticauda colubrina,* and others.

10.5.11 Sea Turtles

Malaysia beaches provide nesting habitats for 4 species of sea turtles. However, the population and landings had been reduced greatly because of egg exploitation, incidental capture in fishing gears, coastal and tourism development, marine pollution and from direct harvesting for their meat. The species that occur in Malaysia are *Lepidochelys olivacea, Eretmochelys imbricata, Chelonia mydas,* and *Dermochelys coriacea.*

10.5.12 Amphibia

In the World, amphibian estimates are found to be around 6,771 species of which 5,966 species of frogs and toads, 242 species of Caecilians and 19 species of Salamanders. The systematic list of Malaysian Amphibia includes more than 203 species comprising of the order Anura includes 107 species in Peninsular Malaysia, 111 species in Sabah, of which 18 are endemic, and Sarawak has more than 130 species. The rest belong to order Caudata and the order Gymnophia (Table 10.19).

10.5.13 Fishes

In the World, it was estimated that there are more than 20,000 fish species and more than 4,000 are found in Malaysian waters. Currently, there are 1,971 species belonging to 704 genera and 186 families of both the freshwater and marine fish. Of these at least 470 are found in freshwater habitats, 81 species in brackish water and 1,400 in marine water. Earlier only 280 species of freshwater fishes were reported in Peninsular Malaysia (Mohsin and Ambak, 1983; Ahmad and Khairul-Adha, 2007), with more than 100 and 200 species in Sabah (Inger and Chin, 1990) and Sarawak (Kottelat and Whitten, 1996), respectively. The figures for Sabah and Sarawak are believed to be underestimates as the two states have been poorly researched. At present, there are 1,751 species of marine and brackish water fish recorded and more than 400 species are recorded in coastal waters and river estuaries and more than 450 species recorded offshores in Sabah and Sarawak alone.

10.5.14 Phytoplanktons

Many species of phytoplanktons are found in the Malaysian seas and among the most common are species of *Chaetoceras, Coscinodiscus,* and *Rhizosolenia.* Diatoms (Bacillariophyceae) constitute the major part of the phytoplanktons.

Table 10.19 The Amphibian Fauna in Malaysia

Family	Number of species
Bufonidae	35
Megophryidae	28
Microhylidae	34
Ranidae	51
Rhacophoridae	48
Ichtyophiidae	7
Total	**203**

10.5.15 Zooplanktons

The largest group of zooplanktons is the crustaceans and of the crustaceans, the copepods are the most abundant, including species of *Undinula, Labidocera, Eucalanus, Centropages, Euchaeta,* and *Oithona,* among others (Othman and Lotfy, 1995).

10.5.16 Hard Corals

Though relatively small in area, Malaysia's coral reefs are high in biodiversity. Together with other Coral Triangle countries such as Indonesia, the Philippines, Timor Leste this region in the Asia Pacific has 500–600 species of hard corals and has more coral species than the Great Barrier Reefs of Australia which has about 70 genera of hard corals genera. In Malaysian waters it has been estimated that there are 35 genera at Pulau Payar, Langkawi, 70 at Pulau Redang, Terengganu, and Pulau Tioman, Pahang, 70 at Pulau Sipadan, Sabah.

10.5.17 Soft Corals

The term soft corals are broadly used for the members of Cnidarians, comprises of sea fans and sea whips. It is estimated that there are more than 200 species of soft corals in Malaysian shallow seawaters. More than half belongs to Alcyoniidae which include the common species of *Sinularia* and *Nephthea.* Others include the species of *Virgularia, Melithaea, Pteroeides, Dendronephthea,* and *Ellisella.*

10.5.18 Cephalopoda

Cephalopods are exclusively distributed in the marine environment, most squids and octopuses inhabit the shallow sea around the continental shelf, while cuttlefish live nearer the coast, especially that of Terengganu, Peninsular Malaysia. The dominant species found are *Loligo chinensis, L. duvaucelli,* and *Spioteuthis lessoniana,* which are commonly popular in local cuisine and as dried sea products. Among the common cuttlefish species are *Sepia aculeata, S. pharaonis,* and *Sepiella inermis.*

10.5.19 Tunicata

There are two groups of barnacles, the acorn barnacles, and the stalked barnacles. The former is predominant in Malaysia, as well as in other countries, attaching itself to rocks and also biofoulers. There is only one stalked barnacle in Malaysia, *Lepas anatifera*, but there are several acorn barnacles including *Balanus amphitrite*, *B. amaryllis*, *Chthanmalus withersi*, *Chelonibia testudinaria*, and *Tetraclita porosa*.

10.5.20 Crabs

There are more than 102 species in 24 genera and four families of freshwater crabs are known in Malaysia, out of the total known 1,000 species. The marine and brackish water crabs are not known.

10.5.21 Mollusca

The molluscs are of great diversity with more than 100,000 species in the world, but only six classes are found in Malaysia. The richness of the species is due to the diverse ecosystems and habitats which include rocky shores, sandy beaches, mud flats, coral reefs, rivers and mangroves, and those living in coral reefs show greater diversity compared with those in other habitats. Among the dominant species in rocky shores are *Saccostrea cucullata* and *Nerita polita*. In muddy habitat including mud flats, *Anadara granosa* and *Pholas* sp. are common. The former is now commercially cultivated in many estuarine areas. The giant clamp, *Charonia tritonis* and *Tridacna squamosa* are now considered an endangered species in Malaysia as a result of exploitive activities to harvest their shells and meat. Today, it is subjected to artificial breeding to ensure their survival in the wild.

10.5.22 Echinodermata

Among the invertebrates, the echinoderms include the starfish (Asteroidea), and sea cucumbers (Holothuroidea), feather stars (Crinoidea), sea urchins (Echinoidea), and brittle stars (Ophiuroidea). They are found from the shallow intertidal areas to the deep ocean floors. The global diversity is estimated to have 6,600 extant species and 13,000 extinct ones. The most common among the starfishes are species of *Fromia, Choriaster, Linckia, Acanthaster,* and *Coscinasterias*. The holuthurians also occur in the rocky and fringes of coral areas and among the common species are *Holothuria atra* and *H. edulis*. For decades the sea cucumbers have been harvested for food and medicine. In Malaysia, the Chinese eat the sea cucumbers as a special cuisine and the Malays extract the sea cucumbers for medicine.

10.5.23 Worms

A highly diversified group of rounds worms occurring in all types of habitat. Globally, 26,646 species are recognized with 8359 parasitic in vertebrates and 10,681 free living, 4105 plant-parasitic worms and 3501 species parasites on invertebrates. In Malaysia, there are seven phyla that include the flatworms, segmented worms and ribbon worms. Though they occur widely they were under-studied, hence we don't know their estimates.

10.5.24 Protozoa

These unicellular organisms are found in various natural conditions and have been reported from freshwater, brackish and marine, soil and moss as well as free-living forms. The world protozoan fauna is represented by 31,250 species of which 10,000 are parasitic. Parasitic protozoans occur in epizoic, luminicolous, coelozoic, histozoic, and coprozoic environs. In Malaysia their number is not known.

10.5.25 Arachnida

The arachnids include spiders, scorpions, ticks, and mites as well as a few smaller groups of pseudoscorpions and phalangids. Globally 42,751 spider species from 110 families have been reported under 3,859 genera. Besides about 30,000 species of Acarina, 2,300 species of Pseudoscorpions, 1,752 species of Scorpions, 1600 species of opilions 1,000 species of Solpugidas and 4 species of King Crabs have been recorded. In Malaysia, the actual number of spiders is not known in Peninsular Malaysia, study of scorpion fauna is limited but the earliest published study was conducted by Tweedie and Harrison (1956), they recorded five scorpion species there. Kovarik (1994, 2000, 2003, 2004, 2005) described *Isometrus zideki* based on specimens collected from Malaysia and Indonesia marked the "re-birth" of research interest in this fauna. Later, several works were published on scorpions in Peninsular Malaysia by the same author (*see* Kovarik, 2003, 2004, 2005a, 2005b). Zhiyong et al. (2011) also contributed to the research of the scorpion fauna in Peninsular Malaysia. The common species include *Nephila maculata, Hippasa holmerae, Thiania bhamoensis, Cyclosa mulmeinensis, Amyciaea lineatipes, Gasteracantha arcuata, Liphistius desultory,* among others. In general, studies on scorpions in Peninsular Malaysia are still limited.

10.5.26 Insecta

The Arthropoda alone includes more than 1 million species, constituting about 80% of the total number of species. The most successful group, Insecta, accounts for about 66% of all animals and the most successful orders are Coleoptera and Lepidoptera. The rainforests of Malaysia are known to support a very high diversity of beetle species (Chung, 2007). As an example, a study of subfam. Galerucinae in a 2 km^2 plot at Danum Valley, Sabah revealed about 200 species. Malaysia has a diverse butterflies and moths, including the most beautiful *Troides brookiana* and the largest moth, *Attacus atlas*. There are more than 1,031 butterfly species, and the biggest are Papilioidae and Nymphalidae which comprise of more than 300 species in Peninsular Malaysia. The others are Lycaenidae and Hesperidae. Moths are more numerous than the butterflies, and the macromoths alone comprise of more than 5,000 species. The common families are Limacodidae, Sphingidae, Saturniidae, Lymantriidae, Nocturidae, and Geometridae. The micromoths are not well-known.

The dragonflies and damselflies (Odonata) are represented by 230 species in Peninsular Malaysia alone (Orr, 2005). Malaysia also has the greatest density of stingless bees, *Trigona,* with more than 35 species. The honeybees such as *Apis dorsata* and *A. cerana* are well-known as they are harvested for honey. Today the honey is produced commercially through bee domestication.

The ant fauna is not known too but the estimate is of the 9,000–10,000 species (Wilson, 1988). Of about 2,300 species of termites in the world, nearly 200 have been recorded in Malaysia. According to Maryati (1995) and Cheng and Kirton (2001) more than 1,200 species of ants have been recorded in Malaysia. Similarly of the 2,500 species of stick and leaf insects in the world, 200–250 species have been recorded in Malaysia. This includes one of the largest stick insects in the world, *Phobaeticus kirbyi* which is about 30 cm long. One of the commonest insects is the cicadas belonging to Homoptera which are represented by more than 104 species in families Cicadidae and Tibicinidae in Peninsular Malaysia. Many species of *Purana, Muda, Platypleura,* and *Nelcyndana* are yet to be named and published.

10.5.27 Animal Breeds in Malaysia

Except for the Kelantan-Kedah breed of cattle that is quite popular in Peninsular Malaysia, the majority of the animal breeds known in Malaysia are the foreign origin.

Acknowledgments

The author wishes to express his thanks to his colleague and students at the Universiti Kebangsaan Malaysia for their constant encouragement and help. The author is indebted to Prof. Dato' Dr. Ismail Sahid for introducing him to the editor.

Keywords

- amphibian fauna
- animal breeds
- faunal resource
- floral diversity

References

Ahmad, A., & Khairul-Adha, A. R., (2007). State of knowledge of freshwater fishes of Malaysia. In: *Status of Biological Diversity in Malaysia and Threats Assessment of Plant Species in Malaysia,* Chua, L. S. L., Kirton, L. G., & Saw, L. G., (eds.). Forestry Research Institute Malaysia, pp. 83–90.

Anderson, J. A. R., (1963). The flora of the peat swamp forests of Sarawak and Brunei including a catalogue of all recorded species of flowering plants, ferns and fern-allies. *Gards. Bull. Singapore, 20,* 131–228.

Ashton, P. S., (1982). Dipterocarpaceae. *Flora Malesiana, Series 1, 92,* 237–552.

Ashton, P. S., (1988). *Manual of the Non-Dipterocarp Trees of Sarawak.* Dewan Bahasa & Pustaka, Sarawak Branch, vol. 2.

Bidin, A. A., & Latiff, A., (1995). The status of terrestrial biodiversity in Malaysia. In: *Prospects in Biodiversity Prospecting,* Zakri, A. H., (ed.). Genetic Society of Malaysia & Universiti Kebangsaan Malaysia, pp. 59–76.

Brunig, E. F., (1974). *Ecological Studies in the Kerangas Forests of Sarawak and Brunei.* Borneo Literature Bureau, Kuching.

Chase, M. W., Cameron, K. M., Barrett, R. L., & Freudenstein, J. V., (2003). *DNA Data and Orchidaceae Systematics: A New Phylogenetic Classification.* Orchid conservation. Natural history Publication (Borneo). Kota Kinabalu, pp. 69–89.

Cheng, S., & Kirton, L. G., (2007). Overview of insect biodiversity research in Peninsular Malaysia. In: *Status of biological diversity in Malaysia and threats assessment of plant species in Malaysia.* Chua, L. S. L., Kirton, L. G., & Saw, L. G., (eds.). Forestry Research Institute Malaysia, pp. 12–128.

Chin, S. C., (1977). The limestone hill flora of Malaya 1. *Gards. Bull. Singapore, 30,* 165–219.

Chuah-Petiot, M. S., (2011). A checklist of Hepaticae and Anthocerotae of Malaysia. *Polish Bot. Journ., 56*(1), 1–44.

Chung, A. Y. C., (2007). An overview of research on beetle diversity and taxonomy in Malaysia. In: *Status of Biological Diversity in Malaysia and Threats Assessment of Plant Species in Malaysia,* Chua, L. S. L., Kirton, L. G., & Saw, L. G., (eds.). Forestry Research Institute Malaysia, pp. 137–148.

Corner, E. J. H., (1978). The freshwater swamp forest of South Johore and Singapore. *Gards. Bull. Singapore, Suppl. 1*, 1–266.

Davison, G. W. H., & Akbar, Z., (2007). The status of mammalian biodiversity in Malaysia. In : *Status of Biological Diversity in Malaysia and Threats Assessment of Plant Species in Malaysia,* Chua, L. S. L., Kirton, L. G., & Saw, L. G., (eds.). Forestry Research Institute Malaysia, pp. 3–27.

Davison, G. W. H., (1995). The birds of Temengor Forest Reserve, Hulu Perak, Malaysia. *Malay. Nat. J., 48,* 371–386.

De Kok, R. P. J., Rusea, G., & Latiff, A., (2009). The genus *Teijsmanniodendron* Koords. (Lamiaceae). *Kew Bulletin, 64,* 587–625.

Faridah-Hanum, I., S., Kamis, K. A., & Hamzah, A., (2009). *Handbook on the Peat Swamp Flora of Peninsular Malaysia.* PSF Technical Series No. 3, UNDP/GEF.

Foxworthy, F. W., (1932). *Dipterocarpaceae of the Malay Peninsula.* Mal. Forest Record No. 10.

Gilliland, H. B., (1971). *A Revised Flora of Malaya. Grasses of Malaya.* Botanic Gardens Singapore, vol. 3.

Holttum, R. E., (1950). The Zingiberaceae of the Malay Peninsula. *Gard. Bull. Singapore, 13*(1), 1–249.

Holttum, R. E., (1951). The Maranthaceae of Malaya. *Gard. Bull. Singapore, 13,* 254–296.

Holttum, R. E., (1953). *A Revised Flora of Malaya.* Orchids of Malaya. Singapore: Pejabat Percetakan Kerajaan, vol. 1.

Holttum, R. E., (1954). *A Revised Flora of Malaya.* Ferns of Malaya. Singapore, Pejabat, Percetakan Kerajaan, vol. 2.

Inger, R. F., & Chin, P. K., (1990). The freshwater fishes of North Borneo. *Fieldiana Zool., 45,* 1–268.

Jaman, R., (2014). *Cyathea bunnemeijeri* v. A new fern record to Peninsular Malaysia. *Malayan Nature Journ., 67*(1), 52–54.

Japar Sidik, B., & Muta Z. H., (2011). Sea grass in Malaysia. In: *Seagrasses Resource Status and Trends in Indonesia, Japan, Malaysia, Thailand and Vietnam,* Ogawa, H., Japar Sidik, B., & Muta Harah, Z., (eds.). Japan Society for the Promotion of Science & Atmosphere and Ocean Research Institute. The University of Tokyo, pp. 22–37.

Jeyarajasingam, A., (2007). As assessment of the current knowledge of Malaysia's avifauna. In: *Status of Biological Diversity in Malaysia and Threats Assessment of Plant Species in Malaysia,* Chua, L. S. L., Kirton, L. G., & Saw, L. G., (eds.). Forestry Research Institute Malaysia. pp. 29.

Keng, H., (1978). *Orders and Families of Malayan Seed Plants.* Singapore University Press.

Kiew, R., (1998). The seed plant flora of Fraser's hill. *Research Pamphlets No. 121.* Forest Research Institute Malaysia, Kuala Lumpur.

Kiew, R., (2005). *Begonias of Peninsular Malaysia.* Natural History Publications (Borneo).

Kiew, R., Chung, R. C. K., Saw, L. G., Soepadmo, E., & Boyce, P. C., (2010). *Flora of Peninsular Malaysia. Series II: Seed Plants. Malayan Forest Records No, 49.* Forest Research Institute Malaysia, Kepong, vol. 1.

Kiew, R., Chung, R. C. K., Saw, L. G., Soepadmo, E., & Boyce, P. C., (2011). *Flora of Peninsular Malaysia. Series II: Seed Plants,* Malayan Forest Record No. 49. Forestry Research Institute Malaysia, vol. 2.

Kiew, R., Sang, J., Repin, R., & Ahmad, J. A., (2015). *A Guide to Begonias of Borneo.* Natural History Publications (Borneo), Kota Kinabalu, Sabah.

Kottelat, M., & Whitten, T., (1996). Freshwater fishes of Sarawak and Brunei Darussalam: A preliminary annotated checklist. *The Sarawak Museum Journ.*, *48*, 228–256.

Kovarik, F., (1994). *Isometrus zideki sp. n.* from Malaysia and Indonesia, and a taxonomic position of *Isometrus formosus*, *I. thurstoni* and *I. sankariensis* (Arachnida: Scorpionida: Buthidae). *Acta Societatis Zoologicae Bohemicae*, *58*, 195–203.

Kovarik, F., (2000). First reports of *Liocheles nigripes* from Indonesia and Malaysia and *Hormiops davidovi* from Malaysia (Scorpiones: Ishnuridae). *Acta Societatis Zoologicae Bohemicae*, *64*, 57–64.

Kovarik, F., (2003). A review of the genus *Isometrus* Ehrenberg, (1828). (Scorpiones: Buthidae) with description of four new species from Asia and Australia. *Euscorpius, 10.*

Kovarik, F., (2004). A review of the genus *Heterometrus* Ehrenberg, 1828, with description of seven new species (Scorpiones, Scorpionidae). *Euscorpius, 15.*

Kovarik, F., (2005). Two new species of the genus *Chaerilus* Simon, (1877). From Malaysia (Scorpiones: Chaerilidae). *Euscorpios, 26.*

Latiff, A., & Faridah-Hanum, I., (2013). Mangroves ecosystem of Malaysia: Status, challenges and management strategies. In: *Mangrove Ecosystems in Asia: Status, Challenges and Management Strategies*, Faridah-Hanum, I., Latiff, A., Khalid, R. H., & Ozturk, M., (eds.), pp. 1–22.

Latiff, A., & Mashhor, M., (2013). *Ex Situ Conservation of Orang Utan.* Bukit Merah. Orang Utan Island Foundation, Perak, Malaysia.

Latiff, A., & Zakri, A. H., (1998). Environmental and conservation issues in Malaysia. In: *Biodiversity Conservation in ASEAN: Emerging Issues and Regional Needs,* Ismail, G., & Mohamed, M., (eds.). Asean Academic Press, pp. 136–162.

Latiff, A., (1997). Malaysia's plant diversity: A disappearing resources? In: *State of the Environment in Malaysia.* Consumers' Association of Penang., pp. 218–228.

Latiff, A., (1998a). Conservation and utilization of botanical diversity in Malaysia. *Malaysian Applied Biology*, *27* (1&2), 1–11.

Latiff, A., (1998b). Problems in Managing Biodiversity in Malaysia. In: *Fluttering Around: Malaysia's Biodiversity Policy,* Singh, G., & Jacob, R., (Comp.). Centre for Environment Technology and Development, Malaysia, pp. 17–28.

Latiff, A., (2001). Biodiversity Policy challenges in Malaysia's. In: *Making Megabiodiversity Meaningful,* Gurmit Singh, K. S., (Comp.). The challenge for Southeast Asian Nations. *A. CETDEM Report,* pp. 10–21.

Latiff, A., (2007). Managing plant resources and conservation of biodiversity in Malaysia. In: *Proceedings of Conference on Natural Resources in the Tropics: Development and Commercialization of Tropical Natural Resources,* Ipor, I. B., et al., (eds.), pp. 1–12.

Latiff, A., (2009). Species diversity and management of threatened plants. In: *Fundamental Science Congress 2009–Accelerating Research Excellence,* Jambari, H. J. Ali, et al., (eds.). Faculty of Science, University Putra Malaysia, pp. 2–13.

Latiff, A., (2010). Biodiversity in Malaysia: The most undervalued natural asset, In: *Biodiversity-Biotechnology: Gateway to Discovery, Sustainable Utilization and Wealth Creation,* Rita, M., et al., (eds.). Sarawak Biodiversity Centre, pp. 13–22.

Latiff, A., (2011). Loss of biodiversity and resources due to forest exploitation and degradation. In: *Proceedings International Symposium on Rehabilitation of Tropical Rainforest Ecosystems,* Nik, M. M., et al., (eds.). University Putra Malaysia. pp. 137–144.

Latiff. A., (2015). *Cyathea arjae* (Cyatheaceae), A new species of dwarf tree-fern from Mount Kinabalu, Sabah, Malaysia. *Sains Malaysiana,* *44* (1), 57–60.

Lee, G. E., Robert, G. S., & Latiff, A., (2013). A catalogue of the Lejeuneaceae of Malaysia. *Malayan Nature Journ.*, *65*(2&3), 81–129.

Lee, G. E., Robert, G. S., Damanhuri, A., & Latiff, A., (2011). Towards a revision of *Lejeunea* (Lejeuneaceae) in Malaysia. *Gards. Bull. Singapore*, *63*(1&2), 163–174.

Marsh, C. W., & Wilson, W. L., (1981). *Survey of Primates in Peninsular Malaysian Forests.* University of Cambridge and University Kebangsaan Malaysia.

Maryati, M., (1995). *Semut UBTP*. University Malaysia Sabah & Dewan Bahasa dan Pustaka (In Malay).

Mat-Salleh, K., & Latiff, A., (2002). *Tumbuhan Ubatan Malaysia*. University Kebangsaan. Malaysia.

Medway, L., (1977). Mammals of Borneo. *Monographs of the Malaysian Branch, Royal Asiatic Society No. 7*, Kuala Lumpur.

Medway, L., (1983). *The Wild Mammals of Malaya (Peninsular Malaysia) and Singapore.* Oxford University Press, Kuala Lumpur.

Meijer, W., & Wood, G. H. S., (1964). *Dipterocarps of Sabah (North Borneo).* Sabah Forest. Record No. 5. Forest Department, Sandakan.

Mohamed, M. A. H., & Tan, B. C., (1988). A checklist of Peninsular Malaya and Singapore. *The Bryologist*, *91*, 24–44.

Mohsin, A. K. M., & Ambak, A., (1983). *Freshwater Fishes of Peninsular Malaysia*. University Pertanian Malaysia, Selangor.

Ng, F. S. P., (1978). *Tree Flora of Malaya*, vol. 3, Kuala Lumpur: Longman, Malaysia.

Ng, F. S. P., (1989). *Tree Flora of Malaya*, vol. 4, Kuala Lumpur: Longman, Malaysia.

Ng, F. S. P., Low, C. M., & Mat Asri, N. S., (1990). Endemic trees of the Malay Peninsula. *Research Pamphlet No. 106*. Forest Research Institute Malaysia.

Ong, P. T., O'Byrne, P., Yong, W. S. Y., & Saw, L. G., (2011). *Wild Orchids of Peninsular Malaysia*. Forest Research Institute Malaysia.

Orr, A. G., (2005). *A Pocket Guide: Dragonflies of Peninsular Malaysia and Singapore*. Natural History Publication (Borneo), Kota Kinabalu.

Othman, B. H. R., & Lotfi, W. N., (1995). The status of marine biodiversity in Malaysia. In: *Prospects in Biodiversity Prospecting*, Zakri, H., (ed.). Genetic Society of Malaysia & University Kebangsaan Malaysia, pp. 77–93.

Parris, B. S., & Latiff, A., (1997). Towards a checklist of pteridophytes in Malaysia. *Mal. Nat. J.*, *50*, 235–280.

Parris, B. S., Kiew, R., Chung, R. C. K., & Saw, L. G., (2013). *Flora of Peninsular Malaysia. Series I: Ferns and Lycophytes,* Malayan Forest Record No. 48. Forestry Research Institute Malaysia, vol. 2.

Parris, B. S., Kiew, R., Chung, R. C. K., Saw, L. G., & Soepadmo, E., (2010). *Flora of Peninsular Malaysia, Series I: Ferns and Lycophytes,* Malayan Forest Record No. 48. Forestry Research Institute Malaysia.

Phang, S. M., Lim, P. E., & Yeong, H. M., (2010). Malaysian seaweed resources in the South China Sea and their potential economic and ecological applications. *Journal of Science and Technology in the Tropics*, *6*(2), 87–109.

Ridley, H. N., (1922–1925). *Flora of the Malay Peninsula*. Reeves, L., & Co., London, vols. 1–5.

Saw, L. G., & Chung, R. C. K., (2007). Towards the flora of Malaysia. In: *Status of Biological Diversity in Malaysia and Threats Assessment of Plant Species in Malaysia*, Chua, L. S. L., Kirton, L. G., & Saw, L. G., (eds.). Forestry Research Institute Malaysia, pp. 211–287.

Seidenfaden, G., & Wood, J. J., (1992). *The Orchids of Peninsular Malaysia and Singapore.* Olsen & Olsen, Fredensborg.

Soepadmo, E., & Wong, K. M., (1995). *Tree Flora of Sabah and Sarawak.* Forestry Department of Sabah, Forestry Department of Sarawak & Forestry Research Institute Malaysia, vol. 1.

Soepadmo, E., Saw, L. G., & Wong, K. M., (1995). *Tree Flora of Sabah and Sarawak.* Forestry Department of Sabah, Forestry Department of Sarawak & Forestry Research Institute Malaysia, vol. 2.

Soepadmo, E., Saw, L. G., Chung, R. C. K., & Kiew, R., (2007). *Tree Flora of Sabah and Sarawak.* Sabah Forestry Department, Forest Research Institute Malaysia & Sarawak Forestry Department, vol. 6.

Soepadmo, E., Saw, L. G., Chung, R. C. K., & Kiew, R., (2014). *Tree Flora of Sabah and Sarawak.* Forestry Department of Sabah, Forestry Department of Sarawak & Forestry Research Institute Malaysia, vol. 8.

Suleiman, M., Akiyama, H., & Tan, B. C., (2006). A revised catalogue of mosses reported from Borneo. *J. Hattori Bot. Lab.*, *99*, 107–183.

Symington, C. F., (1943). (Revised by Ashton, P. S., & Appanah, S., 2004). *Foresters' Manual of Dipterocarps.* Malay. For. Records No. 16. University of Malaya Press, Kuala Lumpur.

Turner, I. M., (1995). A catalogue of the vascular plants of Malaya. *Gards. Bull. Singapore*, *47*(1&2), 1–757.

Tweedie, M. W. F., & Harrison, J. L., (1956). *Malayan Animal Life.* New York, Longsman.

Whitmore, T. C., (1972). *Tree Flora of Malaya*, vol. 1. Longmans, Malaysia.

Whitmore, T. C., (1973). *Tree Flora of Malaya*, vol. 2. Longmans, Malaysia.

Wilson, D. E., & Reeder, D. M., (2005). *Mammal Species of the World: A Taxonomic and Geographic Reference.* Johns Hopkins University Press, USA.

Wilson, E. O., (1988). The current state of biological diversity. In: *Biodiversity,* Wilson, E. O., & Francis, M. P., (eds.). National Academy Press, pp. 3–18.

Wong, M., & Latiff, A., (1994). Rafflesias of Peninsular Malaysia. *Nature Malaysiana*, *19*(3), 84–88.

Yong, K. T., Tan, B. C., Ho, B. C., Ho, Q. Y., & Mohamad, H., (2013). *A Revised Moss Checklist of Peninsular Malaysia and Singapore.* Research Pamphlets No. 133. Forest Research Institute Malaysia.

Zhiyong, D., Yawen, H., Yingliang, W., Zhijian, C., Hui, L., Dahe, J., & Wenxin, L., (2011). The scorpion of Yunnan (China: Updated identification key, new record and redescriptions of *Euscorpiops kubani* and *E. shidian* (Arachnida, Scorpiones). *ZooKeys*, *82*, 1–33.

CHAPTER 11

Biodiversity in Mongolia

GOMBOBAATAR SUNDEV,[1,2] **NATHAN CONABOY,**[3]
URGAMAL MAGSAR,[4] **TERBISH KHAYANKHIRVAA,**[1] and
GANTIGMAA CHULUUNBAATAR[4]

[1]Department of Biology, National University of Mongolia, Mongolia
[2]Mongolian Ornithological Society, Ulaanbaatar, Mongolia, E-mail: gomboo@num.edu.mn; info@mos.mn
[3]Zoological Society of London, UK
[4]Institute of General and Experimental Biology, Mongolian Academy of Sciences, Mongolia

11.1   Introduction

Mongolia is a landlocked country in East Asia, situated between Russia and China. Mongolia is well known for the Empire of the Great Chinggis Khan, born in 1206 in Eastern Mongolia's Onon River basin. The total area of the country is 1,564,116 square kilometers and is the 18th largest and the most sparsely populated country in the world. The total population was 3,130,527 people in February, 2017. Mongolia's administrative units are 1,575 bags (small rural communities), 152 khoroo (sub-districts of the Capital), 9 durags (districts of the Capital), 330 soums (towns) and 21 aimags. Out of the total soums, 51.7% belong to small sum with less than 3,000 people, 30.1% fairly small with 3,000–5,000 inhabitants, 8.5% medium sized with 5,001–7,000 people, and 9.7% large with more than 7,001 inhabitants. A total of 483,000 people live in province centers, 438,700 in soum centers, and 667,100 in the countryside (Tungalag, 2014).

The capital of Mongolia, Ulaanbaatar city, is home to about 45% of the country's population. Mongolia's traditional nomadic and seminomadic lifestyle is practiced by approximately 30% of the population living in rural areas. Many of these families own the majority of Mongolia's 61,549,236 head of livestock (Mongolian Statistical Yearbook, 2015).

Economic activity of Mongolia has traditionally been based only on livestock husbandry and agriculture. In the past, there were small industries in the mining of copper, coal, molybdenum, tin, tungsten, and gold. In recent years, the Mongolian government extensively focused on the development of the mining sector which led to a boom from 2007 (with 9.9% growth) through to 2011 when GDP growth was as high as 17%.

Mongolia has five major high mountains ranges. The highest point in Mongolia is the Khuiten Orgil (peak) in the Altai Tavan Bogd Mountain

Range in the west at 4,374 m above sea level (a.s.l.). The lowest is Höh Lake of Dornod Province in Eastern Mongolia at 560 m a.s.l. Most of the country is hot in the summer and extremely cold in the winter, with January averages dropping as low as −30°C. A vast front of cold air flow comes from Siberia in winter and collects in river valleys and low elevated basins causing very cold temperatures. The country averages 257 cloudless days per a year. Precipitation is higher in the north (average of 200 to 350 millimeters per year) and lower in the south (100 to 200 millimeters).

Mongolia has rich vegetation through a broad range of natural zones and ecosystems. According to Yunatov (1954), Lavrenko (1979), Hilbig (1995), Pavlov et al. (2005), and Chimed-Ochir et al. (2010), there are six major ecosystems including high mountain, forest, steppe, desert, aquatic, and patch. Mongolia has rich biodiversity due to the range of habitats and ecosystems. However, because of its large area and small scientific community, there is still a lot to be understood regarding species composition of invertebrates and microorganisms. However, few authors have published an overview of invertebrate diversity in Mongolia.

11.2 Habitats, Ecozones and Ecosystems

There are multiple approaches for zoning and division based on vegetation, ecosystem, habitats, soil, and fauna. Based on Yunatov (1954), Lavrenko (1979), Hilbig (1995), Pavlov et al. (2005), and other references, Chimed-Ochir et al. (2010) categorized six major ecosystems including high mountain, forest, steppe, desert, aquatic, and patch (Figure 11.1).

The ecosystem division was originally focused on the identification of gaps for biodiversity conservation in Mongolia. These authors' idea of the ecosystem analysis was to define the conservation priority in the protected areas of Mongolia. Based on a gap analysis on the ecosystem (Chimed-Ochir et al., 2010), Protected Areas and Biodiversity in Mongolia, we can conclude that remote Protected Areas with reduced human impacts are effectively well protected. These ecosystems of the Protected Areas are located in true Gobi Desert and high mountain glaciers. Biodiversity protection plan for Protected Areas in Mongolia is very narrowly focused on certain species of wildlife rather than the broader ecosystem. This analysis clearly showed that the future plan of the Protected Areas should focus on the ecosystem level. Each major ecosystem consists of its own unique minor habitats or subtypes of ecosystems.

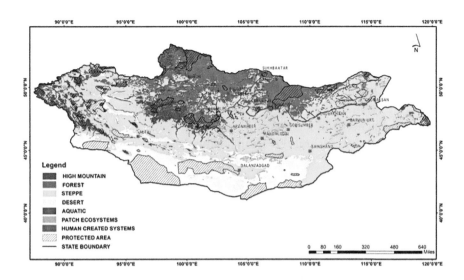

Figure 11.1   Ecosystems of Mongolia.

11.2.1   High Mountain Ecosystem

The high mountain ecosystem contains alpine tundra (1,626,075.2 ha, 1.03% of Mongolian territory), alpine meadow (5,464,041.4 ha, 3.47%), high mountain steppe (4,129,235.3 ha, 2.62%), and subalpine woodland (2,335,709.5 ha, 1.48%) (Chimed-Ochir et al., 2010). The altitude of the ranges varies depending on the origin and formation of any particular mountain range. Hentii Mountain is the lowest range and the peak, Asralt Khairkhan, reaches 2,751 m a.s.l., the average altitude is 2,400–2,500 m a.s.l. (the lowest altitude is 2,200 m in the west and 2,000–2,500 m in the east). Hövsgöl Mountain has glaciers and perpetual snows at its 3,460 m peak, Mönh Saridag. Average altitude in the range varies between 2,800–3,000 m. Khangai Mountain Range shows a typical high mountain characteristic with a high elevated mountain-flat along the 700 km mountain range. The lowest borderline of the high mountain reaches 2,500–2,600 m a.s.l. The peak of the range, Otgontenger, is 4,031 m and covered by perpetual snow. The western range of the Mongol-Altai Mountains is comparatively high where Altai Tavan Bogd reaches 4,653 m with decreased elevation to the east and southeast (from 4,653 to 3,400 m). Many patched glaciers and perpetual snow on the top of the mountain range are visible at great distances from the peak itself. The dominant plants are *Kobresia sibirica, K. bellardi*, and *K. lilifolia* and *Carex melanantha, C.capitata, C.pauciflora*, and *C. angarae. Poa –Festuca* community also occurs in this habitat. Alpine *Kobresia* meadow transforms to breaking rocks and gravels with *Salix-Juniperus-Dryas* and also *Cargana –Sedum-Lagotis-Polygonum* on the upper limit of the alpine belt. Nationally and globally threatened species of birds (Altai Snowcock, Bearded Vulture, Saker Falcon and White-throated Bushchat) and mammals (Siberian Ibex, Argali Sheep, Grey Wolf and Snow Leopard) inhabit this habitat (Figure 11.2).

11.2.2 Forest Ecosystem

In Mongolia, there are two major forest types, boreal coniferous forest (2,785,863.5 ha, 1.77%) and sub-boreal mixed forest (6,738,795.1 ha, 4.27%) (Chimed-Ochir et al., 2010). Large parts of Hentii, Khangai and Hövsgöl Mountain ranges in the northern part of the country are covered with needle-leaf forests. The dominant needle-leaved trees are Larch *Larix sibirica* (72% of total forest area), Siberian Pine *Pinus sibirica* (12%), Scots Pine *Pinus silvestris* (7%), and Siberian Spruce *Picea obovata* (0.25%). Siberian Fir *Abies sibirica* (0.02%) grows only in northwestern Hentii.

Figure 11.2 Habitats of Gobi-Altai High Mountain (Photo credit: Gombobaatar Sundev and Mongolica Publishing).

Broad-leaved forest, aspen forest, elm bush forest, riverine poplar forest, riverine shrubbery, *Halimodendron halodendron* shrubbery, and other forest types grow in major mountain ranges and open areas including steppe and Gobi (Hilbig, 1995). Various forest habitats support a variety of wildlife including many mosses, and lichens, also invertebrates; forest insects in the north of Hentii Mountain range include *Carabus canaliculatus, C. arcensis, C. billbergi, C. hummeli, Pterostichus (Metallophilus) interruptus, P. dauriscus, P. eximius, Cicindela gracilis,* as well as *Xylotrechus rusticus, Monochamus urussovi, Saperda scalaris, Xylotrechus ibex,* and *Xylotrechus hircus* (Gantigmaa et al., 2012; Buyanjargal and Namkhaidorj, 2010), and vertebrates such as amphibian (Far Eastern Tree Frog), reptile (Halys Viper), birds (globally and nationally threatened: Greater Spotted Eagle, Eastern Imperial Eagle, Saker Falcon, Common Pheasant, Tree Pipit, and Saxaul Sparrow), and mammals (globally and nationally threatened: Sable, Grey Wolf, Eurasian Lynx, Musk Deer, Red Deer, and Eurasian Elk).

11.2.3 Steppe Ecosystem

Steppe habitat covers large open areas across the eastern, central, and southern part of Mongolia. There are a number of sub-habitats of steppe including meadow steppe (16,678,504.5 ha, 10.58%), moderate dry steppe (17,183,523.8 ha, 10.90%), dry steppe (23,222,677.3 ha, 14.73%), and desert steppe (30,293,371.9 ha, 19.21%) (Chimed-Ochir et al., 2010). Steppe sub-habitats vary depending on the plant community. Dominant plants are *Stipa, Festuca, Cleistogenes, Agropyron, Koeleria, Poa, Elymus, Helictotrichon,* and *Filifolium* in the ecosystem. Other steppe communities consist of *Stipa-Carex duriuscula, Artemisia frigida-Stipa, Artemisia frigida-Cleistogenes-Stipa, Cleistogenes-Stipa,* and *Stipa-Caragana.* This community can be always changed due to overgrazing of livestock and succession of vegetation (Hilbig, 1995). The steppe ecosystem supports various wildlife such as lichen, moss, fungus (26 species of saprophytic or coprolitic mushrooms) (Kherlenchimeg and Burenbaatar, 2016), invertebrates (dominating families of Carabidae, Cerambycidae, etc.), and vertebrates including amphibian (Mongolian Toad), reptiles (Mongolian Racerunner, Steppes Ratsnake and Halys Viper), birds (globally threatened: Steppe Eagle, Saker Falcon, and Great Bustard), and mammals (nationally and globally threatened: Mongolian Marmot, Long-eared Jerboa, Pallas's Cat, Grey Wolf, Red Deer, and Mongolian White-tailed Gazelle).

11.2.4 *Desert Ecosystem*

According to Chimed-Ochir et al. (2010), the desert ecosystem consists of semi-arid desert (11,641,030.5 ha, 7.38%), true desert (12,101,045.5 ha, 7.67%), and extreme arid desert (55,969,44.9 ha, 3.55%). The desert ecosystem area covers 18.6% of Mongolian territory. The dominant plant community contains low shrubs and semishrubs belonging to Chenopodiaceae (*Anabasis, Haloxylon, Iljinia, Kalidium, Kochia, Nanophyton, Salsola, Sympegma*, etc.), Compositae (*Ajania, Artemisia, Asterothamnus, Brachanthemum*, etc.), Polygonaceae (*Atraphaxis, Calligonum*), Zygophyllaceae (*Nitraria, Zygophyllum*), Tamaricaceae (*Reaumuria, Tamarix*), Fabaceae (*Ammopiptanthus, Caragana, Oxytropis,* etc.), Rosaceae (*Potaninia*), and Convolvulaceae (*Convolvulus*) (Hilbig, 1995). Wild fauna include: amphibian (Mongolian Toad), reptiles (Kaspischer Even-fingered Gecko, Gobi Naked-toed Gecko, Przewalski's Wonder Gecko, Sunwatcher Toad-head Agama, Tuva Toad-head Agama, Mongolian Agama, Multi-oscillated Racerunner, Gobi Racerunner, Variegated Racerunner, Viviparous Lizard, Sand Lizard, Tatar Sand Boa, Slender Racer, Steppe Ribbon Racer, Northern Viperand Meadow Viper), birds (globally and nationally threatened species such as Relict Gull, Short-toed Snake-eagle, Saker Falcon, Houbara Bustard, Mongolian Ground-Jay, and Saxaul Sparrow) and mammals (globally and nationally threatened species: Gobi Bear, Alashan Ground Squirrel, Small Five-Toad Jerboa, Mongolian Three-Toed Jerboa, Tamarisk Gerbil/ Jird, Asiatic Wild Ass, Bactrian Camel, Saiga Antilope, Long-eared Jerboa and Black-tailed Gazelle).

11.2.5 *Aquatic Ecosystem*

The aquatic ecosystem consists of major water resources including perennial rivers and floodplains (7,052,707.9 ha, 4.47%) and lakes (1,288,128.5 ha, 0.82%) (Chimed-Ochir et al., 2010). Based on the report from a water census conducted across Mongolia in 2011, out of 6,646 rivers, about 6,095 were running while 551 rivers were dry. Out of 3,613 lakes, about 3,130 had standing water while 483 had dried. Out of 10,557 springs and wells, about 8,970 had water whereas 1,587 were dry (Gombobaatar et al., 2014). Fresh water ecosystems provide habitat for diverse aquatic wildlife including many species of algae, fish (Taimen, Siberian Sturgeon, Gobi Loach, Dzundarian Dace, Pidschian, Amur Grayling, Hövsgöl Grayling, Lake Osman, Small Osman, Mongolian Grayling and Lenok), amphibains (Siberian Salamander, Mongolian and Pewzow's Toads, Far Eastern Tree, Siberian Wood and

Asiatic Grass Frogs), birds (Pallas's Fish Eagle, Dalmatian Pelican, Baikal Teal, Swan Goose, Lesser White-fronted Goose, Asian Dowitcher, and Reed Parrotbill), and mammals (Eurasian Beaver and Wild Boar *Sus scrofa nigripes*).

11.2.6 Patch Ecosystem

This ecosystem consists of many different habitats and sub-ecosystems. This ecosystem includes intermittent rivers and ephemeral channels (586,282.7 ha, 0.37%), closed depressions and salt banks (3,463,984.5 ha, 2.20%), sand dunes (3,314,182.3 ha, 2.10%), and glaciers (94,907 ha, 0.06%) (Chimed-Ochir et al., 2010).

11.3 Status, Trends, and Threats of Plants

The floristic elements and distribution characteristics also indicate that the climate and environmental conditions, ecological amplitude, and adaptive capacity of the plants are influenced by the floristic origin and spatial patterns of plant diversity. Plant community patterns can be a functional tool for ecological restoration. Therefore, native species should be used for cultivation and regional vegetation should be used during the process of ecological recovery and rehabilitation. Such research will improve our global understanding of the ecology of arid lands.

The first attempt to produce a flora record for this area was that of Maximovich (1859), which, however, was not completed. The research of Flora and Vegetation in Mongolia started in the 19th century by the researchers studying Central Asia. Floristic research relevant to the area of the present-day Mongolian Republic was conducted by Grubov (1955, 1982), Ulziykhutag (1989), Gubanov (1996), Oyuntsetseg and Urgamal (2014), and Urgamal et al. (2014, 2016). The knowledge of the vascular plants of Mongolia has been compiled by Grubov (1982) in the "Keys to Vascular Plants of Mongolia," which was published in Russian and later on translated into English (Grubov, 2001). While Grubov's work includes keys for species identification and information on the distribution of species within Mongolia, Gubanov (1996) published a list of the vascular plant species and their distribution known at that time from Mongolia in his "Conspectus of the Flora of Outer Mongolia (Vascular Plants)."

According to an overview of plant diversity in Mongolia (Urgamal et al., 2016), there are 7,315 native species and infraspecific taxa of plants belonging to 1,522 genera, 392 families, 116 orders 39 classes and 19 phyla.

Among them, there are 574 mushroom species in 156 genera and 56 families, 2003 algae species in 268 genera and 87 families (Tsetsegmaa, 2008), 1,031 lichens species in 207 genera and 63 families (Enkhtuya, 2007), 580 bryophytes species in 208 genera and 74 families, and 3,127 vascular plant species in 683 genera and 112 families (Urgamal et al., 2014 & 2016) (Table 11.1 and Figure 11.3).

Table 11.1   Summary Statistics for the Plant Diversity of Mongolia

Groups	Divisions	Classes	Orders	Families	Genera	Species and infraspecies	% of total
Mushrooms	2	7	20	56	156	574	7.85
Algae	10	13	39	87	268	2,003	27.38
Lichens	1	3	12	63	207	1,031	14.09
Bryophytes	1	2	6	74	208	580	7.93
Vascular plants	5	14	39	112	683	3,127	42.75
Total	19	39	116	392	1,522	7,315	100.0

Figure 11.3   The National flower of Mongolia, *Scabiosa comosa* (Photo credit: Gombobaatar Sundev and Mongolica Publishing).

11.3.1 Fungi

A total of 574 species of mushrooms distributed in 156 genera and 2 division (Ascomycota of 12 species in 4 families) and Basidiomycota (562 species in 50 families) predominate over this groups (Table 11.2).

The largest 5 families with mushroom diversity are Russulaceae (2 genera/39 species), Agaricaceae (14/38), Tricholomataceae (13/34), Polyporaceae (15/27), and Strophariaceae (8/21) and the largest 5 genera are *Russula* (22 species), *Lactarius* (17), *Cortinarius* (17), *Suillus* (9), and *Tricholoma* (9) (Table 11.3).

11.3.2 Algae

Currently, a total of 2,003 species and infraspecific taxa of algae in 268 genera and 10 divisions were reported. The largest 5 families with highest diversity are Naviculaceae (19 genera/156 species), Pinnulariaceae (4/153), Bacillariaceae (4/149), Desmidiaceae (19/125), and Fragilariaceae (15/114)

Table 11.2 Summary Statistics for the Mushrooms of Mongolia

Division	Classes	Orders	Families	Genera	Species and infraspecies	% of total
Ascomycota	3	4	6	10	12	2.09
Basidiomycota	4	16	50	146	562	97.91
Total	**7**	**20**	**56**	**156**	**574**	**100.0**

Table 11.3 The 10 Largest Families and Genera of Mushrooms of Mongolia

Family	Genera	Species	Genera	Species
Russulaceae	2	39	*Russula*	22
Agaricaceae	14	38	*Lactarius*	17
Tricholomataceae	13	34	*Cortinarius*	17
Polyporaceae	15	27	*Suillus*	9
Strophariaceae	8	21	*Tricholoma*	9
Cortinariaceae	2	18	*Clitocybe*	9
Hygrophoraceae	5	15	*Hygrocybe*	7
Marasmiaceae	6	14	*Agaricus*	7
Suillaceae	2	12	*Pleurotus*	6
Hymenochaetaceae	6	11	*Amanita*	6

and largest 5 genera are *Navicula* (127 species), *Pinnularia* (121), *Nitzschia* (98), *Gomphonema* (79), and *Eunotia* (61) (Table 11.4).

11.3.3 Lichens

A total of 1031 species and infraspecific taxa of lichens distributed in 207 genera and 1 division. The largest 10 families with highest diversity are Parmeliaceae (34 genera/176 species), Physciaceae (11/95), Lecanoraceae (9/81), Verrucariaceae (11/70), Acarosporaceae (4/58), and largest 10 genera are *Cladonia* (57 species), *Lecanora* (48), *Peltigera* (25), *Melanelia* (19), and *Usnea* (15) (Table 11.5).

Table 11.4 The 10 Largest Families and Genera of Algae of Mongolia

Family	Genera	Species	Genera	Species
Naviculaceae	19	156	*Navicula*	127
Pinnulariaceae	4	153	*Pinnularia*	121
Bacillariaceae	4	149	*Nitzschia*	98
Desmidiaceae	19	125	*Gomphonema*	79
Fragilariaceae	15	114	*Eunotia*	61
Gomphonemataceae	4	105	*Cymbella*	51
Cymbellaceae	7	102	*Cosmarium*	45
Pleurosigmataceae	4	85	*Stauroneis*	42
Eunotiaceae	1	61	*Surirella*	38
Phormidiaceae	8	61	*Phormidium*	38

Table 11.5 The 10 Largest Families and Genera of Lichens of Mongolia

Family	Genera	Species	Genera	Species
Parmeliaceae	34	176	*Cladonia*	57
Physciaceae	11	95	*Lecanora*	48
Lecanoraceae	9	81	*Peltigera*	25
Verrucariaceae	11	70	*Melanelia*	19
Acarosporaceae	4	58	*Usnea*	15
Teloschictaceae	4	49	*Hypogymnia*	11
Peltigeraceae	3	45	*Xanthoparmelia*	10
Collemataceae	2	32	*Evernia*	7
Bacidiaceae	6	24	*Parmelia*	7
Lichenaceae	13	21	*Flavopunctelia*	2

11.3.4 Bryophytes

Currently, a total of 580 species and infraspecific taxa of mosses in 208 genera and 1 division were recorded. The largest 5 families with highest moss diversity are Pottiaceae (25 genera/82 species), Dicranaceae (16/47), Grimmiaceae (6/37), Bryaceae (4/36), and Amblystegiaceae (13/25) and largest 5 genera are *Bryum* (31 species), *Didymodon* (22), *Grimmia* (18), *Sphagnum* (17), and *Dicranum* (16) (Table 11.6).

11.3.5 Vascular Plants

Since research and educational organizations such as the National University of Mongolia (1942), Mongolian Academy of Sciences (1961), and Institute of Botany (1974) were founded and the Mongolian-Russian Biological Complex Expedition started in 1970, the research of the flora was intensified. The number of known flora species increased to 2,239 in Grubov's key to the vascular plants in 1982 and to 2,823 in Gubanov's conspectus in 1996. Floristic research relevant to the area of the present-day Mongolian Republic (Grubov 1955, 1982; Ulziykhutag 1989; Gubanov 1996). Since then, most notable milestones of the late 21st century were checklists of the Mongolian vascular flora by Sanchir et al. (2004), Kamelin and Ulziykhutag (2005), Ulziykhutag et al. (2009), Ganbold (2010), Dariimaa et al. (2010, 2014), Oyuntsetseg and Urgamal (2014), Urgamal et al. (2013, 2014), Urgamal (2016), and others (Table 11.7).

Table 11.6 The 10 Largest Families and Genera of Mosses of Mongolia

Family	Genera	Species	Genera	Species
Pottiaceae	25	82	*Bryum*	31
Dicranaceae	16	47	*Didymodon*	22
Grimmiaceae	6	37	*Grimmia*	18
Bryaceae	4	36	*Sphagnum*	17
Amblystegiaceae	13	25	*Dicranum*	16
Lophaziaceae	8	23	*Scapania*	13
Hypnaceae	11	22	*Pohlia*	13
Brachythiceae	7	20	*Hypnum*	11
Sphagnaceae	1	17	*Brachythecium*	11
Mniaceae	6	17	*Syntrichia*	10

Table 11.7 Development of the Number of Vascular Plant Taxa Known From
Mongolia

Sources and references	Families	Genera	Species and infraspecific taxa
Grubov (1955).	97	555	1897
Grubov (1982)	103	599	2239
Ulziykhutag (1989)	122	625	2443
Gubanov (1996).	128	662	2823
Sanchir et al. (2004)	128	669.	2871
Kamelin and Ulziykhutag (2005)	130	~700	2930
Ulziykhutag et al. (2009)	134	~700	2946
Dariimaa et al. (2010)	130	–	2950
Virtual Flora of Mongolia (Flora GREIF,Germany, 2013)	128	669	2879
Oyuntsetseg & Urgamal (National convention, 2013)	*112	676	3014
Urgamal, Oyuntsetseg & Nyambayar (2013)	*112	679	3053
Urgamal & Sanchir (as of April 2014)	*112	683	3113
Conspectus of the vascular plants of Mongolia (2014)	*112	683	3127
Urgamal (2016)	^108	683	3127

* According to APG III (2009).
^ According to APG IV (2016).

The last complete update of Mongolian flora appeared in "Conspectus of the Vascular Plants of Mongolia" written by Urgamal et al. (2014). This book includes more than 400 new taxa, added in the last 20 years and about 480 new nomenclatural combinations, made according to the International Code of Nomenclature for algae, fungi, and plants (ICN, 2012). Mongolian flora now includes 3,127 taxa (updated with 131 subspecies and 32 varieties) of vascular plants, distributed over 683 genera, 108 families, 39 orders, 12 classes or clades, 5 divisions, and 3 superclades (*Ferns, Gymnospermae and Angiospermae*) (APG IV, 2016; Urgamal, 2016). Magnoliophyta make up the majority (97.72 %) of the vascular flora recorded (Table 11.8).

The most recent additions to the species, families and genera are Asteraceae (80 species), Fabaceae (54), Brassicaceae (30) families and *Astragalus* (21), *Taraxacum* (20), *Potentilla* (17) genera.

The 10 largest families as well as the 10 largest genera of the Mongolian vascular plant flora are listed in Table 11.9 (Urgamal et al., 2014). The

Table 11.8 Statistics for the Vascular Flora of Mongolia (Urgamal, 2016)

Subkingdom	Division	Class	Order	Family	Genus	Species	% of total
Fern	Lycopodiophyta	2	2	2	4	7	0.22
	Pteridophyta	3	3	6	18	42	1.34
Gymnospermae	Pinophyta	1	1	2	5	13	0.41
	Gnetophyta	1	1	1	1	9	0.28
Angiospermae	Magnoliophyta	5	32	97	655	3,056	97.72
	Total	**12**	**39**	**112**	**683**	**3,127**	**100.0**

Table 11.9 The 10 Largest Genera and Families to the Vascular Flora

Family	Species	% of total	Genus	Species	% of total
Asteraceae	478	15.28	*Astragalus*	132	4.22
Fabaceae	356	11.38	*Artemisia*	104	3.32
Poaceae	259	8.28	*Oxytropis*	99	3.16
Rosaceae	161	5.14	*Carex*	92	2.94
Brassicaceae	160	5.11	*Potentilla*	73	2.33
Ranunculaceae	138	4.41	*Taraxacum*	57	1.82
Cyperaceae	132	4.22	*Saussurea*	53	1.69
Amaranthaceae	105	3.35	*Allium*	52	1.66
Lamiaceae	103	3.29	*Salix*	43	1.37
Caryophyllaceae	97	3.10	*Pedicularis*	36	1.15

Asteraceae are by far the largest family with 478 species, followed by the Fabaceae (356 species), Poaceae (259), Rosaceae (161), and Brassicaceae (160). The largest 5 genera are *Astragalus* (132 species), *Artemisia* (104), *Oxytropis* (99), *Carex* (92), and *Potentilla* (73).

11.3.6 Conservation Measures

Currently, there are 151 (4.82% of Mongolian flora) known endemic species of vascular plants in Mongolia, 458 species of subendemic, 133 species of very rare, 356 species of rare, 35 species of alien plants, 438 species of antropophilus plants, 143 species of aquatic plants, 70 species of relict plants, 47 species of wilding crops, 135 species in Mongolian Red Book, 148 species of Mongolian Plant Red List, and 8 species of in II annex of CITES. A total of 164 species are endemic to Mongolia including 1 species of mushroom, 5 species of algae, 5 species of lichens, and 153 species of vascular plants (Urgamal et al., 2016) (Table 11.10). The percentages of endemic species (POES) for each group in the country are 0.1% for mushrooms, algae, lichens and bryophytes and 2.1% for vascular plants (Urgamal et al., 2016).

11.4 Status, Trends, and Threats of Animals

11.4.1 Insects

11.4.1.1 Species Composition and Status

Insects are spread throughout the vast regions of forest steppe (cool temperate forests, alpine mountains, meadows, high mountain steppe, cool coniferous forests, and mixed forests), steppe (arid steppe, semi-arid steppe, meadows, and desert steppe), and desert ecosystems as well as wetlands, lakes, rivers and springs are yet to be fully researched. According to the data at the Entomology Laboratory of the Mongolian Academy of Sciences, there are about 13,000 species and 26 orders of insects known in Mongolia (Table 11.11). However, the biology and ecology of most of these insects have not been extensively studied.

Table 11.10 Plant Conservation Groups to the Vascular Plants to the Flora

Groups	Families	Genera	Species	Sources
Endemic plants	25	64	151	Urgamal et al., 2015
Subendemic plants	39	157	458	Urgamal et al., 2014
Very rare plants	56	108	133	Law of Natural Plants, 1995
Rare plants	64	211	356	Law of Natural Plants, 1995
Alien plants	14	33	35	Urgamal et al., 2014
Antropophilus plants	49	212	438	Tserenbaljid, 2002
Aquatic plants	31	50	143	Dulmaa, 2003
Relict plants	23	36	70	Urgamal et al., 2014
Wilding crops	12	39	47	Urgamal et al., 2014
Mongolian Red Book	49	103	135	Mongolian Red Book, 2013
Mongolian Plant Red List	53	116	148	Nyambayar et al., 2011
Appendix II of CITES	2	6	8	CITES, 2008

Table 11.11   Species Composition of Insect Groups in Mongolia

Name of the groups	Numbers of species	Name of the groups	Numbers of species
Springtail (Collembola)	48	Thrips (Thysanoptera)	87
Bristletails (Thysanura)	10	Beetles (Coleoptera)	>2,900
Dragonfly (Odonata)	68	Dobsonfly (Megaloptera)	2
Mayfly (Ephemeroptera)	96	Twisted winged-parasites (Strepsiptera)	2
Stonefly (Plecoptera)	53	Snakefly (Rhaphidioptera)	7
Earwig (Dermaptera)	5	Net winged insects (Neuroptera)	86
Cockroaches (Blattoptera)	3	Scorpion fly (Mecoptera)	1
Mantis (Mantodea)	5	Caddisfly (Trichoptera)	198
Grasshoppers and crickets (Orthoptera)	152	Butterflyand moth (Lepidoptera)	>1,550
Rubbed wings (Psocoptera)	39	Fly (Diptera)	>2,400
Mallophaga (Malloptera)	148	Fleas (Siphonaptera)	134
Sucking louse (Phthiraptera)	19	Membranous wings (Hymenoptera)	2,100
True bugs and homopterans (Hemiptera)	>1,500		

Joint long-term ecological studies by the University of Goettingen, Germany and the National University of Mongolia have been conducted at Yeröö River, Northern Hentii since 1996. The study results on stonefly (Plecoptera), butterfly and moth (Lepidoptera), grasshopper (Orthoptera), true bug (Hymenoptera), dragonfly (Odonata), caddisfly (Trichoptera), mayfly (Ephemeroptera), dobsonfly (Megaloptera), fly (Diptera) and beetle (Coleoptera) including 154 species of butterfly and 285 species of moths were published in many journals (Gantigmaa, 2005).

Recent extensive surveys on some groups of insects determined 34 species belonging to *Colletes* genus of bees (Apoidea) (Proshchalykin and Kuhlmann, 2015) and 100 species belonging to 26 genera in 4 subfamilies of wasps (Hymenoptera, Vespidae) (Buyanjargal, 2016). Over the last decade an American and Mongolian joint team conducted field surveys on aquatic invertebrates in rivers, springs, lakes and ponds in the Central Asian Internal and Arctic Ocean Drainage Basins. During this expedition American, Mongolian and Russian scientists recorded 96 species of mayfly belonging to 29 genera and 14 families in Mongolia (Soldan et al., 2009). This project also extend their survey topics to fly (Diptera) and described a new species of fly in Mongolia (Gelhausat et al., 2000; Gelhaus and Podenas, 2006). An updated checklist of the 198 caddisfly species belonging to 69 genera in 16 families has been produced (Chuluunbat et al., 2016). Some results of the taxonomic studies on several insect groups in Mongolia, such as ground beetle (Carabidae), water scavenger beetles (Helophoridae), lamellicorn (Scarabaeidae), predaceous diving beetles (Dytiscidae) and darkling beetles (Tenebrionidae) are summarized in Table 11.12 (Anichtchenko et al., 2007–2017; Enkhnasan, 2007).

A detailed study on forest insects conducted in North Hentii Mountain Range has resulted in illustrated distribution maps of *Carabus canaliculatus, C. arcensi, C. billbergi, C/Morphocarabus hummeli, Pterostichus/Metallophilus interruptus, Pterostichus dauricus, P. eximius, Cicindela gracilis,* (Carabidae, Coleoptera), *Xylotrechus rusticus, Monochamus urussovi, Saperda scalaris, Xylotrechus ibex,* and *Xylotrechus hircus* (Cerambycidae, Coleoptera) (Gantigmaa et al., 2012; Buyanjargal and Namkhaidorj, 2010).

During the study on insects and invertebrates in the Gobi Desert, species such as *Bryodema gebleri, Bryodema mongolicum, Calliptamus barbarus, Compsorhipis bryodemoides, C. davidiana, Gryllus desertus, Leptopternis gracilis, Mongolotmethis kozlovi, Zichya baranovi,* and *Zichya* sp. were found in large numbers. Field surveys on arachnids (Arachnida), crustaceans (Branchiopoda), annelid worms (Clitellata), rotifers (Eurotatoria), gastropods (Gastropoda) and leeches (Hirudinea) resulted in the documentation of 30 species belonging to 9 classes in Khanbogd Mountain, Khanbogd

Table 11.12 Diversity of Intensively Studied Insect Groups in Mongolia

S.No.	Name of insect orders	Number of species
1	Beetles (Coleoptera): Carabidae	372
2	Tenebrionidae	202
3	Scarabaeidae	171
4	Dytiscidae	>88
5	Helophoridae	~25
6	Hydraenidae	18
7	Haliplidae	17
8	Gyrinidae	7
9	Hydrophilidae	14
10	Cerambycidae	134

soum, Galba Gobi, Undain River and Nariin Sukhait valley and Nemetgei Mountain Range in the Gobi Desert (Gantigmaa et al., 2012). Arachnid surveys were conducted near Dayan Nuur of Bayan-Ölgii Province and found 53 species from 22 genera of 11 families of which 9 species (*Larinioides cornutus* (Araneidae), *Phrurolithus festivus* (Corinnidae), *Acantholycosa altaiensis*, *Alopecosa pictilis*, *Pardosa albomaculata*, *P. palustrus*, *P. tesquorumoides*, *P. uintana* (Lycosidae) and *Xysticus obscurus* (Thomisidae) were new for Mongolian fauna (Azjargal, 2013).

11.4.1.2 Threats and Conservation Issues

The development of infrastructure, mining, and exploration activities are one of the threats to various insect species. Habitat degradation and loss caused by overgrazing in the Gobi and other desert regions is also occurring (Dash et al., 2003).

There is no significant use of insects in traditional or modern Mongolian culture and so there is a very limited direct use of insects. A limited number of insects are used, for example, the Darkling Beetle (*Blaps* sp., Tenebrionidae), which secretes a bio-lipid for self-protection. This secretion is used for traditional medicines and treatments (Gunbilig et al., 2009; Aldarmaa et al., 2010). For respiratory illnesses, medicinal treatments derived from *Bryodema, Angaracris,* and *Caliptamus* species of grasshoppers are used (Ganbold, 1993). In recent years, taxonomical and biological studies of insects have been intensively performed in association with advanced technology for controlling pest insects and biological technology.

11.4.2 Fish

11.4.2.1 Species Composition and Status

There are 64 species of fish recorded in Mongolia (Ocock et al., 2006). These species inhabit three different river drainage basins across Mongolia—The Pacific, Central Asian Internal, and Arctic Ocean drainage basins. Of these three, the Pacific drainage basin has the greatest diversity of fish. The Central Asian Internal drainage basin has unique fish species due to it being significantly isolated for a long period of time. The Mongolian Law on Fauna (2000) covers fish species and lists the Amur Sturgeon (*Acipenser schrenckii*) and Tench (*Tinca tinca*) as a very rare and the eastern Brook Lamprey (*Lethenteron reissneri*) and Taimen (*Hucho taimen*) as rare (in accordance with Government decree number 7, Appendix 1 passed in 2012).

Mongolian and international experts assessed all occurring fish species in Mongolia using the IUCN criteria and categories (IUCN, 2003) in 2006. According to the assessment of 64 species of fish, 16 species are listed as Not Applicable due to lack of data and information. While a quarter of Mongolian fish species are considered to be Least Concern, 23% are considered as Threatened (categories Critically Endangered, Endangered and Vulnerable). A further 6% are listed as Near Threatened (NT) (Ocock et al., 2006) (Figure 11.4).

Based on the assessment, one species is listed as Critically Endangered (Siberian Sturgeon *Acipenser baerii*), 6 species listed as Endangered (Gobi Loach *Barbatula dgebuadzei*, Dzungarian Dace *Leuciscus dzungaricus*, Pidschian *Coregonus pidschian*, Amur Grayling *Thymallus grubei*, Hövsgöl Grayling *T. nigrescens*, Taimen *Hucho taimen*), 4 species listed as Vulnerable (Lake Osman *Oreoleuciscus angusticephalus*, Small Osman *Oreoleuciscus humilis*, Mongolian Grayling *Thymallus brevirostris* and Lenok *Brachymystax lenok*), and 3 species listed as Near Threatened (Amur Bitterling *Acheilognathus asmussi*, Ide *Leuciscus idus*, and Arctic Grayling *Thymallus arcticus*) (Ocock et al., 2006).

The Amur River basin in the Pacific drainage basin has the highest levels of species richness, particularly around the Onon River basin and Buir Lake, where up to 31 species are present. The next ranked area of species richness

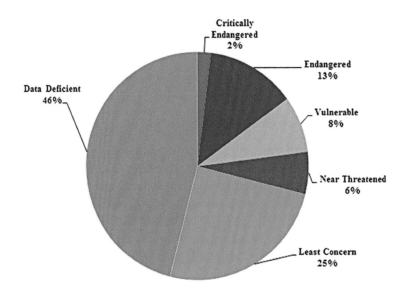

Figure 11.4 Status of fish in Mongolia.

is the Arctic Ocean drainage basin. Western Mongolia and the Gobi Valley of the Lakes in the Central Asian Internal drainage basin have the lowest richness of fishes. However, the area contains a significant number of threatened species of fish, such as Mongolian Grayling, Lake or Potanin's Osman, Dzungarian Dace, and Gobi Loach.

11.4.2.2   Threats and Conservation Issues

Ocock et al. (2006) showed that intentional mortality was the primary threat to the fish of Mongolia. Fishing is a dominant threat to fishes because high levels of unregulated and illegal fishing are being carried out across Mongolia. Fishing has become more popular in recent years among amateurs. Fishing near towns and cities has decreased the stock of Taimen, Lenok, Grayling, and Whitefish. There is a lack of monitoring in the implementation of the law on illegal fishing, therefore, it is difficult to evaluate the damages accurately. There is a view of threats indicated that local demand plays only a small part in the commercial nature of the threat, with most of the fishes sold to foreign markets, predominantly China but also Russia (Ocock et al., 2006).

Habitat degradation caused by gold mining and overgrazing is the primary threat to some threatened species and is the secondary threat for over half of the threatened species. Water pollution associated with mining, livestock grazing, and untreated sewage, was further considered as the third most dominant threat to all fishes. Due to mining and industrial operations, the Tuul River near Zaamar has been polluted, and 70 cm of slime and mud have settled on the bottom, slowing the flow. This has resulted in the extinction of rhyofite insects such as Plecoptera and Ephemeroptera in the river. There are also fewer observations of Ephemerella and Baetis insects. Such changes to the Tuul River are affecting the breeding habits of Taimen, Grayling and Lenok. During a monitoring study at Zaamar, not a single fish from the species mentioned above were caught (Gombobaatar et al., 2014).

Climate change is the primary threat to two threatened species and represents a significant future threat for all fish species in Mongolia. Many rivers, lakes, and springs have dried out due to the climate change and global warming, thus having a negative impact on the distribution and population of fish.

The tourism sector is developing rapidly in Mongolia with a trend towards fishing. Today, tourists from all over the globe are coming to fish in Mongolia, and the number of tour operators offering such packages has increased. The combined threats of fishing, global warming and pollution pose significant challenges for the survival of Mongolian fish (Ocock et al., 2006).

In order to reduce illegal fishing, private sectors and NGO's have established fish breeding facilities. The Mongolian Pisciculture Center was established under the decree of the government to conserve the fish stock and increase the population. This will ensure the sustainable growth of the populations (Gombobaatar et al., 2014).

11.4.3 Amphibians

11.4.3.1 Species composition and status

Currently, there are 6 species of 4 families of 2 orders of amphibians in Mongolia (Table 11.13 and Figure 11.5).

Comparatively, low numbers of amphibian species in Mongolia are a result of the geographical location and continental climate of this country. These few species of amphibians are distributed throughout Mongolia. However, in the arid Gobi regions or the cool Altai mountain region, there are very few amphibian species. Most species of amphibians are found in the north and northeastern regions such as the Khangai Mountain Range, the Hövsgöl, Mongol Daguur Steppe, Middle Khalkh Steppe, Eastern Mongolia, Ikh Khyangan Mountain Range, and at greatest concentrations in the Hentii Mountain Range. The Hentii Mountain Range has the highest greatest number of amphibian species in Mongolia (Terbish et al., 2006) (Figure 11.6). The majority of amphibian species are assessed as Vulnerable (67%). This includes Pevzow's Toad (*Bufo pewzowi*) and the Siberian Salamander (*Salamandrella keyserlingii*). The remaining 33% are categorized as Least Concern (Terbish et al., 2006).

Table 11.13 Species Composition of Amphibians

Order	Family	Species
Caudata	Hynobiidae	Siberian Salamander (*Salamandrella keyserlingii*)
Anura	Bufonidae	Mongolian Toad (*Bufo raddei*)
		Pewzow's Toad (*Bufo pewzovi*)
	Hylidae	Far Eastern Tree Frog (*Hyla japonica*)
	Ranidae	Siberian Wood Frog (*Rana amurensis*)
		Asiatic Grass Frog (*Rana chensinensis*)

Figure 11.5 Mongolian Toad (*Bufo raddei*). (Photo credit: Gombobaatar Sundev and Mongolica Publishing).

11.4.3.2 Threats and Conservation Issues

Impacts from climate change, global warming, and human activities such as pollution, disturbance, overgrazing, and infrastructure development are threatening the populations of amphibians in Mongolia. Clear evidence of these impacts is the decline of the population of Siberian salamander in the Tuul River basin near Ulaanbaatar, where the Mongolian Toad has not been recorded in recent years. The drought of large wetlands including Taatsiin Tsagaan, Ulaan, and Adgiin Tsagaan Lakes in the Gobi became a cause of the decline in the population and distribution of the Mongolian Toad. The same situation was documented in the Orkhon-Selenge basin, where almost 67% of the amphibians in Mongolia inhabit. Researchers of a Mongolian-Russian joint biological expedition studied two water reservoirs near Shaamar, where many amphibians are found, from 1983–1984. The population of the 4 species at these water reservoirs was 880 per 1,000 m² and the juvenile population was 10,800 (Borkin et al., 1988). However, 24 years later, in

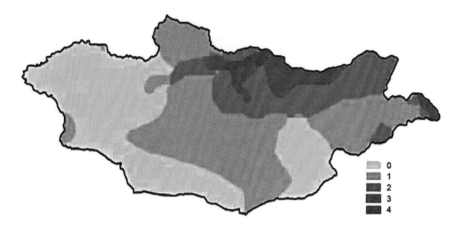

Figure 11.6 Species richness of amphibians in Mongolia (Terbish et al., 2006).

2008, a repeat study at the site showed that the first group of lakes had dried, therefore, no signs of amphibians inhabiting the area were found. There are also impacts of water pollutants. An albino species was found amongst the Siberian Wood Frog population at Shatan River which pours into Kharaa at Batsumber Sum of Töv Province (Munkhbaatar, 2008). The pH level of the marshes of these species inhabiting is 8.3. The same situation was observed at Mankhaadai Lake in Dadal Sum of Hentii Province (Munkhbaatar and Terbish, 2008).

11.4.4 Reptiles

11.4.4.1 Species Composition and Status

Based on the research studies (Terbish et al., 2006), there are 22 species of 7 families of 2 sub-order reptiles in Mongolia (Table 11.14 and Figure 11.7).

True lizards (Lacertidae) and snakes (Colubridae) make up 55% of reptile species in Mongolia. Reptiles in Mongolia are mainly distributed in the Gobi region, with warmer and arid climate. National and international experts assessed all reptile species in Mongolia based on IUCN criteria and categories in Mongolia in 2006. The assessment results show that Gobi Naked-toad Gecko and Adder were listed as Vulnerable (Terbish et al., 2006). Przewalski's Wonder Gecko, Mongolian Agama, Tatary Sand Boa, Slender Racer, and European Grass Snake were categorized as Near Threatened. However,

Table 11.14 Species Composition of Reptiles in Mongolia

Suborder	Family	Species
Lizard (Lacertilia)	Geckos (Gekkonidae)	1. Kaspischer Even-fingered Gecko or Squeaky Pygmy Gecko (*Alsophylax pipiens*)
		2. Gobi Naked-toed Gecko (*Cyrtopodion elongatus*)
		3. Przewalski's Wonder Gecko or Plate-tailed Gecko (*Teratoscincus przewalskii*)
	Agamids (Agamidae)	4. Sunwatcher Toad-head Agama (*Phrynocephalus helioscopus*)
		5. Tuva toad-head Agama (*Phrynocephalus versicolor*)
		6. Mongolian Agama or Mongolian Rock Agama (*Paralaudakia stoliczkana*)
	True lizards (Lacertidae)	7. Mongolian Racerunner (*Eremias argus*)
		8. Stepperunner or Arguta (*Eremias arguta*)
		9. Multi-oscillated Racerunner (*Eremias multiocellata*)
		10. Gobi Racerunner (*Eremias przewalskii*)
		11. Variegated Racerunner (*Eremias vermiculata*)
		12. Viviparous Lizard or Common Lizard (*Zootoca vivipara*)
		13. Sand Lizard (*Lacerta agilis*)
Snakes (Serpentes)	Non-venomous snakes (Boidae)	14. Tatar Sand Boa (*Eryx tataricus*)
	Black Snakes (Colubridae)	15. Slender Racer (*Coluber spinalis*)
		16. Steppes Ratsnake (*Elaphe dione*)
		17. Amur Ratsnake (*Elaphe schrenckii*)
		18. European Grass Snake (*Natrix natrix*)
		19. Steppe Ribbon Racer (*Psammophis lineolatus*)
	Pit vipers (Crotalidae)	20. Halys Pit Viper (*Gloydius halys*)
	Venomous Snakes (Viperidae)	21. Northern Viper (*Vipera berus*)
		22. Meadow Viper (*Vipera ursinii*)

due to insufficient information, Stepperunner, Sand Lizard, and Amur Rat Snake were listed as Data Deficient.

Reptile species richness is unequal across the country with a majority of the species occurring either in the southwest or south of the country (Figure 11.8) (Terbish et al., 2006). Relatively, large numbers of species occur in the west

Figure 11.7 Mongolian Racerunner (*Eremias argus*). (Photo credit: Gombobaatar Sundev and Mongolica Publishing).

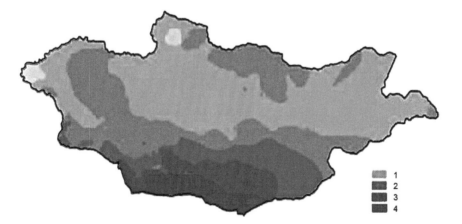

Figure 11.8 Species richness of reptiles in Mongolia (Terbish et al., 2006).

(Great Lakes Depression and the western Khangai Mountain Range) of the country. The greatest species richness areas are the Trans-Altai Gobi Desert, the Gobi-Altai Mountain Range, Alashan Gobi Desert, and the Eastern Gobi.

11.4.4.2   Threats and Conservation Issues

There are almost no instances of using reptiles for food, medicine, commercial or trade purposes in Mongolia. Therefore, there is very little direct anthropogenic threat to reptiles in the country. However, there are few indirect threats to the habitat, distribution, and population of reptiles caused by human activities. Habitat degradation and loss caused by overgrazing, desertification, wildfires, exploration for natural resources, and land erosion are threats, although there is a very little information on how these threats are affecting the reptiles. Many reptile species occur in southern Mongolia, where droughts and dry conditions commonly occur. However, it remains unclear if these represent natural environmental changes or are driven by anthropogenic activity (Terbish et al., 2006).

A total of 6 species of reptiles (Gobi Naked-toed Gecko, Przewalski's Wonder Gecko, Stepperunner, Tatar Sand Boa, Slender Racer, and Northern Viper) are listed in the Mongolian rare species list (Government decree number 7, appendix 1, 2012) as well as the Mongolian red list. These species make up 27.2% of the total reptiles in Mongolia. A detailed study on the distribution, biology, ecology, population dynamics of reptiles in Mongolia and implementing a monitoring system is vital to the conservation of reptile species in this country.

11.4.5   Birds

11.4.5.1   Species Composition and Status

In Mongolia, 491 species of birds are recorded (Gombobaatar and Bayanmunkh, 2016). Out of those species, 6% are winter visitors, 11% vagrants, 32% breeding visitors, 40% passage migrant, and 11% resident breeder. From this calculation, we can conclude that Mongolia is an important breeding and stopover site for many migrants from Siberian breeding grounds to South-East Asian wintering ground.

The National Bird Red List Workshop was organized in 2009, where many issues facing bird species found in Mongolia were discussed. National and international researchers and experts assessed every bird species using the IUCN Red List criteria. Based on this assessment, 9.3% of species are categorized as Threatened. This also includes species that are near threatened. Out of all the species, 7.7% are Data Deficient, and 83% are Least Concern, while 18.3% are Not Applicable by IUCN Red List criteria (Table 11.15).

Table 11.15 Threatened Species of Birds in Mongolia

Critically Endangered (0.5%)	Endangered (1.5%)	Vulnerable (3.2%)	Near Threatened (4.1%)
Dalmatian Pelican (*Pelecanus crispus*), Siberian Crane (*Grus leucogeranus*)	White-headed Duck (*Oxyura leucocephala*), Relict Gull (*Larus relictus*), Greater Spotted Eagle (*Aquila clanga*), Pallas's Fish-eagle (*Haliaeetus leucoryphus*), Short-toed Snake-eagle (*Circaetus gallicus*), Reed Parrotbill (*Paradoxornis heudei*)	Lesser White-fronted Goose (*Anser erythropus*), Baikal Teal (*Anas formosa*), Ferruginous Duck (*Aythya nyroca*), Lammergeier (*Gypaetus barbatus*), Eastern Imperial Eagle (*Aquila heliaca*), Saker Falcon (*Falco cherrug*), White-naped Crane (*Grus vipio*), Hooded Crane (*Grus monacha*), Asian Dowitcher (*Limnodromus semipalmatus*), Great Bustard (*Otis tarda*), Houbara Bustard (*Chlamydotis undulata*), Mongolian Ground-jay (*Podoces hendersoni*)	Great Bittern (*Botaurus stellaris*), Little Bittern (*Ixobrychus minutes*), Purple Heron (*Ardea purpurea*), Greater White-fronted Goose (*Anser albifrons*), Swan Goose (*Anser cygnoides*), Mute Swan (*Cygnus olor*), Falcated Duck (*Anas falcata*), White-tailed Eagle (*Haliaeetus albicilla*), Altai Snowcock (*Tetraogallus altaicus*), Common Pheasant (*Phasianus colchicus*), Common Crane (*Grus grus*), Tree Pipit (*Anthus trivialis*), White-throated Bushchat (*Saxicola insignis*), Saxaul Sparrow (*Passer ammodendri*), Yellow-breasted Bunting (*Emberiza aureola*), Ochre-rumped Bunting (*Emberiza yessoensis*)

Globally threatened species include Dalmatian Pelican, Greater White-fronted Goose, Swan Goose, White-headed Duck, Baikal Teal, Greater Spotted Eagle, Pallas's Fish Eagle, Eastern Imperial Eagle, Saker Falcon, Siberian Crane, White-naped Crane, Hooded Crane, Great Bustard, Houbara Bustard, Relict Gull, White-throated Bushchat, and Yellow-breasted Bunting.

The species richness of Mongolian birds varies depending on the regions and habitats. Landscapes such as Mongol Daguur, Dornod Mongol Steppe, Ikh Khyangan, Khangai, Great Lakes Depression, Darkhad Valley, and Hövsgöl Lake have high species richness; whereas Gobi districts and desert steppe landscapes are low species richness. Species richness is high in most parts of Mongolia because it is on the junction of Central, Western, and Eastern Asian Flyways (Figure 11.9) (Gombobaatar et al., 2011, 2011a).

11.4.5.2   Threats and Conservation Issues

Threats were evaluated using the IUCN Red List threat criteria as well as conservation status criteria. Based on the evaluation, the largest threats to

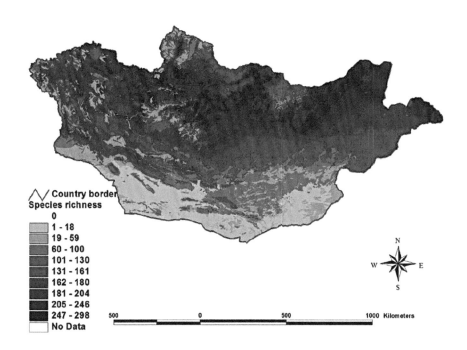

Figure 11.9   Species richness of birds of Mongolia.

Mongolian birds were habitat loss and degradation at 38.1%, human disturbance at 13.6%, pollution at 11%, and changes in native species dynamics at
10.7% (Gombobaatar et al., 2011, 2011a).

Habitat loss and degradation due to mining and infrastructure developments, the establishment of tourist camps, and spread of wildfires in the fall
and autumn must be considered for the conservation actions. Altai Snowcock, Black Kite, and Mongolian Ground-jay are being hunted for exhibit
purposes as well as medicinal uses.

A national and international research team in 2010 found that many
birds are fatally electrocuted as a result of the design and construction of
15 Kilovolt power lines. To alleviate this threat, organizations such as the
Mongolian Ornithological Society, Ministry of Environment and Tourism,
National University of Mongolia, and the Mongolian Academy of Sciences
have been collaborating on action plans (Gombobaatar et al., 2011). Since
1994, Saker Falcons have been exported to Arabian countries for scientific
and commercial purposes. The former Ministry of Culture, Sport and Tourism together with scientists from the Mongolian Ornithological Society and
National University of Mongolia were successful in petitioning to proclaim
the falcon as the national bird of Mongolia in 2012 (Figure 11.10).

Figure 11.10   The National bird of Mongolia, Saker Falcon (*Falco cherrug*).
(Photo credit: Gombobaatar Sundev and Mongolica Publishing).

11.4.6 Mammals

11.4.6.1 Species Composition and Status

There are 128 species of mammal species recorded (Clark et al., 2006) (Figure 11.11).

In 2006, national and international experts and mammal biologists gathered together in Ulaanbaatar, Mongolia and assessed 128 species of mammals in Mongolia. The assessement result showed that 17% of mammal species are categorized as threatened (Table 11.16)

The Asiatic Wild Dog is regionally extinct. Almost 6% of the population is categorized as Near Threatened (Eurasian Red Squirrel (*Sciurus vulgaris*), Eurasian Lynx (*Lynx lynx*), Pallas's Cat (*Otocolobus manul*), Grey Wolf (*Canis lupus*), Corsac Fox (*Vulpes corsac*), Red Fox (*Vulpes vulpes*), Wild Boar (*Sus scrofa*), and Siberian Ibex (*Capra sibirica*)) (Clark et al., 2006).

Mammalian species richness is the highest in Mongolia's northern regions, with the greatest number of species in Hövsgöl and Hentii Mountain Ranges, and northern parts of Khangai Mountain Range (Clark et al.,

Figure 11.11 Przewalski's Wild Horse (*Equus przewalski*). (Photo credit: Gombobaatar Sundev and Mongolica Publishing).

Table 11.16 Threatened Species of Mammals in Mongolia

Critically Endangered (2%)	Endangered (11%)	Vulnerable (3%)
Gobi Bear (*Ursus arctos gobiensis*), Przewalski's Wild Horse (*Equus przewalski*), Red Deer (*Cervus elaphus*)	Mongolian Marmot (*Marmota sibirica*), Eurasian Beaver (*Castor fiber*), Alashan Ground Squirrel (*Spermophilus alashanicus*), Small Five-toad Jerboa (*Allactaga elater*), Mongolian Three-toed Jerboa (*Stylodipus sungorus*), Tamarisk Gerbil/Jird (*Meriones tamariscinus*), Snow Leopard (*Uncia uncia*), Asiatic Wild Ass (*Equus hemionus*), Bactrian Camel (*Camelus bactrianus ferus*), Argali Sheep (*Ovis ammon*), Mongolian White-tailed Gazelle (*Procapra gutturosa*), Saiga Antilope (*Saiga tatarica*), Musk Deer (*Moschus moschiferus*), Eurasian Elk (*Alces alces*)	Long-eared Jerboa (*Euchoreutes naso*), Sable (*Martes zibellina*), Black-tailed Gazelle (*Gazella subgutturosa*)

2006). Species richness tends to decrease from north to south Mongolia, with lowest species richness in southern and southeastern parts of the country, particularly Gobi-Altai Mountain Range, Alashan Gobi Desert and southern parts of Eastern Mongolia.

11.4.6.2   Threats and Conservation Issues

The assessment process identified the main activities or processes driving the decline of species and the direct threats causing these declines (e.g., loss of habitat or intentional mortality). Intentional mortality caused by hunting is the primary threat identified for more than half (57%) of Mongolia's threatened mammals. This is a particularly serious threat for ungulates such as Red Deer, Argali Sheep, Mongolian White-tailed Gazelle, and Musk Deer, as well as species hunted for their furs, such as Mongolian Marmot and Snow Leopard. Habitat degradation is believed to result from increasing numbers of livestock grazing. When combined, habitat loss, fragmentation, and degradation constitute a secondary threat to Mongolia's threatened mammals (Clark et al., 2006). Habitat loss is another serious threat to many species, including artiodactyls, commonly caused by increasing resource extraction such as mining. Other forms of resource extraction, such as logging and clear-felling of forests also threaten species such as Eurasian Beaver, Sable, and rodents. Climate change represents a threat to several taxonomic groups, including rodents, artiodactyls and carnivores, and is highlighted as an important threat for the future. Natural threats such as desertification, water shortage, and land degradation are affecting the landscape, which is changing year-to-year.

11.5   The National Biodiversity Strategic Action Plan

The first review of the National Biodiversity Action Plan in 2000 indicated that the NBAP was only partially filling its objectives due to there being issues with implementation; a lack of coordination, monitoring, and management meant there was a lack of detailed awareness of overall implementation of projects. The main result of the evaluations shows that the NBAP fulfilled its main objectives and it made its potential contribution to protect biodiversity in Mongolia. According to the evaluation, 46 activities were 90–100% complete, 39 activities 70–80% complete and 3 activities (0–20% completed) were ongoing or had just started being implemented. This indicated that 96.5% of the NBAP had been achieved since 1996 (Adiya et al., 2010).

In 2014, national experts concluded all achievements of the National Biodiversity Action Plan since 2000 and completed the 5th National Report.

Mongolia's 5th National Report follows the structure outlined by the guidelines of the CBD Secretariat. The main body of information for the report comes from: (a) written contributions from Mongolian taxonomic experts in the given fields in part one of the reports; (b) Mongolia's current NBAP; (c) written contributions and reports from Mongolian government ministries; (d) written contributions from NGOs and NFP organisations conducting work in fields of biodiversity conservation and development in Mongolia. Previous Mongolia's NBAP consists of 21 overall targets with a total of 87 activities, 96% of which have been carried out (Gombobaatar et al., 2014).

According to the assessment of the threats to the biodiversity of Mongolia, several points are highlighted in the following subsections.

11.5.1 Climate Change

The annual average temperature in Mongolia increased by 2.14°C between 1940 and 2008. Global averages for the period 1880–2012 increased by 0.85°C. The area of glaciers in Mongolia decreased by 12.3% during 1940–1990, 9.8% in 1990–2000 and 11.7% in 2000–2010, totaling 27.8% loss in the past 70 years with an accelerated loss noted in the past decade. These factors indicate that climate change is occurring rapidly in Mongolia, melting glaciers and permafrost (Gombobaatar et al., 2014).

11.5.2 Habitat Degradation

Due to the change in livestock herd composition, the goat population has increased rapidly to occupy 40% of herds and there is an excess of 32.5% or 16 million head of sheep over the advised national herd quota. This has significantly contributed to overgrazing and pasture land degradation. Forested areas have also seen a decline in size.

11.5.3 Desertification

The distribution of areas considered to suffer from severe and very severe desertification changed drastically, with new additions to the category of very severe. Calculations based on the methods outlined in the UN's Convention to Combat Desertification showed that approximately 90% of pasture land in Mongolia lies within a vulnerable region that is susceptible to desertification and land degradation. Overgrazing has become the main source of desertification and land degradation in Mongolia.

11.5.4 Pollution

Due to a growth in mining operations near large riverbeds, such as Ork-hon and Selenge, pollution has become a significant problem. The breeding migration route of the Omul fish has been shortened because of pollution caused by mining operations (Gombobaatar et al., 2014).

The Ministry of Environment and Tourism (formerly Ministry of Environment Green Development and Tourism) with collaborating partners completed the National Biodiversity Strategic Action Plan in 2014. Mongolia's National Biodiversity Strategic Action Plan has the vision of guaranteeing all citizens' "right to a healthy and safe environment and to be protected against environmental pollution and ecological imbalance" as defined by the Constitution of Mongolia. The Action Plan has 14 goals, 29 objectives, and 74 outputs within the frame of four strategies to ensure the conservation and sustainable use of Mongolia's biological diversity until 2025.

- Strategy 1: Increase awareness and knowledge of Biodiversity conservation and sustainable use among both decision makers and the general public (2 goals, 4 objectives, and 9 outputs).
- Strategy 2: Develop and implement science-based policy on conservation and sustainable use of biological resources (5 goals, 12 objectives, and 34 outputs).
- Strategy 3: Sustainable Use of Biodiversity (3 goals, 5 objectives, and 14 outputs).
- Strategy 4: Improve policies and legal environment for conservation and use of biological diversity and ecological services (4 goals, 8 objectives, and 17 outputs) (Batbold et al., 2015).

11.6 Endemics

More than 140 endemic vascular plant species are listed in Mongolia (Grubov, 1976, 1984, 1989; Gubanov, 1996; Ulziykhutag, 1989; Pyak et al.2008; Ganbold, 2010; Dariimaa, 2014; Urgamal et al., 2014) (Table 11.17).

At present, total of 151 species (included 14 subspecies) are endemic (4.82%) belonging to 64 genera and 25 families to Mongolia (Urgamal et al., 2015). The most of families and genera richest in endemic species are Fabaceae (46 species), Asteraceae (32 species), Rosaceae (17) families and *Astragalus* (21), *Oxytropis* (17), and *Potentilla* (12) genera (Table 11.18).

The highest species richness of endemics occur in Mongol-Altai (61 species), Khangai (45) and Gobi-Altai (44) regions (Table 11.19).

Table 11.17 The Numbers of the Endemic Vascular Plant Species Taxa Known
From Mongolia

Sources and references	Families	Genera	Species
Grubov (1976)	–	–	77
Ulziykhutag (1981)	–	–	74
Ulziykhutag (1989)	–	–	145
Gubanov (1996)	–	–	143
Urgamal et al. (2014)	26	65	153
Urgamal et al. (2015)	25	64	151

The endemic gymnosperms are not represented in the flora of Mongolia. The regional differences in endemic species richness are in part due to differences in habitat characteristics and diversity between the individual regions. Areas of greatest species richness is concentrated in the middle altitudinal belts of the mountain systems of Mongol-Altai, Gobi-Altai, and Khangai.

A complete list of the endemic species of animals has not been analyzed and published yet. This is a field that national and international experts should focus on near future. Of the insect species, *Compsorhipis bryodemoides*, *Mongolotmethis kozlovi*, and *Zichya baranovi* occur in Southern Mongolia as endemics.

11.7 Protected Areas

For Mongolia, the issue of special protected areas has been the core of government policy and it is the basis to ensure preservation and sustainable use of natural and cultural heritage and to maintain a sustainable ecological condition. Usually these protected areas are preferred habitats for rare species of animals and plants. Under the relevant laws of Mongolia, the special protected areas are divided into four categories: Strictly Protected Areas, National Parks, Nature Reserves, and Natural Monuments. As of 2017, there are 99 special protected areas covering a total of 27.2 million hectares of land in Mongolia, consisting of 20 strictly protected areas (12.4 million hectares), 32 National parks (11.88 million hectares), 34 Natural reserves (2.77 million hectares), and 13 Monuments (0.12 million hectares). As such, Mongolia has designated 17% of its territory as specially protected areas (Myagmarsuren and Namkhai, 2012) (Figure 11.12).

The Government of Mongolia has been taking step by step actions to include areas, which are crucial for maintaining the ecological integrity of

Table 11.18 The Endemic Species in the Genera and Families to the Flora

Family name	Genera number	Species number
Fabaceae	6	46
Asteraceae	13	32
Rosaceae	4	17
Brassicaceae	6	8
Ranunculaceae	6	7
Lamiaceae	5	6
Papaveraceae	2	6
Caryophyllaceae	3	5
Juncaceae	2	3
Plumbaginaceae	1	3
Amaranthaceae	1	2
Solanaceae	2	2
Plantaginaceae	1	2
Apocynaceae	1	1
Bignoniaceae	1	1
Campanulaceae	1	1
Caprifoliaceae	1	1
Cleomaceae	1	1
Euphorbiaceae	1	1
Gentianaceae.	1	1
Orobanchaceae	1	1
Polygonaceae	1	1
Rubiaceae	1	1
Scrophulariaceae	1	1
Zygophyllaceae	1	1
Total	**64**	**151**

regional and global importance, in the World Heritage sites and the World Network of Biosphere Reserves. As for Mongolia, the following six areas covering a total of 7,120,378 hectares of land are included in the World Network of Biosphere Reserves of UNESCO's Man and the Biosphere Programme (Table 11.20).

The protected area surrounding Uvs Lake basin has been registered as a World Heritage site, while Orkhonii Höndii rock paintings in the Altai

Table 11.19   The Richness of Endemic Species Distributed in the 16 Phyto-
Geographical Regions

Phyto-geographical region name	Total species	Percentage of total flora	Rank
Hövsgöl	15	0.47	VIII–IX
Hentii	15	0.47	VIII–IX
Khangai	45	1.43	II
Mongol Daguur	13	0.41	XI–XII
Foothills of Great Khyangan	4	0.12	XVI
Khovd	18	0.57	VI
Mongol-Altai	61	1.95	I
Middle Khalkh	12	0.38	XIII–XIV
Eastern Mongolia	13	0.41	XI–XII
Great Lakes Depression	21	0.70	V
Valley of the Lakes	12	0.38	XIII–XIV
Eastern Gobi	16	0.51	VII
Gobi-Altai	44	1.40	III
Dzungarian Gobi	32	1.02	IV
Trans-Altai Gobi	6	0.19	XV
Alashan Gobi	14	0.44	X

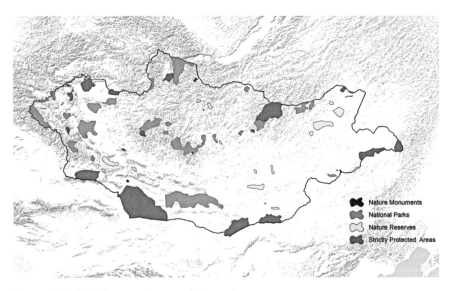

Figure 11.12   Protected areas of Mongolia.

Table 11.20 Protected Areas of Mongolia Included in the Man and the Biosphere Programme

Name of the protected area	Year	Area by ha
Great Gobi Strictly Protected Area	1990	5,560,412
Bogd Khan mountain Strictly Protected Area	1996	41,322
Uvs Nuur Strictly Protected Area	1997	771,700
Khustai Nuruu National Park	2004	48,884
Eastern Mongolian Steppe Strictly Protected Area	2005	589,906
Mongol Daguur Strictly Protected Area	2007	108,154

Mountains and Burkhan Khaldun Uul have been registered as cultural heritage sites under UNESCO's Convention Concerning the Protection of the World Cultural and Natural Heritage. Furthermore, seven areas in Mongolia, namely Mongol Daguur, Ganga Lake, Terhiin Tsagaan Lake, Uvs Lake, Khar-Us Lake, Airag Lake, and Achit Lake where water-birds occur in great numbers have been included in the list of the Ramsar Convention on Wetlands (Gombobaatar, 2012).

11.8 Main Directions of Government Policy Regarding Biodiversity

Mongolia has been implementing its government policy regarding biodiversity through relevant laws, regulations, and other national programs. For instance, the state policy directed towards the protection of biological diversity has been reflected in more than 30 laws, including the Law on Environmental protection (1995), the Law on Forests (1995), the Law on Natural Plants (1995), the Law on the Fees for Use of Natural Plants (1995), the Law on Payments for Hunting Resource Use and on Fees for Hunting and Trapping Authorization (1995), the Law on Protection of Plants (1996), and the Law on Fauna (2000). Furthermore, the adaptation of specialized legislature, such as the Mongolia Comprehensive National Development Strategy-based on the Millennium Development Goals (2008) approved by the Parliament of Mongolia, "National Program on Biodiversity" (2015), "Government Policy on Ecology" (1997), "National Program on Public Ecological Education" (1997), "Action Program for Conservation and Proper Use of Rare Plants of Mongolia" (2002),"Snow Leopard Conservation Policy of Mongolia" (2005), "National Program on Wild Mountain Sheep Conservation" (2002), and "National Action Plan for the Protection of Endangered and Threatened Species" (2011) have established a favorable legal environment.

The policy on biodiversity conservation is aimed at preserving pristine natural environments, protecting natural and cultural heritage and ensuring appropriate use, rehabilitation and just allocation of resources. The policy is therefore in line with sustainable economic development and the improvement of people's livelihoods.

The Mongolian Government is paying close attention to protect flora and fauna from adverse effects of climate change, natural disasters, human activities and diseases, and to maintain their gene pools, herd composition, natural habitat and preferred areas. In addition to protecting threatened species in their natural habitats and increasing their abundance, issues related to domestication and cultivation of extremely rare plants and economically beneficial plants and expanding the area of occurrence have been included in the abovementioned government policy documents. The sustainable harvesting of plant and animal species for use in international markets will be an essential part of the government policy on the protection and proper use of Mongolia's biological diversity.

11.9 Livestock and Its Products

Mongolia has a long tradition of breeding free ranging livestock. Most morphological and physiological features of the livestock have developed in the direction of adaptation to the harsh continental weather conditions. Thus their productivity is not as high as livestock breeds in farms in other countries. For Mongolia, livestock husbandry is fundamental for the provision of food and many raw materials for the national economy, contributing 20% of the total GDP. The livestock sector is responsible for almost 90% of total agricultural production and comprises more than 10% of all exports. The national census of livestock in 2016 revealed the number of camels has reached 401,300 head (368,000 head in 2015), horses 3,634,900 head (329,530 head in 2015), cow 4,080,300 head (3,780,400 head in 2015), goats 25,572,200 head (23,592,900 head in 2015) and sheep 27,853,400 head (24,943,100 head in 2015) (Mongolian Statistical Yearbook, 2015).

The main products of the livestock sector are meat, milk and skin from all livestock, wool from camel and sheep, and cashmere from goat. In 2014, 765,400,000 liters of milk and 238,200 tonnes of meat, and 8,000 tonnes of goat cashmere were produced in Mongolia (https://www.mofa.gov.mn).

11.10 Mining Activity

As of 2012, 73.7% of land was being used for agriculture, farming, and grazing, 0.44% for urbanized cities and settlements, 0.27% for infrastructure, 9.11% forested, 0.43% water bodies, and 16% for strictly national government purposes. Out of this, land use for mining purposes is drawing attention. Although mining is beneficial to a country's development, it is one of the main causes of land degradation. Exploration projects, preparation of building materials, road construction, geological studies, construction of buildings and structures are impacting the landscape by stripping the soil, building up dirt piles and digging trenches, causing technical soil degradation. Of the 9,856.3 ha of land that has been degraded, 699.4 ha is the result of geological exploration activities, 8,028.6 ha due to mineral exploitation, 205 ha by activities for defense purposes, 125.5 ha by construction, engineering grids and lines, and 797.2 ha by infrastructure developments.

Acknowledgment

We would like to express our thanks to Mrs. Sh. Myagmarsuren and Mrs. M. Munkhjargal (Steppe Forward Programme at the National University of Mongolia and ZSL Mongolia), and Mrs. E. Unurjargal (Mongolian Ornithological Society) for their help for gathering data and information on this paper. Our thanks go to the Dr. S. Shar (National University of Mongolia) for his contribution to the mammal session, N. Kherlenchimeg (Mongolian Academy of Sciences) for her help for fungus text, and D. Batbold (WWF Mongolia) for his encouragement and some data for the NBSAP.

Without help from the Mongolian Ornithological Society and Mongolica Publishing, we would not have been able to include wonderful images of Mongolian wildlife and habitats, and so we thank them for this and the inclusion of data on birds and other species included in this text.

Keywords

- endemics
- protected areas
- status
- strategic action plan
- threats
- trends

References

Adiya, Y., Enkhbileg, D., & Ariundayar, B., (2010). *Evaluation of National Programme on Conservation of Biodiversity Mongolia*. Ulaanbaatar, Mongolia.

Aldarmaa, J., Dagvatseren, B., Enkhtur, E., Khishigjargal, L., & Naratsetseg, D., (2010). Bio preparation of Stink Beetle (*Blaps femoralis*) pharmacological effects. *Scientific Proceedings of the Mongolian Academy of Sciences, 1*, 3–17.

Anichtchenko, A., et al., (2007–2017). *Carabidae of the World*. http://www.carabidae.org (accessed on 14 February, 2017).

APG (Angiosperm Phylogeny Group), (2016). An update of the angiosperm phylogeny group classification for the orders and families of flowering plants: APG IV. *Bot. J. Linn. Soc., 181* (1), 1–20.

Azjargal, J., (2013). *Community Studies of True Spider (Araneae), Master Thesis*, National University of Mongolia. Ulaanbaatar, Mongolia, 1–75.

Batbold, D., Batkhuyag, B., Ganzorig, B., Munkhnast, D., Munkhchuluun, B., Onon, Yo., Purevdorj, S., Sumiya, E., Selenge, G., Chimeddorj, B., & Enkhbayar, N., (2015). *National Biodiversity Program* (2015–2025). Ministry of Environment, Green Development and Tourism and WWF Mongolia. Ulaanbaatar, Mongolia, 1–57.

Batsaikhan, N., Samiya, R., Shar, S., Lkhagvasuren, D., & King, S. R. B., (2014). *A Field Guide to the Mammals of Mongolia,* 2nd edition. National University of Mongolia. Ulaanbaatar, Mongolia.

Borkin, L. Y., Borobieva, E. I., Darevskii, I. S., Kuzmin, S. L., Munkhbayar, K., & Semenov, D. V., (1988). *Amphibians and Reptiles of the People's Republic of Mongolia*. Nauka, Russia, 48.

Buyanjargal, B., & Namkhaidorj, B., (2010). Habitats of various species of longhorn beetles *(Cerambycidae). Proceedings of the Seminar of Agriculture and Biotechnology of Khurel Togoot*. Ulaanbaatar, Mongolia, 27–33.

Buyanjargal, B., (2016). *Fauna and Ecology of Vespid Wasps (Hymenoptera, Vespidae) of Orkhon - Selenge Depression in Mongolia*: *Dissertation of PhD Degree*, 1–199.

Chimed-Ochir, B., Hertzman, T., Batsaikhan, N., Batbold, D., Sanjmyatav, D., Onon, Y., & Munkhchuluun, B., (2010). *Filling the Gaps to Protect the Biodiversity of Mongolia*. WWF Mongolia, Ulaanbaatar, Mongolia, 1–134.

Chuluunbat, S., Morse, J. C., & Sosorburam, B., (2016). Caddisflies of Mongolia: Distribution and diversity. *Zoosymposia, 10*, 96–116.

Clark, E. L., Munkhbat, J., Dulamtseren, S., Baillie, J. E. M., Batsaikhan, N., Samiya, R., & Stubbe, M., (2006). *Mongolian Red List of Mammals. Regional Red List Series,* Zoological Society of London, London, vol. 1.

Dariimaa, S., (2014). *Flora of Mongolia*. (Asteraceae, *Cichorioideae – Cardioide*). Ulaanbaatar, Mongolia, "Bembi san" Press, vol. 14a.

Dariimaa, S., Kamelin, R. V., Ulziykhutag, N., & Ganbold, E., (2010). The flora of Mongolia: Results of floristical researches during the last years. *Proceedings of the International Conference"Ecological Consequences of Biosphere Processes in the Ecotone Zone of Southern Siberia and Central Asia."* Ulaanbaatar, Mongolia, *1*, 55–58. (in Mongolian).

Dash, D., Jalbaa, K., Khaulenbek, A., & Mandakh, N., (2003). *Scientific Background of Ecosystem Protection and Restoration in the Gobi and Grassland Region*. Desertification

Research Center, Geo-Ecological Institute, Mongolian Academy of Sciences. Ulaanbaatar, Mongolia.

Dulmaa, A., (2003). The research news of the water vascular plants of Mongolia II. *News of the Mongolian Academy of Sciences, 4*(169), 22–35.

Enkhnasan, D., (2007). *Dytiscidae of Mongolia*, Masters Thesis at the National University of Mongolia, Ulaanbaatar, Mongolia.

Enkhtuya, O., (2007). *The Checklist of Lichens of Mongolia*. Institute of Botany of the Mongolian Academy of Sciences. Ulaanbaatar, Mongolia, 1–73.

Ganbold, E., (2010). Flora of the Northern Mongolia. *Series of the Russian Mongolian Complex Biological Expedition,* Tom LIII. Moscow.

Gantigmaa, C., (2005). *Butterfly Communities in the Natural Landscape of West Khentii, Northern Mongolia: Diversity and Conservation Value*. PhD thesis, Dissertation in biology, University of Goettingen, 1–126.

Gantigmaa, C., Bayartogtokh, B., Aibek, U., Altanchimeg, D., & Enkhnasan, D., (2012). *Report of the Management of the Research on Distribution of Insects and Invertebrates in Mongolia*. Ulaanbaatar, Mongolia.

Gelhaus, J., & Podenas, S., (2006). The diversity and distribution of crane flies (Insecta: Diptera: Tipulidae) in the Khovsgol Lake watershed, northern Mongolia. *The Geology, Biodiversity and Ecology of Lake Hovsgol*, Mongolia, Backhuys Publishing, USA.

Gelhaus, J., Podenas, S., & Brodo, F., (2000). New and poorly known species of long-palped crane flies (Diptera: Tipulidae) from Mongolia. *Scientific Proceedings of the Mongolian Academy of Natural Sciences, 150*, 145–157.

Gombobaatar, S., & Bayanmunkh, D., (2016). *An Annotated Bird List of Mongolia*. Mongolica Publishing and Mongolian Ornithological Society. Ulaanbaatar, Mongolia, 1–99.

Gombobaatar, S., (2012). *National Report on the Convention on Wetlands of International Importance, Especially as Waterfowl Habitat – Ramsar Convention*. Ulaanbaatar, Mongolia, 1–44.

Gombobaatar, S., Brown, H. J., Sumiya, D., Tseveenmyadag, N., Boldbaatar, S., Baillie, J. E. M., Batbayar, G., Monks, E. M., & Stubbe, M., (2011). *Summary Conservation Action Plans for Mongolian Birds*. Regional red list series, Zoological Society of London, Mongolian Ornithological Society and National University of Mongolia, Ulaanbaatar, Mongolia, vol. 8, 1–145.

Gombobaatar, S., Monks, E. M., Seidler, R., Sumiya, D., Tseveenmyadag, N., Bayarkhuu, S., Baillie, J. E. M., Boldbaatar, S., & Uuganbayar, C., (2011). *Mongolian Red List for Birds. Regional Red List Series, Birds*. Zoological Society of London, National University of Mongolia and Mongolian Ornithological Society, vol. 7, 1–1036.

Gombobaatar, S., Myagmarsuren, S., Conaboy, N., & Munkhjargal, M., (2014). *The 5ᵗʰ National Report of the Convention on Biological Diversity, Mongolia*. Ministry of Environment and Green Development, Steppe Forward Programme, National University of Mongolia and Zoological Society of London. Ulaanbaatar, Mongolia, 1–142.

Grubov, V. I., (1955). *Conspectus of Flora of People's Republic of Mongolia*. Series of Mongolian Commission of Academy of Sciences of USSR. Moscow-Leningrad, vol. 67.

Grubov, V. I., (1976). Sixth addition to the check-list of flora of Mongolian people republic. *Journal of Botany, 61*, 1751–1753.

Grubov, V. I., (1982). *Key To The Vascular Plants of Mongolia*. Leningrad, Nauka.

Grubov, V. I., (1984). Endemic species of Mongolia. *News on Systematics of Vascular Plants, 21*, 202–220.

Grubov, V. I., (1989). Endemism of the Flora in Mongolia. Exploration into the Biological Resources of Mongolia, Halle (Saale), *6*, 87–90.

Grubov, V. I., (2001). *Key to the Vascular Plants of Mongolia*. Science Publishers.

Gubanov, I. A., (1996). *Conspectus of the Flora of Outer Mongolia*. Moscow, "Valang" Press.

Gunbilig, D., & Boland, W., (2009). Defensive Agents of *Blaps femoralis*, A traditional Mongolian Medicinal Insect. *Scientia Pharmaceutica, 77*, 597–604.

Hilbig, W., (1995). *The Vegetation of Mongolia*. SPB Academic Publishing. The Netherlands, 1–258.

http://www.carabidae.org (accessed on 14 February, 2017).

http://www.cbd.int/doc/world/mn/mn-nr-04-en.pdf (accessed on 14 February, 2017).

https://www.mofa.gov.mn (accessed on 11 February, 2017).

https://www.state.gov/r/pa/ei/bgn/2779.htm (accessed on 10 February, 2017).

IUCN, (2003). *IUCN Red List Categories*. Prepared by the IUCN Species Survival Commission. IUCN, Gland, Switzerland.

Kamelin, R. V., & Ulziikhutag, N., (2005). New data of vascular plants in the Mongolian flora. *In the International Conference on Ecosystem of Mongolia*. Ulaanbaatar, Mongolia, 104–106.

Kherlenchimeg, N., & Burenbaatar, G., (2016). *A Handbook of Mushrooms of Mongolia*. Mongolian Academy of Sciences. Ulaanbaatar, Mongolia.

Lavrenko, E. M., (1995). *Vegetation Map of MPR*. Scale 1:1, 500, 000. Moscow, Russia, (1979). *Law of Natural Plants*. Ulaanbaatar. Mongolia.

Maximowicz, K. J., (1859). Primitiae florae Amurensis. *Mem. Acad. Imp. Sci. St.-Petersbourg Divers Savans, 9*, 1–467.

Mongolian Statistical Yearbook, (2015). National Statistical Office of Mongolia. Ulaanbaatar, Mongolia.

Munkhbaatar, M., & Terbish, K., (2008). Reptiles and Amphibians in Onon-Balj Nature Reserve. *In the Proceedings of the Workshop of the Nature and Sustainable Development in Onon River Basin*. Ulaanbaatar, Mongolia, 37–44.

Munkhbaatar, M., (2008). Albinism detected from Siberian wood frog in Mongolia (*Rana amurensis* Boulenger, (1886), Ranidae). *Bulletin of the Mongolian Academy of Sciences, 2*(188), 39–43.

Myagmarsuren, D., & Namkhai, A., (2012). *Special Protected Areas of Mongolia. Munkhiin Useg, Ulaanbaatar, Mongolia, 1–407.*

Nyambayar, D., Oyuntsetseg, B., Tungalag, R., Jamsran, T., Sanchir, C., Bachman, S., Soninkhishig, N., Gombobaatar, S., & Baillie, J. E. M., (2011). *Mongolian Red List and Conservation Action Plans of Plants (Part 1)*. Regional Red List Series, Zoological Society of London, National University of Mongolia. Ulaanbaatar, Mongolia, vol. 9.

Ocock, J., Baasanjav, G., Baillie, J. E. M., Erbenebat, M., Kottelat, M., Mendsaikhan, B., & Smith, K., (2006). *Mongolian Red List of Fishes*, vol. 3. Regional Red List Series, Zoological Society of London, London.

Oyuntsetseg, B., & Urgamal, M., (2014). *Vascular Plants: Convention on Biological Diversity the 5th National Report of Mongolia*. Ulaanbaatar, Mongolia, 21–24.

Pavlov, D. S., Galbaatar, T., Kamelin, R. V., & Ulziykhutag, N., (2005). *Ecosystems of Mongolia Atlas*. Joint Russian and Mongolian Complex Biological Expedition of the Russian Academy of Sciences and Institute of Botany of the Mongolian Academy of Sciences, Moscow, 1–48.

Proshchalykin, M. Y., & Kuhlmann, M., (2015). The bees of the genus *Colletes* Latreille (Hymenoptera, Colletidae) collected by the Soviet-Mongolian expeditions (1967–1982). *Far Eastern Entomologist, 296*, 1–18.

Pyak, A. I., Shaw, S. C., Ebel, A. L., Zverev, A. A., Hodgson, J. G., Wheeler, B. D., Gaston, K. J., Morenko, M. O., Revushkin, A. S., Kotukhov, Y. A., & Oyunchimeg, D., (2008). *Endemic Plants of the Altai Mountain Country*. WILD Guides Press.

Sanchir, C., Batkhuu, J., & Munkhbaatar, A., (2004). New species in the Middle Khalkha steppe region. *Scientific Proceedings of the Institute of Botany, Mongolian Academy of Sciences, 14*, 6–8.

Shiirevdamba, T., (ed.). (2013). *Mongolian Red Book*. Ministry of nature and environment. Ulaanbaatar, Mongolia.

Soldan, T., Enktaivan, S., & Godunko, R. J., (2009). Commented checklist of mayflies (Ephemeroptera) of Mongolia. *Aquatic Insects, 31*, 653–670.

Terbish, K., Munkhbayar, K., Clark, E. L., Munkhbat, J., Monks, E. M., Munkhbaatar, M., Baillie, J. E. M., Borkin, L., Batsaikhan, N., Samiya, R., & Semenov, D. V., (2006). *Mongolian Red List of Reptiles and Amphibians*, vol. 5. Regional Red List Series, Zoological Society of London, London.

Tserenbaljid, G., (2002). *Colour Atlas of Antropophilus Plants Seeds of Mongolia*. Admon press, Ulaanbaatar, Mongolia, 1–251.

Tsetsegmaa, D., (2008). *An Overview of Algae of Mongolia*. Institute of Botany of the Mongolian Academy of Sciences. Ulaanbaatar, Mongolia, 1–120.

Tungalag, O., (2014). *History and Development of Administration of Mongolia*. Ulaanbaatar, Mongolia.

Ulziykhutag, N., (1989). *Overview of the Flora of Mongolia*. State Publishing, Ulaanbaatar, Mongolia.

Ulziykhutag, N., Dugarjav, C., Tuvshintogtokh, I., Mandakh, B., Narantuya, N., Tsooj, S., Urgamal, M., & Munkhjargal, B., (2009). Introduction of Institute of Botany, Mongolian Academy of Sciences in 35 years, *Scientific Proceedings of the Institute of Botany of the Mongolian Academy of Sciences, 21*, 10 28.

Urgamal, M., & Sanchir, C., (2015). Preliminary analysis of the vascular flora of Mongolia. In: *Commemoration of the 45th Anniversary of the Joint Russian-Mongolian Complex Biological Expedition, RAS and MAS, and 50th Anniversary of the Institute of the General and Experimental Biology, MAS. Proceedings of International Conference of Ecosystems of the Central Asia under Current Conditions of Socio-economic Development*, Ulaanbaatar, Mongolia, 262–264.

Urgamal, M., (2016). Renewed taxonomical macrosystem of flora in Mongolia. *Scientific Proceedings of the Institute of General and Experimental Biology, 32*, 100–106.

Urgamal, M., Bat-Enerel, B., & Oyuntsetseg, B., (2015). Contribution of the Endemic Plants of Mongolia. In: *Commemoration of the 45th Anniversary of the Joint Russian-Mongolian Complex Biological Expedition, RAS and MAS, and 50th Anniversary of the Institute of the General and Experimental Biology, MAS. Proceedings of International Conference of Ecosystems of the Central Asia under Current Conditions of Socio-economic Development*, Ulaanbaatar, Mongolia, 265–271.

Urgamal, M., Enkhtuya, O., Kherlenchimeg, N., Enkhjargal, E., Bukhchuluun, T., Buren-baatar, G., & Javkhlan, S., (2016). Current overview of plant diversity in Mongolia. *News of the Mongolian Academy of Sciences, 3*(219), 86–94.

Urgamal, M., Oyuntsetseg, B., & Nyambayar, D., (2013). Synopsis and recent additions to the flora of Mongolia. *Scientific Proceedings of the Institute of Botany of the Mongolian Academy of Sciences, 25*, 53–72.

Urgamal, M., Oyuntsetseg, B., Nyambayar, D., & Dulamsuren, C., (2014). *Conspectus of the Vascular Plants of Mongolia,* Mongolian Academy of Sciences and National University of Mongolia, Ulaanbaatar, Mongolia, 1–333.

Urgamal, M., Oyuntsetseg, B., Nyambayar, D., & Dulamsuren, C., (2014). *Conspectus of the Vascular Plants of Mongolia,* Sanchir, C., & Jamsran, T., (eds.). Admon Press, Ulaan-baatar, Mongolia, 1–334.

US Department of State (2015). U.S. Relations with Mongolia. Bureau of East Asian and Pacific Affairs.

Yunatov, A. A., (1954). *Forage Plants of Pastures and Meadows of the MPR.* USSR Academy of Sciences Publishing House. Moscow-Leningrad, Russia, 1–352.

CHAPTER 12

Biodiversity in Myanmar

NAING ZAW HTUN

Deputy Director, Nature and Wildlife Conservation Division, Forest Department, Ministry of Natural Resources and Environmental Conservation, The Republic of the Union of Myanmar, E-mail: nzhtun@gmail.com

12.1 Introduction

Myanmar is the largest country in mainland Southeast Asia with a land area of 676,577 km^2, bordered by Bangladesh and India to the northwest, the People's Republic of China to the northeast and the Lao PDR and Thailand to the southeast (Figure 12.1). The Bay of Bengal and Andaman Sea lie to the south and west. More than 40% of Myanmar is mountainous. Prominent mountain chains include an extension of the eastern Himalaya, the Chin Hills, the Western Plateau/Rakhine Yoma, Bago Yoma, the Eastern Plateau/Shan Plateau and the Taninthayi Range. The Ayeyawady, Thanlwin/Salween, Chindwin, Sittaung, and Kaladan are Myanmar's major rivers.

The country has three seasons: wet (from mid-May to mid-October), cold (from early November to late February), and dry (from March to mid-May). Temperature, precipitation, and humidity vary greatly; from the Taninthayi coast which receives about 5,000 mm of rain annually to the arid Central Dry Zone in the central plains which receive only 500–750 mm of rain a year (Figure 12.2). This diverse topography and climatic conditions create numerous different ecosystems and support an incredibly wide range of associated species.

12.2 Overview of Biodiversity of Myanmar

Myanmar is situated at the transition zone between different biogeographic regions: in the north, Indochina, the Indian sub-continent, and Eurasia; in the south, Taninthayi forests cover the northern section of the transition between Indochina and Sundaic ecological zones. These transitional zones produce unique and diverse species assemblages. The region's most intact lowland Sundaic forests are found in Myanmar, along with patchy but regionally significant areas of dry deciduous forest. Birds that migrate on both the Central Asian and East Asian Flyways rest at globally important wetlands in the country. Myanmar contains almost 10% of global turtle and tortoise diversity,

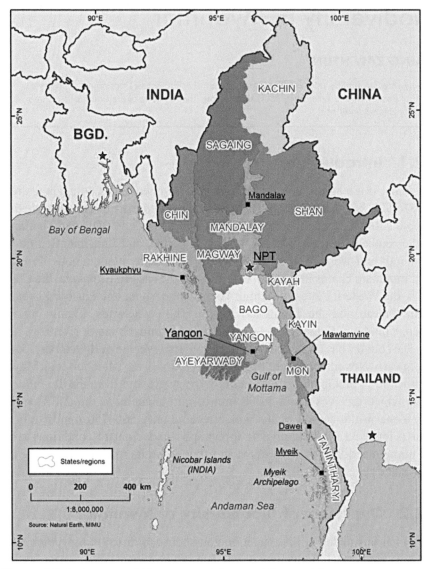

Figure 12.1 Location of Myanmar and State and Region Administrative Boundaries (Source: https://www.cbd.int/doc/world/mm/mm-nbsap-v2-en.pdf).

including seven endemic species. Some regions and taxa are relatively under-studied, and surveys continue to identify new endemic species and range extensions of globally threatened species. Ongoing surveys are also developing a better understanding of the distribution and status of these species.

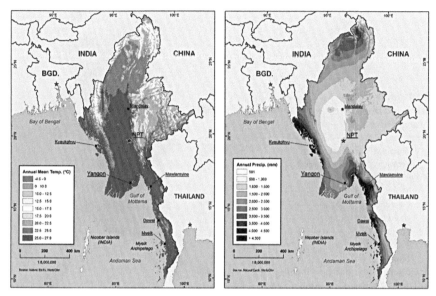

Figure 12.2 Annual Mean Temperature and Annual Precipitation in Myanmar (Source: https://www.cbd.int/doc/world/mm/mm-nbsap-v2-en.pdf).

12.3 Ecosystem Diversity

12.3.1 Forest Ecosystems

Forests constitute the dominant ecosystem in Myanmar, with 45% of the country ecologically classified as forest (FAO, 2015). Furthermore, as a result of a wide altitudinal range, with the corresponding variation in climatic conditions, the country supports a range of forest types and vegetation zones. Broadly speaking, forests in Myanmar can be categorized into the types as shown in Figure 12.3. These include the extensive teak forests for which Myanmar is renowned. In addition, one of the largest homogenous bamboo stands in the world is found in Rakhine State, covering an area of over 7,770 km².

12.3.2 Freshwater Ecosystems

Myanmar supports a diversity of freshwater ecosystems, from fast-flowing mountain streams to wide, slow-flowing lowland rivers, as well as lakes and wetlands. These rivers, lakes, and wetlands provide enormous economic and cultural values. The Salween and the Ayeyawady Rivers are some of

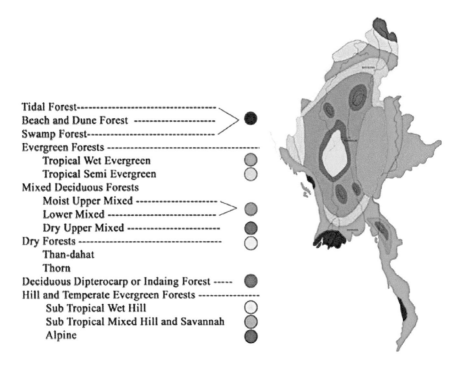

Tidal Forest--
Beach and Dune Forest ------------------------ ●
Swamp Forest--
Evergreen Forests --------------------------------------
    Tropical Wet Evergreen   ◯
    Tropical Semi Evergreen   ◯
Mixed Deciduous Forests
    Moist Upper Mixed ----------------------
    Lower Mixed --------------------------- ●
    Dry Upper Mixed ----------------------- ●
Dry Forests ------------------------------------ ◯
    Than-dahat
    Thorn
Deciduous Dipterocarp or Indaing Forest ----- ●
Hill and Temperate Evergreen Forests ----------------
    Sub Tropical Wet Hill   ◯
    Sub Tropical Mixed Hill and Savannah ◯
    Alpine   ●

Figure 12.3 Major Vegetation Types of Myanmar (Source: Adapted from Kress et al., 2003).

the most intact major rivers in Asia, providing livelihoods to the people living along their banks and rich with historical and cultural significance. The Chindwin River flows through Hukaung Valley and creates one of the largest seasonally flooded grasslands of the region. Indawgyi Lake is the largest freshwater lake in Myanmar, hosting globally significant aggregations of water birds and providing livelihoods for people who fish and grow unique varieties of rice around the lake.

12.3.2.1 Rivers

Myanmar is endowed with tremendous inland water resources in the form of rivers, streams, and springs (Figure 12.4). Major rivers include the 1,800 km-long Ayeyawady River which arises from the confluence of the N'mai Kha and Mali Kha Rivers. The Chindwin River, with headwaters in the northwestern hills, is the main tributary of the Ayeyawady. The Sittaung River starts in the hills southeast of Mandalay, and the Thanlwin River,

Figure 12.4 Major Rivers of Myanmar (Source: https://www.cbd.int/doc/world/mm/mm-nbsap-v2-en.pdf).

the last undammed river, races through deep gorges in the Shan Plateau. The Kaladan River is formed by tributaries discharging from the Arakan Mountains.

12.3.2.2   Lakes

Myanmar contains several large lakes, which provide critical habitat for a range of species and a source of livelihood for local residents. Indawgyi Lake in Kachin State is the largest, with around 12,000 hectares of open water. The lake provides habitat for numerous endangered species and for globally significant aggregations of migratory water birds. Inlay Lake on the Shan plateau is the most famous lake in Myanmar, known for its floating gardens and the leg-rowing Intha people who live around the lake. The country also contains numerous small and medium-sized lakes, including glacial lakes in the north that are crucial sources of freshwater. Lakes within urban areas provide freshwater, flood control, and opportunities for recreation.

12.3.3   Coastal and Marine Ecosystems

Myanmar has a large marine territory. The coastline stretches from the Naf River, the dividing line between Bangladesh and Myanmar, to Kawthaung at the border with Thailand, 2,831 km to the south (Figure 12.5). Along the southern coastline, the Myeik Archipelago is made up of more than 800 islands. The continental shelf covers 225,000 km^2, and the Exclusive Economic Zone covers 512,000 km^2. Coastal areas also include 5,000 km^2 of brackish and freshwater swampland that provides essential ecological habitat for spawning and as a nursery and feeding ground for fish, prawns and other aquatic fauna and flora of economic and ecological importance. Mangroves are found in many coastal regions, particularly near estuaries in Rakhine State, Taninthayi Region, and Ayeyawady Region. Other coastal habitats include intertidal mud and sand flats, which are very important for migratory water birds, as well as sand dunes and beach forest. The Gulf of Mottama contains one of the largest intertidal mudflats in the world and is thought to be key for the survival of the critically endangered spoon-billed sandpiper.

12.3.4   Mountain Ecosystems

Forty-two percent of Myanmar is mountainous and these areas form some of the most important landscapes in terms of biological, cultural, traditional and ethnic diversity and identity (Figure 12.6). Mountainous areas are also important for the country's economy, providing most of the fresh water for the country. In addition, three-quarters of Myanmar's 132 Key Biodiversity Areas (KBAs), areas identified as being particularly important

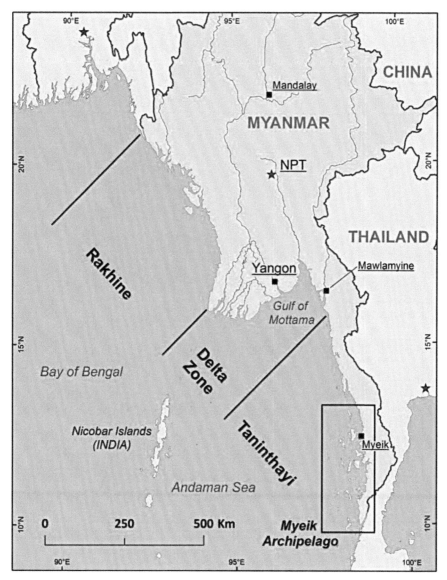

Figure 12.5 Coastal Areas of Myanmar (Source: https://www.cbd.int/doc/world/mm/mm-nbsap-v2-en.pdf).

for biodiversity, are located in mountainous areas, and are home to several endemic and globally important species. Major mountain ranges in Myanmar are shown in Table 12.1. In the far north, with an elevation of 5,881 m, Hkakaborazi is the highest peak as well as part of the only permanently

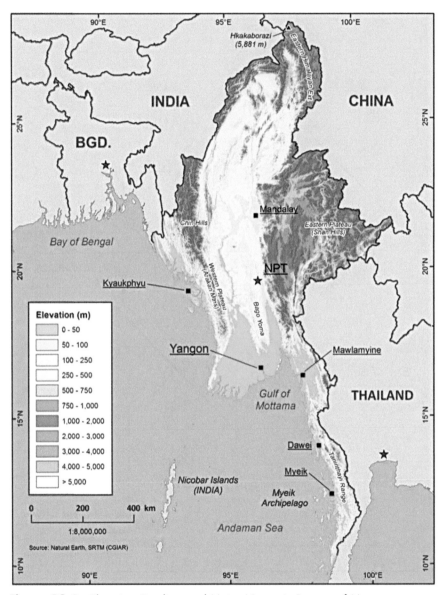

Figure 12.6 Elevation Gradient and Major Mountain Ranges of Myanmar
(Source: https://www.cbd.int/doc/world/mm/mm-nbsap-v2-en.pdf).

snow-capped mountain range in the Indo-Burma region. These mountain ranges are home to diverse ethnicities who practice traditional upland cultivation systems and are dotted with peaks and caves of cultural and historical importance.

Table 12.1 Major Mountain Ranges in Myanmar

Mountain Range	Location	Notable Features
Eastern Himalayan Extension	Northern part of country, eastern extent of the Himalayan range	This range contains the highest peaks in Southeast Asia, unique forest habitat and rich bird biodiversity
Chin Hills	Western part of Myanmar, extending to south of India	Natmataung National Park, in the Chin Hills, is an alpine island that is home to rich plant diversity and an endemic bird species.
Western Plateau/ Rakhine Yoma	Between the Ayeyawady River and Bay of Bengal	Acting as a barrier to the monsoon, western slopes of the Rakhine Yoma can receive 1 m of precipitation per month in the wet season. This range extends under water to the south and later emerges to form the Nicobar Islands
Bago Yoma	Between the Ayeyawady and Thanlwin Rivers	Largely forested, these mountains supply many reservoirs and provide habitat to a resident elephant population. The Bago Yoma is a historically important source of high-quality teak.
Eastern Plateau	Northeast, bordering with China, Laos and Thailand	The Shan Hills cover mountainous Shan State with forest, karst formations, and agricultural land.
Taninthayi Range	South, bordering with Thailand	Covering the northern transition zone between Indochina and Sundaic zones, this range is home to a variety of threatened species. PAs in Thailand are connected to this forest complex across the border.

(Source: https://www.cbd.int/doc/world/mm/mm-nbsap-v2-en.pdf)

Karst formations can be found in Taninthayi Region, Karin State, Shan State, and stretching along the upper Ayeyawady River in Kachin State. Karst formations are home to species with severely restricted ranges, some of which are confined to a single cave or peak. This high rate of endemism makes Karst systems particularly important for biodiversity conservation. Limestone quarrying for cement production threatens Karst ecosystems.

12.3.5 Agricultural Ecosystem

Myanmar has six major agro-ecological zones corresponding to topographical variation and climatic conditions (Table 12.2).

Table 12.2 Major Agro-Ecological Zones of Myanmar

Name	Geographical description	Administrative units	Main agricultural crops
A. Bago and Kachin riparian areas and floodplains	Upper Delta, Kachin Plain, flat plains adjacent to Ayeyawady and Sittaung rivers; moderate rainfall (1,000–2,500 mm)	Ayeyawady Region, Kachin State, Sagaing Region, Mandalay Region and Bago Region	Rice, pulses, oilseeds, sugarcane, tobacco and alluvial/island cultivation
B. Central Dry Zone	Central Dry Zone, rainfall less than 1,000 mm, highest temperatures in summer, flat plains, some areas with rolling hills	Magway Region, Mandalay Region, and Sagaing Region	Upland crops, oilseeds, pulses, rice, cotton, irrigated agriculture and alluvial/island cultivation
C. Delta and Coastal Lowland	Delta, lowland and coastal river outlets and estuaries; heavy rainfall (more than 2,500 mm)	Ayeyawady Region, Yangon Region, Bago Region, Mon State, Taninthayi Region and Rakhine State	Rice, pulses, oilseeds and nipa palm
D. Kachin and Coastal Upland	Mountainous, sloping land; heavy rainfall (more than 2,500 mm)	Kachin State, Rakhine State, Taninthayi Region, Mon State, Karin State, Kayah State, Yangon Region and Bago Region	Orchards, plantation crops, and upland agriculture
E. North, East and West Hills	Hilly areas, uneven topography, moderate to heavy rainfall, sloping land	Kachin State, Chin State, and Shan State	Upland crops, shifting cultivation and fruit trees
F. Upper, Lower Myanmar and Shan Plain	Upper and lower plains outside of central dry zone, Shan Plain	Sagaing Region, Kachin State, Shan State, Bago Region, Magway Region, Mandalay Region and Yangon Region	Upland crops, oilseeds, pulses, vegetables and wheat

Source: Adapted from FAO/WFP (2009).

12.4 Wild Species Diversity

12.4.1 Plants

Myanmar's variations in latitude, altitude, and climate create a variety of habitats and support correspondingly rich plant biodiversity. To date, more than 18,000 plant species have been recorded in Myanmar. These include more than 800 orchid species, 80 bamboo species, numerous rattan species, and more than 800 medicinal plant species. However, there are large research and information gaps for several species groups. On-going collaborative botanical surveys by the Wildlife Conservation Society (WCS; USA), National Institute of Biological Resources (NIBR; Republic of Korea), Institute of Botany, Chinese Academy of Sciences (IBCAS) and Xishuangbanna Tropical Botanical Garden (XTBG) (People's Republic of China), and Makino Botanical Garden (MBK; Japan) will likely identify additional plant species, including endemic species. Enhanced coordination of these efforts is required.

There are 61 globally threatened plant species known to occur in Myanmar. Of these, 16 are assessed on the IUCN Red List of Threatened Species (hereafter referred to as the "Red List") as Critically Endangered (CR), 24 as Endangered (EN), and 21 as Vulnerable (VU). The main threats to plant species in Myanmar are overexploitation by legal and illegal logging, conversion to agriculture—especially commercial plantations, and degradation and fragmentation from road construction and small scale agriculture. Illegal logging for valuable timber species is a driver of deforestation. Rosewood species (Padauk, *Pterocarpus macrocarpus* and Tamalan, *Dalbergia oliveri*) are highly valued and increasingly sold illegally across the border as rosewood supplies are exhausted in neighboring countries. Orchids are also threatened by unregulated collection and sale across the borders.

12.4.2 Mammals

Nearly 300 mammal species have been recorded in Myanmar, but a number of these have not been sighted in recent years, including the Sumatran rhinoceros (*Dicerorhinus sumatrensis*), Javan rhinoceros (*Rhinoceros sondaicus*), and Indian water buffalo (*Bubalus arnee*). Myanmar is a home to the Western Hoolock gibbon (*Hoolock hoolock*), Eastern Hoolock gibbon (*Hooloc kleuconedys*) and Myanmar snub-nosed monkey (*Rhinopithecus strykeri*), discovered in the mountains near the Chinese border in northeast

Kachin State in 2010. There are 47 globally threatened mammal species in Myanmar; five Critically Endangered, 17 Endangered and 25 Vulnerable.

Two large mammals, the Asian elephant (*Elephas maximus*) and tiger (*Panthera tigris*) are threatened, mainly due to illegal trafficking, and their populations are thought to be decreasing. Black musk deer (*Moschus fuscus*), sun bear (*Helarctos malayans*), Malayan pangolin (*Manis javanica*), and Chinese pangolin (*Manis pentadactyla*) are also severely threatened by illegal trafficking. On the other hand, camera trap surveys have shown that Htamathi Wildlife Sanctuary, and the proposed Taninthayi, Lenya and Lenya (extension) National Parks are home to a considerable number of tigers and prey species, as well as the Asian elephant.

The population of the Irrawaddy dolphin (*Orcaella brevirostris*) has been decreasing, mainly due to destructive electrofishing practices. Another large aquatic mammal, the dugong (*Dugong dugon*), has been sighted in the Myeik Archipelago and off the Rakhine coast.

Myanmar is also a home to at least five endemic mammal species, including: Anthony's pipistrelle (*Hypsugo anthonyi*), Joffre's pipistrelle (*Hypsugo joffrei*), Myanmar pipistrelle (*Hypsugo lophurus*) and the Popa soft-furred rat *(Millardia kathleenae)*. Dry mixed deciduous forests in Myanmar are home to the largest remaining population of the endangered Eld's deer (*Rucervus eldii*).

12.4.3 Avifauna

Myanmar is recognized as having possibly the greatest diversity of bird species in Southeast Asia, with at least 1,096 avifauna species recorded including 6 endemic species and 46 bird species listed on the Red List. Although some of these species have not been recorded for decades they may be present in low numbers. Jerdon's babbler *(Chrysomma altirostre)*, was rediscovered in grassland near Yangon in 2014, with the first recorded sighting in 73 years.

Bird species endemic to Myanmar include Jerdon'sminivet (*Pericrocotus albifrons*), hooded treepie (*Crypsirina cucullata*), Burmese bush lark (*Mirafra microptera*), Burmese tit (*Aegithalos sharpie*), white-throated babbler (*Turdoides gularis*) and white-browed nuthatch (*Sitta victoriae*).

Of the 45 globally threatened bird species in Myanmar, eight are listed as Critically Endangered. Of these, five have globally significant populations which depend on the country as a critical refuge or wintering area. These include the white-bellied heron (*Ardea insignis*), spoon-billed sandpiper (*Calidris pygmaea)*, white-rumped vulture (*Gyps bengalensis*), slender-billed vulture (*Gyps tenuirostris*), and red-headed vulture (*Sarcogyps calvus*). Myanmar is a home to the bulk of the world's population of Gurney's pitta (*Pitta gurneyi*), an endangered species, which, outside of Myanmar, is only known from very small populations in southern Thailand.

12.4.4 Herpetofauna

Ongoing surveys indicate that Myanmar hosts a high diversity of reptiles and amphibians. Myanmar has an exceptional turtle and tortoise diversity, with seven endemic species. A herpetofauna survey, jointly conducted between 1999 and 2010 by the Forest Department (FD) and the California Academy of Sciences (CAS), marked an initial effort to understand diversity in Myanmar and subsequent surveys have filled in gaps and discovered new species.

Twenty-one reptile species and three amphibian species endemic to Myanmar have been recorded, including the Burmese frog-faced softshell turtle (*Chitra vandijki*), Myanmar star tortoise (*Geochelone platynota*), Rakhine forest turtle (*Heosemys depressa)*, Burmese roofed turtle (*Batagur trivittata)*, Myanmar flapshell turtle (*Lissemys scutata*), Burmese-eyed turtle (*Morenia ocellata*) and Burmese peacock softshell turtle (*Nilssonia formosa*). Wildlife trafficking and consumption are major threats to these species. *Geochelone platynota* is considered functionally extinct in the wild, and conservation efforts focus on assurance colonies and reintroduction. The status of several species including *Manouria emys*, *Manouria impressa*, *Batagur baska*, Gharial crocodile (*Gavialis gangeticus*), *Crocodylus palustris* and *Tomistoma schlegelii* remains poorly understood.

12.4.5 Invertebrates

Invertebrates are one of the least studied taxa in Myanmar. A joint study by FD and Smithsonian Institution identified 1,197 butterflies in Myanmar, about 12% of the global total, which makes Myanmar the fifth richest country in the world in terms of butterfly diversity. This also includes six of the rarest known butterfly species in the world (Table 12.3). The diversity of other invertebrate species such as beetles, bees, and spiders are largely unknown.

12.4.6 Freshwater Fish

Freshwater fish is one of the least studied fauna in Southeast Asia. Nevertheless, Myanmar is already known to be rich in freshwater fish species, with

Table 12.3 Rare Butterfly Species Found in Myanmar

Scientific Name	Common Name
Parnassius imperator	Apollo
Troides helenacerberus	Common birdwing
Troides aeacus praecox	Golden birdwing
Bhutanitis ledderdalii	Bhutan glory
Teinopalpus imperialis	Kaiser
Euthalia phemius phemius	White edge baron
Euthalia phemius	White-edged blue baron

(Source: https://www.cbd.int/doc/world/mm/mm-nbsap-v2-en.pdf).

520 species recorded, including a number of endemic species (Fish Base, 2015). Recent studies conducted by FD and Fauna & Flora International (FFI) revealed some species new to science (*Lepidocephalichthys* spp., *Acanthocobitis* spp. and *Physoschistura* spp. from Indawgyi Lake). Freshwater endemic fish species in Myanmar are presented in Table 12.4. Notable areas for endemic freshwater species are Inlay Lake and Indawgyi Lake.

12.4.7 Marine Fauna

Myanmar has a long coastline and large marine territory. Its marine resources play an important role in the country's development. A growing understanding of coral reef resilience and species composition is helping to identify key areas for conservation. The initial result of a marine ecosystem survey by the Research Vessel RV Fridtjof Nansen conducted November–December 2013 indicated that the maximum sustained yield (MSY) in Myanmar's marine territory has been significantly reduced compared to the MSY calculated in the early 1980s. With the exception of marine fish species, the majority of the data is collected from the Myeik Archipelago.

In total, 17,043 species are recorded in Myanmar (Table 12.5), and a comprehensive survey will discover more new species for the world or new record for the country.

12.5 Domesticated Biodiversity

12.5.1 Crops

Plants play a vital role for the survival of human society. Plant Genetic Resources (PGR) provide enormous potential for food security, biofuel and

Table 12.4 Endemic Freshwater Fish Species in Myanmar

No.	Species	No.	Species	No.	Species
1	*Akysis pictus*	21	*Garra poecilura*	41	*Neolissochilus blythii*
2	*Akysis prashadi*	22	*Garra propulvinus*	42	*Neolissochilus compressus*
3	*Caragobius burmanicus*	23	*Garra rakhinica*	43	*Neolissochilus stevensonii*
4	*Chaca burmensis*	24	*Garra spilota*	44	*Olyra burmanica*
5	*Channa harcourtbutleri*	25	*Garra vittatula*	45	*Osteochilus sondhii*
6	*Clupisoma prateri*	26	*Gonialosa modesta*	46	*Parasphaerichthys ocellatus*
7	*Cyprinus intha*	27	*Gonialosa whiteheadi*	47	*Physoschistura brunneana*
8	*Danio choprae*	28	*Gudusia variegata*	48	*Physoschistura rivulicola*
9	*Danio erythromicron*	29	*Hemibagrus peguensis*	49	*Physoschistura shanensis*
10	*Danio nigrofasciatus*	30	*Hemibagrus variegatus*	50	*Proeutropiichthys macropthalmos*
11	*Devario auropurpureus*	31	*Homaloptera rupicola*	51	*Pseudolaguvia tuberculata*
12	*Devario sondhii*	32	*Ilisha novacula*	52	*Puntius burmanicus*
13	*Devario spinosus*	33	*Labeo stolizkae*	53	*Sawbware splendens*
14	*Esomus ahli*	34	*Macrognathus caudiocellatus*	54	*Schistura acuticephalus*
15	*Esomus altus*	35	*Mastacembelus oatesii*	55	*Sicamugil hamiltonii*
16	*Exostoma berdmorei*	36	*Microdevario gatesi*	56	*Toxotes blythii*
17	*Exostoma stuarti*	37	*Microphis dunckeri*	57	*Trichogaster labiosa*
18	*Garra flavatra*	38	*Microrasbora rubescens*	58	*Yunnanilus brevis*
19	*Garra gravelyi*	39	*Mystus leucophasis*		
20	*Garra nigricollis*	40	*Mystus rufescens*		

(Source: https://www.cbd.int/doc/world/mm/mm-nbsap-v2-en.pdf).

biopharmaceutical production and play a critical role in adaptation to climate change. More than 60 different crops are grown in the country and they can be grouped into seven categories as follows:

Table 12.5 Recorded Species of Myanmar

Group	Total no. of species	Threatened			
		Critically endangered	Endangered	Vulnerable	Total
Plants	11,824	16	24	23	63
Mammals	252	5	17	25	47
Birds	1096	8	12	25	45
Reptiles	291	-	-	-	-
Amphibians	119	-	-	-	-
Butterflies	1197	-	-	-	-
Freshwater fishes	525	-	-	-	-
Marine fishes	578	-	-	-	-
Phytoplankton	136	-	-	-	-
Zooplankton	150	-	-	-	-
Meroplankton	47	-	-	-	-
Seagrass	12	-	-	-	-
Seaweed	38	-	-	-	-
Gastropods (molluscs)	50	-	-	-	-
Bivalves (molluscs)	41	-	-	-	-
Crab (crustacean)	42	-	-	-	-
Coral	287	-	-	-	-
Marine invertebrates	230	-	-	-	-
Sharks	57	-	-	-	-
Rays	71	-	-	-	-
Total	**17,043**	**29**	**53**	**73**	**155**

- Cereals: Rice, wheat, maize, and millet.
- Oil seeds: Groundnut, sesame, sunflower, and mustard.
- Pulses: Black gram, green gram, butter bean, red bean, pigeon pea, chickpea, cowpea and soybean, etc.
- Industrial crops: Cotton, sugar cane, tobacco, rubber, and jute.
- Culinary crops: Potato, onion, chilli, vegetables, and spices.
- Plantation crops: Tea, coffee, coconut, banana, oil palm, toddy palm and other fruits.
- Other crops: Other crops that are not listed in the above groups.

Inter- and intraspecific genetic variations are also observed among crops sown nationwide, especially for rice, maize, sorghum, millet, sesame, groundnut, ginger, turmeric, custard apple, okra, chilli, pepper, tomato, citrus, water melon, mango, jack-fruit, banana, and medicinal plants.

Myanmar is also home to important crop species such as rice, mango, banana and sugarcane. Wild relatives and local landraces (varieties developed through traditional breeding methods and adapted to local conditions) of these cultivated crops are also found in Myanmar. According to genetic, geographical and molecular studies, Myanmar is believed to be in the center of diversity of cultivated rice, *O. sativa* subsp. *indica*. Several wild legume species related to cultivated mung bean, black gram and azuki bean are distributed in different ecosystems of Myanmar, including coastal sandy soils, lime stone hills and high lands of Shan state. These wild legume species could provide useful genes for legume crop improvement. Moreover, several lesser used plant species are grown and used by diverse ethnic groups in Myanmar.

Recognizing the great value of PGR and the increasing threat of the loss of plant genetic diversity from natural habitats and farm lands, the seed bank of the Ministry of Agriculture and Irrigation (MOAI) has made efforts to collect and conserve the agro-biodiversity of Myanmar. Currently, the seed bank is conserving more than 12,000 accessions of important crops in Myanmar (Table 12.6).

12.5.2 Livestock

The genetic variations of livestock in Myanmar are still largely unknown. Some livestock breeds are common across the country but some are much more localized. For example, mithun (*Bos frontalis*) are bred only in Chin State. Mithun are semi-domesticated cattle that play an important role in the day to day socio-economic life of the local tribal population. The Department of Animal Biotechnology of Kyauk Se Technical University has initiated systematic mithun breeding to maintain the declining population. Myanmar Myin (horse) and Inbinwa chicken are considered at risk because of a population decrease nationwide. The major livestock breeds in Myanmar are presented in Table 12.7.

12.6 Biodiversity Conservation

In Myanmar, the Kings initiated biodiversity conservation as early as 1752 when teak was proclaimed a Royal Tree. The vicinity of the King's palace

Table 12.6 Plant Genetic Resources Conserved by the Myanmar Seed Bank

Crop species	Number of accessions	Crop species	Number of accessions
Rice	7,367	Maize	100
Wild rice	184	Wheat	1,607
Black gram	128	Sorghum	219
Chick pea	617	Millets	123
Pigeon pea	143	Sesame	37
Green gram	189	Groundnut	665
Cow pea	181	Niger	1
Soybean	80	Safflower	1
Lima bean	66	Jute	42
Kidney bean	69	Vegetables	109
Wild *Vigna* spp.	101	**Total**	12,029

(Source: https://www.cbd.int/doc/world/mm/mm-nbsap-v2-en.pdf)

was declared a refuge area for the wild animals in the city of Yadanapon (now Mandalay) in 1850. The Elephant Preservation Act was enacted in 1879, and amended in 1883. The first Game Sanctuaries were established in 1911, but ratified protected areas were not set up until 1920. The Burma Wildlife Protection Act was imposed in 1936. Nature and Wildlife Conservation Division (NWCD) was created within the FD when the "Nature Conservation and National Parks Project (NCNPP)" was implemented from 1981 to 1984. Protected Area System (PAS) management was introduced since then and the Protection of Wildlife and Protected Areas Law was enacted in 1994 by replacing the Burma Wildlife Protection Act 1936. A policy target was set by the Myanmar Forest Policy of 1995 that protected area coverage must be at least 5% of the total land area of the country. In 2000, the 30-years Forestry Sector Master Plan adjusted this target to 10% of the total land area. Currently, Myanmar has established 39 PAs with a cover of 5.75% of the country's land area (Figure 12.7).

Among the 39 current PAs, 7 have been recognized as Asean Heritage Parks (AHPs), namely, Hkakaborazi National Park, Indawgyi Wildlife Sanctuary, Alaungdaw Kathapa National Park, Inlay Lake Wildlife Sanctuary, Meinmahla Kyun Wildlife Sanctuary, Lampi Marine National Park and Natmataung National Park (Figure 12.8). AHPs are recognized for their particular biodiversity value or uniqueness within ASEAN countries. Myanmar also has three Ramsar Sites, Moeyungi Wetland Sanctuary, Indawgyi Wildlife Sanctuary and Meinmahla Kyun Wildlife Sanctuary, and one Man and

Table 12.7 Major Livestock Breeds in Myanmar

Species	Scientific Name	Local Name	Region/Location
Cattle	*Bos indicus*	PyaScin, Shwe Ni, Shan Nwa, Katonwa, Kyaukphyu	Mandalay, Magway, Sagaing, Shan, Kayin, Rakhine
Mythun	*Bos frontalis*	NwaNauk	Chin
Buffalo	*Bubalus bubalis*	Myanmar Kywe, Shan Kywe	Ayeyawady, Sagaing, Shan
Horse	*Equus caballus*	Myanmar Myin, Shan Myin	Magway, Mandalay, Sagaing, Shan
Ass	*Equus asinus*	Myanmar Mye	Shan
Pig	*Sus domesticus*	Bo cake, Chin wet	Badoung, Akhar, Wet taungMagway, Mandalay, Sagaing, Shan
Sheep	*Ovis aries*	Myanmar Thoe, Karla Thoe	Magway, Mandalay, Sagaing
Goat	*Capra hircus*	Seik Ni, Jade Ni, NyaungOo, Htain San, HkwaySeik	Magway, Mandalay, Sagaing, Rakhine
Chicken	*Gallus gallus*	TaikKyet, TainyinKyet, KyetLada, InbinwaKyet	Widespread
Turkey	*Meleagris gallopavo*	Kyet Sin	Widespread
Duck	*Anas platyrhynchos*	Khayan Be, Taw Be	Widespread
Duck, Muscovy	*Cairina moschata*	Mandarli	Widespread
Goose	*Anser cygnoides*	Ngan	Widespread
Quail	*Coturnix* spp.	Ngown	Widespread

(Source: https://www.cbd.int/doc/world/mm/mm-nbsap-v2-en.pdf).

Biosphere Reserve (MAB) of the UNESCO, Inlay Lake Wildlife Sanctuary. Myanmar is preparing a nomination dossier for the inscription of Hkakaborazi Landscape to the UNESCO's World Heritage List.

12.7   Threats to Biodiversity

In Myanmar, ecosystems, and biodiversity face threats from a range of underlying causes. More work is needed to fully understand the forces driving biodiversity loss. However, past research can give adequate insight into

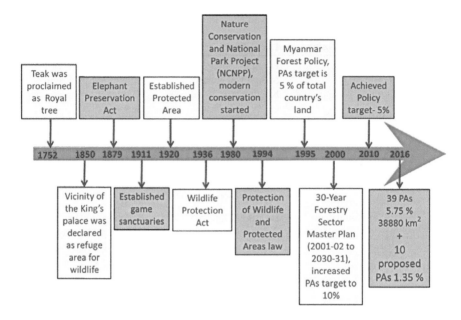

Figure 12.7 History of biodiversity conservation in Myanmar.

these pressures. These major driving forces and threats to biodiversity are discussed in the following subsections.

12.7.1 Land Uses

Myanmar is a largely agricultural county. However, most cultivation employs techniques that can significantly degrade the natural environment. In particular, shifting cultivation in upland areas, over-abstraction of groundwater, and uncontrolled pesticide and herbicide use all negatively affect ecosystems and biodiversity.

Available site-specific data shows that ecosystems and biodiversity are deteriorating mainly due to unsustainable human activities. The clearance of natural forests for agricultural expansion, both for smallholders and industrial agriculture, is leading to habitat loss for endangered species including the tiger (*Panthera tigris*) and Asian elephant (*Elephas maximus*).

Small-scale gold mining is a major polluting industry in the headwaters of many of Myanmar's rivers, ecosystems and biodiversity more widely, and negatively affects aquatic diversity and human health. This is resulting in the deterioration of aquatic ecosystems, aquatic biodiversity, and human health. These effects, in turn, result in chronic, negative impacts to the livelihoods

Protected Areas in Myanmar

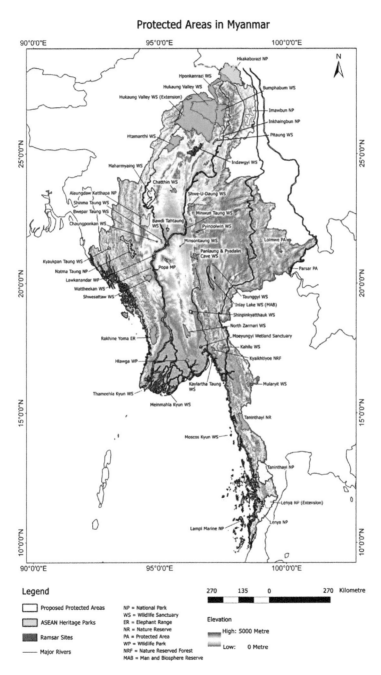

Figure 12.8   Location of Established and Proposed Protected Areas in Myanmar
(Source: https://www.cbd.int/doc/world/mm/mm-nbsap-v2-en.pdf).

of rural communities. Hardest hit are landless, poor, rural or otherwise disadvantaged people who rely on natural resources for subsistence.

Such unsound land-use practices are severely threatening the environment and associated biodiversity, both directly and indirectly, while also worsening the socio-economic situation of local communities, particularly by damaging agricultural land in downstream areas. Consequently, this increases reliance on natural forests, further increasing pressures on ecosystems and biodiversity.

12.7.2 Illegal Wildlife Hunting and Trade

The trade in endangered wildlife is one of the greatest threats to biodiversity in Myanmar. As commercially valuable wildlife species have been wiped out in neighboring countries, Myanmar has increasingly become a source of wildlife products. Particularly vulnerable are the country's endemic species, especially freshwater turtles and tortoises. By monitoring wildlife products in Mong La in Shan State since 2006, TRAFFIC has documented the significant trade in elephants, Asiatic bears, sun bears, tigers, leopards, snow leopards, cloud leopards, turtles, tortoises, and pangolins from Myanmar to its neighbors.

12.7.3 Invasive Alien Species

Little is known about the status of invasive alien species in Myanmar, but a few IAS have been observed in the country, introduced primarily by water, air and/or land transport. Trans-boundary movement of IAS is potentially high along the Myanmar's international borders with India, Bangladesh, China, Laos, and Thailand. IAS can also be introduced unintentionally by tourists or through the transport of cargo or movement of pets, plant parts, seeds and biological residues. Some IAS may be intentionally imported for use in research, manufacture of medicine or ornamental and industrial uses. Though the impact of IAS has not been comprehensively assessed, some impacts on wetland ecosystems and natural forests have been observed.

12.7.4 Threats to Agro-Biodiversity

A variety of human activities threaten the agro-biodiversity of Myanmar. Factors such as the replacement of local landraces with modern varieties, agricultural area expansion, overgrazing, dams and canal construction and urbanization are major threats to the agro-biodiversity of Myanmar.

Compounding these, climate change further threatens the future biodiversity of the country. More efforts are needed to survey and inventory plant genetic resources (PGRs) in Myanmar since the previous collection missions have focused only on local areas or specific target crops. Financial constraints and a lack of well-trained persons for eco-geographic studies of PGRs are also major constraints to surveying and inventorying agrobiodiversity in Myanmar. Large areas of the country remain to be explored, especially in the remote areas of the far north, highlands and border areas where indigenous plant diversity and farmers' knowledge have not yet been well documented.

12.7.5 The Underlying Factors of Threats to Biodiversity

The underlying factors of threats to ecosystems and biodiversity can be stated as given in the following subsections.

12.7.5.1 Poverty

The majority of Myanmar's population is poor, mainly relying on farming and natural resources for subsistence livelihoods. In particular, shifting cultivation and hunting in mountainous areas due to poverty decrease biodiversity. Although Myanmar has made notable progress in reducing poverty, 26% of the population still remains below the poverty line (Ministry of National Planning and Economic Development, 2013). Unsustainable, natural resource-dependent livelihoods are strongly correlated with extreme poverty. The poorest citizens are highly dependent on natural resources, particularly in upland areas. In many cases, use of natural resources by rural communities is potentially sustainable. However, various factors, including external economic forces, population growth, and loss of access to land, can lead to unsustainable levels of natural resource use, and degradation and loss of natural habitats. These problems have been worsened by decades of armed conflict in some, resulting in thousands of people abandoning their land. Poverty and land degradation in the uplands of Myanmar are linked in a mutually reinforcing cycle that is difficult to break. Short-term benefits are given priority in development projects, and the neglect of ecosystem and biodiversity values leads to failure to meet conservation targets. This seriously impacts ecosystems and biodiversity, particularly when funding for restoration or rehabilitation is lacking or insufficient.

12.7.5.2 Economic Growth and Increasing Consumption

While poverty is one major underlying factor for biodiversity loss, economic growth and increasing consumption are likely to be another in Myanmar, as they are throughout the World. Due to improvements in the political situation in Myanmar, foreign direct investment is now rising, resulting in economic growth. This could lead to increased use of natural resources. For example, increased construction work increases extraction of natural resources such as sand. In the short term, economic growth will likely increase pressure on the country's natural resources. In the long term, it may also offer additional resources for biodiversity conservation by lifting people out of poverty and providing increased funding for enforcement and education.

12.7.5.3 Increased Natural Resources Demand from Neighboring Countries

Enhanced logging regulations and an expansion of the illegal wildlife market in neighboring countries have increased pressure on Myanmar's natural forests and biodiversity. Most documented seizures of illegal timber and wildlife trade take place close to or en route to international borders.

12.7.5.4 Limited Environmental Safeguards

After enacting the Environmental Conservation Law on 30 March 2012, environmental safeguards are now required for development activities. However, much improvement is needed on the implementation and enforcement of this requirement.

12.7.5.5 Undervaluation of Ecosystem, Ecosystem services and Biodiversity in Development Planning

Globally, market prices tend to reflect only the direct use values of natural resources, ignoring indirect contributions or inherent value. For this reason, natural resources tend to be severely undervalued. This is broadly the case in Myanmar, where decisions about natural resource use are typically based only on direct use values, such as timber or hydroelectric revenues. Generally, it is perceived that the immediate benefits of exploiting a natural resource are more attractive than the long-term benefits accrued from conservation of a resource, such as watershed protection, soil erosion control or other ecological services.

12.7.5.6   Limited Grassroots Support for Conservation

Although in general, the people of Myanmar are supportive of conservation objectives, rural people living in close proximity to protected areas may not be supportive of conservation efforts and protected area management. Reasons for this may include low awareness about the objectives or value of conservation, lack of mechanisms for local communities to benefit from PAs, and limited opportunities for grassroots participation in conservation activities.

12.7.6   Climate Change Vulnerability

There have been no comprehensive studies on the impacts of climate change on biodiversity in Myanmar, but the country is likely to face the impacts of rising global average temperatures in several areas. The Projected Climate Change and Vulnerability report for Myanmar (2001–2100) predicts the following:

- a general increase in temperature across the country, particularly from December to May, with the Central and Northern regions experiencing the greatest increases;
- an increase in clear sky days, exacerbating drought periods;
- an increase in rainfall variability during the rainy season including an increase across the country from March to November (particularly in Northern Myanmar), and decrease between December and February;
- an increase in the risk of flooding resulting from a late onset and early withdrawal of monsoon events;
- an increase in the occurrence and intensity of extreme weather events, including cyclones/strong winds, flood/storm surges, intense rains, extreme high temperatures and drought.

It is currently difficult to predict detailed future national climate patterns due to a lack of localized data, but preparations nonetheless required to mitigate or adapt to the broad trends that are expected. Some researchers have hypothesized that the dry zone of Myanmar is migrating slowly to the southeast and more comprehensive study and monitoring are urgently needed.

12.8   Conclusion

Myanmar's PAs are vital to sustaining the biodiversity and ecosystem services that underpin sustainable development, poverty reduction, climate stability, and natural disaster reduction. Despite their importance, there are still many challenges to managing these areas. In particular, the PA network has insufficient funding to ensure its effective management. This is starting to change. Over the past few years, both government and international funding for PAs have increased significantly. However, adequate funding remains a key constraint. Though PAs cannot offer full achievement for the biodiversity conservation mainly due to the numerous constraints that have to be addressed immediately, some studies revealed that forests within the PAs has been more effectively conserved than forests outside the PAs (Htun et al., 2010; Songer et al., 2009). These findings indicate that Myanmar's PAs are not merely 'Paper Park' as claimed by others (Aung, 2007; Rao et al., 2002), and Myanmar's PAs achieve a degree of forest and its diversity conservation though they are managing in the given inadequate resources.

Keywords

- conservation
- domesticated biodiversity
- threats
- wild species

References

Aung, U. M., (2007). Policy and practice in Myanmar's protected area system. *J. Environmental Management, 84*(2), 188–203.

Bryant, R. L., (1996). *The Political Ecology of Forestry in Burma.* University of Hawaii Press, Honolulu.

CBD, *A New Era of Living in Harmony with Nature is Born at the Nagoya Biodiversity Summit.* Available online at: http://www.cbd.int/doc/press/2010/pr-2010-10-29-cop-10-en.pdf.

FAO, (2010). *Global Forest Resources Assessment, Country report: Myanmar.* FRA2010/141, Rome, Italy. Available online at: http://www.fao.org/docrep/013/ al576E/al576E.pdf.

FishBase, (2015). Retrieved from http://www.fishbase.org/search.php.

Htun, N. Z., Mizoue, N., Kajisa, T., Yosida, S., (2010). Deforestation and forest degradation as measures of Popa Mountain Park (Myanmar) effectiveness. *Environmental Conservation, 36,* 218–224.

Ministry of National Planning and Economic Development, (2013). *Millennium Development Goals Report.* Republic of the Union of Myanmar, Nay Pyi Taw.

Myanmar's National Biodiversity Strategy and Action Plan (2015–2020). Available online at: https: //www.cbd.int/countries/?country=mm.

Myanmar's Fifth National Report to the Convention on Biological Diversity, (2013). Available online at: https: //www.cbd.int/countries/?country=mm.

Myers, N., Mittermeier, R. A., Mittermeier, C. G., Fonseca, G. A. B., Kent, J., (2000). Biodiversity hotspots for conservation priorities. *Nature, 403,* 853–858.

Rao, M., Rabinowitz, A., Khaing, S. T., (2002). Status review of the protected area system in Myanmar, with recommendations for conservation planning. *Conservation Biology*, *16*, 360–368.

Songer, M., Aung, M., Senior, B., DeFrics, R., Leimgruber, P., (2009). Spatial and temporal deforestation dynamics in protected and unprotected dry forests: A case study from Myanmar. (Burma). *Biodiversity Conservation*, *18*, 1001–1018.

UN-REDD Programme, (2010). *The United Nations Collaborative Programme on Reducing Emissions from Deforestation and Forest Degradation in Developing Countries.* Available online at: http://www.un-redd.org/AboutREDD/tabid/582/language/en-US/Default.aspx.

WCMC, World Conservation Monitoring Centre. Protected areas database in 2011.

Plate 12.1 Ecosystem diversity in Myanmar; from top left to bottom right:
Mountain glacier; Hill forest; Mountain lake; Wetland; Mangrove;
Coral; Mixed deciduous forest and Pine forest (Photo credits: Shijo
Onishi; Forest Department; WCS; FFI; OIKOS).

Plate 12.2 Wildlife diversity in Myanmar; from top left to bottom right: Tiger (*Panthera tigris*); Asian elephant (*Elephas maximus*); Surus crane (Grusantigones); Myanmar golden deer (*Cervus eldithamin*); Leopard (*Pantera pardus*); Asiatic black bear (*Selenarctos thibetanus*); Estuarine crocodile (*Crocodylus porosus*) and Sea turtles (Photo credits: Shijo Onishi; Forest Department; FFI; WCS).

Plate 12.3   Plant diversity in Myanmar; from top left to bottom right: Teak (*Tectona grandis*); Htan (*Borassus flabellifer*); Orchid (*Dendrobium fimbriatum*); Black Orchid (*Paphiopedilum wardii* Summerh); Kanyin-byu (*Dipterocarpus alatus*); Taung-zalat-ni (*Rhododendron arboretum*); Dani (*Nypa fruticans*) and Pan-padauk (*Pterocarpus indicus*). (Photo credits: Forest Department; Saw Lwin; Kyaw Win Maung; ZENG Xiang-Le, XTBG; Naing Zaw Htun; Shijo Onishi).

Plate 12.4 Major causes of biodiversity loss in Myanmar; from top left to bottom right: Habitat loss; Shifting cultivation; Illegal poaching; Wildlife trafficking; Illegal logging; Fuelwood collection; Forest fire and Clearing for plantation (Photo credits: Forest Department; WCS; Naing Zaw Htun).

CHAPTER 13

Biodiversity in Nepal

KRISHNA K. SHRESTHA[1] and SIDDHARTHA B. BAJRACHARYA[2]

[1]Former Head, Central Department of Botany, Tribhuvan University, Kirtipur, Kathmandu, Nepal
E-mail: kkshrestha123@gmail.com
[2]National Trust for Nature Conservation (NTNC), Khumaltar, Lalitpur, Nepal,
E-mail: sid.bajracharya@gmail.com

13.1 Introduction

The biological diversity or biodiversity refers to the diversity of living organisms (microorganisms, plants, and animals) existing in the world. The three elements of biodiversity such as taxonomic diversity (or species diversity), ecological diversity (or ecosystem diversity), and morphological diversity (or genetic diversity) are linked with each other. Understanding of the integral relationship between species, ecosystems, populations, and genes in the Earth are vital issues and a global concern. Therefore, the United Nations Assembly has declared May 22nd as the "World Biodiversity Day" or the "International Day of Biodiversity," and the "Decade of Biodiversity" (2010–2020) to recognize the importance of biodiversity, and to promote public awareness on biodiversity issues.

Although Nepal occupies only 0.1% of the earth surface area, it is one of the biodiversity-rich countries. Nepal stands 27th position in the world, 10th position in Asia, and the second position in South Asia (Groombridge and Jenkins, 2002; MoFSC, 2014b; Shrestha, 2015). For example, among the South Asian countries, India has highest number of species of flowering plants (18,664 species), followed by Nepal (6,973 species), Bhutan (5,468 species), Bangladesh (5,000 species), Pakistan (4,950 species), and Sri Lanka (3,314 species); whereas, Maldives has only 260 species of flowering plants (Groombridge and Jenkins, 2002; Shrestha, 2016).

One of the reasons for the biological richness in the country is due to the extreme altitudinal gradient of Nepal's topography, which allows the occurrence of a wide variety of climatic condition from tropical in the lowland Tarai to nival zone in the High Himalaya. Ranging from as low as 60 m in the lowland Tarai (Kachanakalan, Jhapa, East Nepal) above mean sea level to the highest point on earth, the Mt. Everest (8,848 m), all within a distance of 150 km, Nepal comprises diverse habitat for flora and fauna. The shape of Nepal is long, roughly rectangular, with eastern line shorter than on the west.

It has a length of about 885 km., parallel to the Himalayan axis from the east to west, and an average width of 193 km. It has a total area of 147,181 km² and lies between the latitudes 26° 22' to 30° 27'N and the longitudes 80° 04' and 88° 12'E. The country is landlocked and lies between China on the North and India on the East, South, and West. The major components of land area in Nepal comprise forests (29%), shrubland (10.6%), grassland (12%), water bodies (2.6%), cultivated land (21%), and 7% land is non-cultivated (MoFSC, 2014b). However, the forest and shrubland area in Nepal has been increased to 40.36% and 4.38%, respectively (DFS, 2015).

13.2 Ecosystem Diversity in Nepal

Nepal is divided into seven provinces (previously under five development regions) and 77 districts, of which 20 districts lie in Tarai. Ecologically, Nepal is divided into three regions: the Tarai and Siwalik region (below 1000 m), Mid hill region (1000–3000 m), and the Mountain region (above 3000 m). However, Land Resource Mapping Project (LRMP) divided Nepal into five physical zones (LRMP 1986) namely, Tarai (below 500 m), Siwaliks (500–1000 m), Mid hills (1000–3000 m), High mountains (3000–5000 m), and High Himalaya (above 5,000 m).

Nepal comprises 118 ecosystems, with 112 forest ecosystems, 4 cultivation (agriculture) ecosystems, one water body ecosystem, and one glacier/snow/rock ecosystem (BPP, 1995). Lowland Tarai (60–500 m) comprises 14 ecosystems, 12 ecosystems in Siwaliks (500–1000m), 53 ecosystems in Mid hills (1000–3000 m), and 38 ecosystems in High hills (above 3000 m), whereas aquatic ecosystem occurs in all physiographic zones except Siwaliks (BPP, 1995).

There are nine global ecoregions in Nepal namely, Himalayan subtropical broad-leaved forest, Himalayan subtropical pine forest, Eastern Himalayan broad-leaved forest, Western Himalayan broad-leaved forest, Eastern Himalayan subalpine conifer forest, Western Himalayan subalpine conifer forest, Tarai-Duar savannah and grassland, Eastern Himalayan alpine shrub and meadows, and Western Himalayan alpine shrub and meadows (Olson et al., 2001; Wickamanayake et al.,2002; MoFSC 2009; Miehe et al., 2015).

Using floristic and phytogeographical data, Stearn (1960) divided Nepal into three regions (Western, Central, and Eastern Nepal), which is equivalent to Karnali, Gandaki, and Koshi sections of Nepal (Hara et al., 1978).

1. *Western Nepal*: from Kumaon frontier (border of NW India) to 83°E (Karnali basin),

2. *Central Nepal*: from 83° E to 86° 30 E (Gandaki basin), and
3. *Eastern Nepal*: from 86° 30 E to Sikkim (border of NE India) frontier (Koshi basin).

13.2.1 Forest Ecosystems

The vegetation of Nepal has been variously presented based on altitudinal belts, climatic zones, humidity types, plant life forms, and human impacts. Stainton (1972) recognized 35 different forest types in four major categories, based on altitude and climate, which are still widely accepted in Nepal.

a. Tropical and subtropical (10 forest types: Sal forest to Chir pine forest),
b. Temperate and alpine broadleaved (with 11 forest types: Oak forest to Birch forest),
c. Temperate and alpine conifer (8 forest types: Fir forest to Larch forest), and
d. Minor temperate and alpine associations (6 forest types: from Alder wood to Juniper forest).

Recently, Miehe et al. (2015) classified the forest types into three major categories (i.e., vegetation types of the southern slope of the Himalayas, vegetation types of the inner valleys and vegetation types of Arid Zone), based on altitudinal belts, climatic zones, humidity types, plant life forms, and human impact, which include 55 primary forest types (Shrestha et al., 2018) as discussed in the following subsections.

13.2.1.1 Vegetation and Forest Types

1. Vegetation types in Southern slopes of Himalayas
   a. Tropical belt (5 types): *Shorea robusta* forest; *Terminalia* and *Anogeissus* forest; Riverine grassland; *Dalbergia sissoo-Acacia catechu* riverine forest; and *Bombax* riverine forest.
   b. Subtropical belt (7 types): *Schima-Castanopsis* forest; *Quercus lanata* forest; *Pinus roxburghii* forest; *Toona ciliata-Albizia julibrissin* riverine forest; *Alnus nepalensis* riverine forest; *Euphorbia royleana* grasslands; and Thickets and pastures.
   c. Cloud forest belt (15 types): *Quercus lamellosa* forest; *Lithocarpus pachyphylla* forest; *Quercus floribunda* forest; *Quercus semecarpifolia* forest; *Tsuga dumosa* forest; *Rhododendron*

    arboreum forest; *Rhododendron hodgsonii* forest; *Abies specta-bilis* forest; *Abies densa* forest; *Juniperus recurva* forest; *Juniperus recurva* thickets; *Rhododendron* thickets; Bamboo thickets; Tall forb communities; and Upper tree-line.

    d.   Alpine belt (3 types): *Rhododendron* dwarf thickets; *Kobresia nepalensis* mats; and Crustose lichen covers of rock walls.

2.   Vegetation types in the Inner Valleys

    a.   Subtropical belt (1 type): *Olea europea* subsp. *cuspidata* woodlands.

    b.   Temperate belt (13 types): *Cedrus deodara* forest; *Aesculus-Acer* forest; *Pinus wallichiana* forest; *Picea smithiana* forest; *Abies pindrow* forest; *Betula utilis* forest; *Cupressus torulosa* forest; *Juniperus indica* forest; *Larix* forest; *Hippophae* riverine woodlands; *Caragana sukiensis* thickets; *Rhododendron lepidotum* shrublands; and *Rosa-Berberis-Cotoneaster* shrublands.

    c.   Alpine belt (5 types): *Juniperus squamata* dwarf shrublands; *Kobresia pygmaea* dwarf mats; High alpine cushion communities; *Hippophae tibetana* riverine dwarf thickets; and Pioneer plant successions in glacial forelands.

3.   Vegetation types in Arid Zone (6 types)

Drought line of forests; *Caragana gerardiana* open dwarf shrublands; *Caragana versicolor* open dwarf shrublands; Alpine steppe; Salt meadows; and Plant communities of wastelands.

Major components of biodiversity such as trees, shrubs, and climbers, as well as mammals and birds in three major physiographic zones of Nepal are summarized in the following subsection.

13.2.1.2 Tarai and Siwalik (60–1000 m): Tropical Zone

Lowland Tarai comprises 23 ecosystems, with 1, 885 species of angiosperms, 648 species of birds and 68 species of mammals (BPP, 1995; Bhuju et al., 2007).

- **Trees:** *Acacia catechu, Butea monosperma, Cassia fistula, Dalbergia sissoo* (Fabaceae), *Haldina cordifolia* (Rubiaceae), *Bombax ceiba* (Malvaceae), *Lagerstroemia parviflora* (Lythraceae), *Mallotus phillipinensis, Trewia nudiflora* (Euphorbiaceae), *Shorea robusta* (Dipterocarpaceae), *Syzygium cumini* (Myrtaceae), *Terminalia belerica, T. chebula* and *T. tomentosa* (Combretaceae).

- **Shrubs and climbers:** *Bauhinia vahlii, Butea buteiformis, Mimosa rubicaulis* (Fabaceae), *Clerodendrum viscosum, Colebrookea oppositifolia* (Lamiaceae), *Dioscorea deltoidea* (Dioscoreaceae), *Murraya koenigii* (Rutaceae), *Mussaenda macrophylla, Randia uliginosa* (Rubiaceae), *Thespesia lampas* (Malvaceae), and *Ziziphus mauritiana* (Rhamnaceae).
- **Mammals:** Royal Bengal Tiger (*Panthera tigris*), Greater one-horned Rhinoceros (*Rhinoceros unicornis*), Asiatic Elephant (*Elephas maximus*), Black buck (*Antelope cervicapra*), Gaur (*Bos gaurus*), Dolphin (*Platanista gangatica*), Sloth Bear (*Melursus ursinus*), Swamp dear (*Cervus duvaucelli*), Four-horned Antelope (*Tetracerus quadricornis*), and Wild water-buffalo (*Bubalus arnee*).
- **Birds:** Bar-headed geese (*Anser indicus*), Mallard (*Anas platyrhynchos*), Bengal florican (*Houbaropsis bengalensis*), Black-headed cuckooshrike (*Coracina melanoptera*), Asian Openbill (*Anastomus oscitans*), Dusky eagle owl (*Bubo coromandus*), Ruddy Shelduck (*Tadorna ferruginea*), Great Slaty Woodpecker (*Mulleripicus pulverulentus*), Steppe Eagle (*Aquila nipalensis*), and White rumped Vulture (*Gyps bengalensis*).

13.2.1.3 Mid Hills/Middle Mountains (1000–3000 m): Subtropical to Temperate Zone

The mid hills comprises 52 types of ecosystems, with 3,364 species of flowering plants, 493 species of bryophytes, 272 species of pteridophytes, and 16 species of gymnosperms (BPP, 1995). Similarly, this zone is enriched with 557 species of butterflies, 76 species of fish, 691 species of birds, and 110 species of mammals (MoFSC, 2002; Bhuju et al., 2007).

- **Trees:** *Acer oblongum* (Sapindaceae), *Alnus nepalensis* (Betulaceae), *Castanopsis indica, C. tribuloides, Quercus glauca* (Fagaceae), *Cinnamomum tamala* (Lauraceae), *Engelhardtia spicata, Juglans regia* (Juglandaceae), *Lyonia ovalifolia, Rhododendron arboreum* (Ericaceae), *Magnolia champaca, Magnolia hodgsonii* (Magnoliaceae), *Pinus roxburghii* (Pinaceae), *Schima wallichii* (Theaceae), and *Taxus wallichiana* (Taxaceae).
- **Shrubs and climbers:** *Astilbe rivularis* (Saxifragaceae), *Berberis asiatica, Mahonia napaulensis* (Berberidaceae), *Cissampelos pareira* (Menispermaceae), *Coriaria nepalensis* (Coriariaceae), *Gaultheria fragrantissima* (Ericaceae), *Maesa chisia* (Primulaceae), *Prinsepia*

utilis, Rubus ellipticus (Roscaeae), *Tetrastigma serrulatum* (Vitaceae), and *Viburnum mullaha* (Adoxaceae).

- **Mammals:** Clouded leopard (*Neofelis nebulosa*), Common Leopard (*Panthera pardus*), Himalayan Black bear (*Ursus thibetanus*), Red Panda (*Ailurus fulgens*), Indian Pangolin (*Manis crassicaudata*), Chinese Pangolin (*Manis pentadactyla*), Common Goral (*Naemorhedus goral*), Himalayan Serow (*Capricornis thar*), Common Langur (*Semnopithecus entellus*), Jungle Cat (*Felis chaus*), and Wild boar (*Sus scrofa*).

- **Birds:** Red-billed Blue Magpie (*Urocissa erythrorhyncha*), Scarlet Minivet (*Pericrocotus flammeus*), Spinny Babbler (*Turdoides nepalensis*), Yellow Bellied Fantail (*Rhipidura hypoxantha*), Chesnut-bellied Rock Thrush (*Monticola rufiventris*), White-capped Water Redstart (*Chaimarrornis leucocephalus*), Himalayan Bulbul (*Pycnonotus leucogenys*), Grey-headed Warbler (*Seicercus xanthoschistos*), Rufous Sibia (*Heterophasia capistrata*), and Green-tailed Sunbird (*Aethopyga nipalensis*).

13.2.1.4   High Hills/High Mountains and High Himalaya (Above 3000 m): Subalpine to Nival Zone

The high hills comprise 38 ecosystems, with 1,645 species of flowering plants in subalpine zone (3000–4000 m), and 1,071 species in the alpine and nival zone (above 4000 m).

- **Trees:** *Abies spectabilis, Cedrus deodara. Picea smithiana, Pinus wallichiana, Tsuga dumosa* (Pinaceae), *Betula utilis* (Betulaceae), *Juniperus indica, J. recurva* (Cupressaceae), *Larix griffithiana* (Pinaceae), *Lithocarpus pachyphylla, Quercus semecarpifolia* (Fagaceae), *Rhododendron campanulatum,* and *R. hodgsonii* (Ericaceae).

- **Shrubs and climbers:** *Berberis mucrifolia* (Berberidaceae), *Caragana versicolor* (Fabaceae), *Clematis acuminata* (Ranunculaceae), *Codonopsis convolvulacea* (Campanulaceae), *Cotoneaster microphyllus* (Rosaceae), *Ephedra gerardiana* (Ephedraceae), *Hippophae tibetana* (Elaeagnaceae), *Lonicera myrtillus* (Caprifoliaceae), *Rhododendron anthopogon, R. lepidotum* (Ericaceae), and *Ribes himalense* (Grossulariaceae).

- **Mammals:** Blue sheep (*Pseudois nayur*), Himalayan goral (*Naemorhedus goral*), Himalayan Tahr (*Hemitragus jemlahicus*), Musk deer (*Moschus chrysogaster*), Himalayan Serow (*Capricornis thar*), Snow

leopard (*Panthera uncia*), Tibetan wolf (*Canis lupus*), Pallas's Cat (*Felis manul*), Lynx (*Felis lynx*), and Tibetan Wild Ass (*Equus kiang*).
- **Birds:** Cheer Pheasant (*Catreus wallichii*), Himalayan Monal (*Lophophorus impejanus*), Satyr Tragopan (*Tragopan satyra*), Lammegeier (*Gypaetus barbatus*), Himalayan Griffin (*Gyps himalayensis*), Golden Eagle (*Aquila chrysaetos*), Grey-backed Shrike (*Lanius tephronotus*), Common kingfisher (*Alcedo atthis*), Collared Grosbeak (*Mycerobas affinis*), and Eurasian woodcock (*Scolopax rusticola*).

13.2.2 Rangeland Ecosystems

Rangeland ecosystems in Nepal are comprised of grasslands, pastures, and shrublands that cover about 1.7 million hectares or nearly 12% of the country's land area (MoFSC, 2014b). A significant rangeland area is located in the high mountains. The country's rangeland ecosystems are broadly grouped into five categories: (i) tropical savannas, (ii) subtropical rangelands, (iii) temperate rangelands, (iv) subalpine rangelands, and (v) alpine meadows. Among these five categories, the alpine meadows are exceptionally rich in floral diversity, including numerous species of colorful flowers of alpine herbs, such as *Caltha*, *Potentilla*, *Primula*, *Ranunculus*, etc. Varied associations of *Rhododendron* spp. and *Juniperus* spp. are found scattered across the meadows.

Many of the species found in these rangelands are endemic to Nepal and others are high-value medicinal and aromatic plants, such as *Dactylorhiza hatagirea* (Orchidaceae), *Nardostachys jatamansi* (Caprifoliaceae), *Neopicrorhiza scrophulariiflora* (Plantaginaceae), and *Ophiocordyceps sinensis* (Ophiocordycipataceae), These grasslands are also home to endangered snow leopard, Himalayan goral (*Naemorhedus baileyi*), serow (*Capricornis sumatraensis*), and Himalayan tahr (*Hemitragus jemlahicus*) (MoFSC, 2014b).

13.2.3 Wetland Ecosystems

Wetland ecosystems of Nepal fall into two broad categories: (i) natural wetlands, comprising of lakes and ponds, riverine floodplains, swamps and marshes, and (ii) man-made wetlands, including water reservoirs, ponds, and deep-water paddy fields. Nearly half (45%) of the wetlands are in High Himalaya. This is mainly due to a large number of glaciers and glacial lakes in the Himalayan region. Nepalese wetlands have very high ecological significance, as they harbor many threatened species of

flora and fauna and serve as resting places for many migratory and globally threatened birds (MoFSC, 2014b). Similarly, aquatic, semiaquatic,
and marshy plants, such as *Acorus calamus* (Acoraceae), *Hydrilla verti-
cillata* (Hydrocharitaceae), *Eichhornia crassipes, Monochoria vaginalis*
(Pontederiaceae), *Nasturtium officinale* (Brassicaceae), *Nelumbo nucifera*
(Nelumbonaceae), *Nymphaea nouchali* (Nymphaeaceae), *Potamogeton
crispus* (Potamogetonaceae), and *Trapa quadrispinosa* (Lythraceae), are
the characteristic flora in the wetlands.

The wetland ecosystems offer excellent habitats to at least 230 indigenous species of fish belonging to 104 genera of high economic, environmental and academic value (Rajbanshi, 2013). The wetlands also have high
cultural and economic significance. Many ethnic groups are dependent on
wetlands for their livelihoods. The wide variety of plants and animals that
the wetlands support provide a wide range of goods and services as well
as income-generating opportunities for the local people. Loss and degradation of these vital natural resources during the last few decades has severely
affected these relationships. The wide variety of plants and animals that the
wetlands support provide a wide range of goods and services as well as
income-generating opportunities for the local people. Loss and degradation of these vital natural resources during the last few decades has severely
affected these relationships (MoFSC, 2014b).

13.2.4 *Agroecosystems (Agrobiodiversity)*

The diverse climatic and topographic conditions have favored for maximum diversity of agricultural crops, their wild relatives, and animal species
in Nepal. Crops, livestock, and forests are the three major components of
the country's complex farming systems. The traditional farming systems,
which use local knowledge and experiences and vary across the country, are
assumed to have a great role in maintaining the agricultural diversity. Forests are integral components of such systems (MoFSC, 2014b). The agro-
ecosystem is considered as the integral part of the local livelihood for their
existence and the main source of national economy. Some common species
of agricultural crops in the agro-ecosystems are categorized below (Chaudhary, 1998; Bhuju et al., 2007):

- **Cereals (Poaceae):** *Avena sativa* (Oats, 'Jau'), *Eleusine coracana*
  (millet, 'Kodo'), *Hordeum vulgare* (barley, 'Jau'), *Oryza sativa* (rice;
  'Dhaan'), *Sorghum vulgare* (Great millet, 'Junelo'), *Triticum aestivum*
  (wheat, 'Gahun'), and *Zea mays* (maize, 'Makai').

- **Legumes (Fabaceae):** *Cajanus cajan* (Pigeon pea, 'Arhar'), *Cicer arietinum* (Chickpea 'Chana'), *Lens culinaris* (Lentil, 'Musuro'), *Vicia faba* (Broadbean, 'Bakula'), *Vigna radiata* (Urd bean, Kalo maas).
- **Vegetables:** *Brassica oleracea* (Cabbage, Brassicaceae), *Lablab purpureus* (Beans, Fabaceae), *Lycopersicon esculentum* (Tomato, Solanaceae), and *Solanum tuberosum* (Potato, Solanaceae), etc.
- **Fruits:** *Citrus reticulata* (Orange, Rutaceae), *Malus domestica* (Apple, Rosaceae), *Mangifera indica* (Mango, Anacardiaceae), *Pyrus communis* (Pear, Rosaceae), and *Vitis vinifera* (Grape, Vitaceae), etc.
- **Oil crops:** *Brassica juncea* (Mustard, Brassicaceae), *Glycine max* (Soybean, Fabaceae), *Guizotia abyssinica* (Niger, Asteraceae), and *Helianthus annuus* (Sunflower, Asteraceae).
- **Beverage crops:** *Camellia sinensis* (Tea, Theaceae), *Cocos nucifera* (Coconut, Palmae), *Coffea arabica* (Coffee, Rubiaceae), and *Hippophae salicifolia* (Seabuck thorn, Elaeagnaceae).
- **Cash crops:** *Beta vulgaris* (Sugar beet, Amaranthaceae), *Corchorus capsularis* (Jute, Malvaceae), *Gossypium hirsutum* (Cotton, Malvaceae), and *Saccharum officinarum* (Sugarcane, Poaceae), etc.

13.3 Species Diversity in Nepal

Botanical explorations in Nepal were initiated during 1802–1803 by a British Naturalist Sir Francis Buchanan-Hamilton, who traveled to Nepal via Raxaul to Kathmandu. During his stay in Kathmandu, he had collected several hundred specimens of flowering plants from around the Kathmandu Valley, and described nearly 1200 species, of which 800 species are new to science. Thus Buchanan-Hamilton is well known as "Father of Nepalese Botany." Later, Edward Gardner (1818), Nathalien Wallich (1820–1821) visited Central Nepal, especially in and around Kathmandu Valley, followed by Sir J.D. Hooker (1848) explored Eastern Nepal, bordering Eastern India, and J.F. Duthie (1884) in Far-west Nepal, bordering Northwest India (Press and Shrestha, 1999; Rajbhandari, 2016).

Since 1950s, not only the British explorers, but explorers from Japan, USA, France, Switzerland, India, Russia, and many other countries have collected thousands of herbarium specimens from different physiographic zones of Nepal. Major plant collectors in Nepal include O. Polunin, J.D.A. Stainton, H. Hara, H. Kanai, D. Nicolson, J.F. Dobremez, C. Grey-Wilson, M. Suzuki, M. Watson, etc. Their collections are deposited in the world leading Herbaria of the United Kingdom (BM, E, K), Japan (TI, KYO, TNS),

USA (GH, US), France (P, GRE), Switzerland (G, G-DC), India (CAL, DD), Russia (LE), China (PE, KUN), etc. (Shrestha, 2015; Rajbhandari, 2016).

Brian Hodgson, a British resident, was a pioneer Zoologist who visited Nepal during 1820, but extended his research during his second visit to Nepal in 1824, and later in 1833. He described numerous species of mammals and birds from the Himalayas, especially from around the Kathmandu Valley. During his 20 years of research, he discovered 39 species of mammals and 124 species of birds from the Himalaya as new to Science. Hodgson donated all his collections (10,499 specimens) and 2000 drawings of birds, mammals, and fish to the British Museum, London. Due to his incredible contribution Brian Hudgson is considered as the "Father of Nepalese Zoology."

American ornithologist Dr. S.D. Ripley did a survey and wrote the book 'The Spiny Babler and Bird Hunting in Nepal' in 1948. Various other scientific expeditions to explore animals such as the British Expedition (British Museum) in 1924, 1954, 1961, and 1962; the German Scientific Expedition led by Prof. Walter Hellmich in 1960–1965; Canadian Nepal Expedition in 1967; Swiss Entomological Expedition in 1968; and Japanese Expedition to Nepal Himalayas in 1952. Robert Fleming and R. Fleming, Jr. made an ornithological exploration in Nepal and published the 'Birds of Nepal' in 1976 (Fleming et al., 1976).

The number of species belonging to plant kingdom, animal kingdom, and fungi in Nepal is believed to be well documented, whereas, the microorganisms, especially Monera (Bacteria, and Archaea), Protoctista or Protista (Algae, diatoms, and protozoa) are poorly documented so far. The list of Bacteria, Viruses, Algae, and Fungi (Mycota), described from Nepal, and its comparison with the described and estimated species in the world is presented in Table 13.1.

13.3.1 Plant Diversity in Nepal

The non-flowering plant species in Nepal is about 5000 species, including 807 species of Algae (Baral, 1995; Prasad, 2013), 212 species of Fungi (Kost and Adhikari, 2015), 850 species of Lichens (Miehe et al., 2015), 1,150 species of Bryophytes (Pradhan and Joshi, 2008; Pradhan, 2016), and 534 species of Pteridophytes (Thapa, 2002; Fraser-Jenkins et al., 2015). The documentation of the non-flowering plants in Nepal is based on various publications till 2009 (Kunwar et al., 2010; Shrestha 2016; Shrestha et al., 2018).

In terms of flowering plant diversity, Nepal stands 10th position in Asia and second position in South Asia. It has been estimated that Nepal comprises about 7,500 species of vascular plants, of which 7,000 species belongs to flowering plants (Hara et al., 1978; Groombridge and Jenkins 2002). So far 28 species of gymnosperms, including 20 indigenous species, have been reported from Nepal (Bista, 2006). Based on various publications related to Flora of Nepal, 6653 species of flowering plants have been reported so far (Hara et al., 1978, 1982; Hara and Williams, 1979; Press et al., 2000; Bista et al. 2001; Kunwar et al., 2010), and may go up to 7,000 species in future (Table 13.2).

The top 10 largest families of flowering plants in Nepal are: Orchidaceae (452 species; Rajbhandari, 2015), Asteraceae or Compositae (395 species), Poaceae or Gramineae (366 species), Fabaceae or Leguminosae (304 species), Rosaceae (192 species), Cyperaceae (191 species), Scrophulariaceae *sensu lato* (167 species), Lamiaceae or Labiatae (150 species), Ranunculaceae (143 species) and Apiaceae or Umbelliferae with 123 species (Press et al., 2000; Rajbhandari, 2016).

Similarly, the top 10 largest genera of flowering plants in Nepal are: *Saxifraga* (Saxifragaceae) with 84 species, *Pedicularis* (Orobanchaceae) 72 species, *Carex* (Cyperaceae) 66 species, *Primula* (Primulaceae) 62 species, *Impatiens* (Balsaminaceae) 49 species, *Gentiana* (Gentianaceae) 47 species, *Corydalis* (Papaveraceae) 41 species, *Juncus* (Juncaceae), 40 species, *Saussurea* (Asteraceae) 39 species, and *Ficus* (Moraceae) with 35 species (Press et al. 2000; Shrestha 2016).

Publication of Flora of Nepal is progressing ahead. The Royal Botanic Garden Edinburgh (UK), University of Tokyo (Japan), Nepal Academy of Science & Technology (NAST) in collaboration with Ministry of Forest & Soil Conservation and Tribhuvan University (Nepal) are planning to publish Flora of Nepal (incorporating vascular plants) in 10 volumes, and so far Vol. 3 (Magnoliaceae-Rosaceae) has been published (Watson et al., 2011), and remaining volumes are in progress. In addition, a Companion Volume to the Flora of Nepal is also published (Miehe at al., 2015; Shrestha et al., 2018).

13.3.1.1 Endemic Flora

Endemism is the non-random distribution congruence among different taxa in an area, whereas endemic plants are those which have a very narrow distributional range (Hobohm and Tucker, 2014). Initial documentation reveals 246 species of endemic flowering plants in Nepal (Shrestha and Joshi 1996). It was estimated that Nepal comprises 315 species of endemic plants, especially the flowering plants (Groombridge and Jenkins, 2002). A recent publication on the updated list of endemic plants of Nepal has cataloged 324 species of flowering plants, representing 45 families and 132 genera (Rajbhandari et al., 2016).

Table 13.1 Species Diversity of Microorganisms and Allied Group in Nepal

Organisms (*Kingdom*)	Species in Nepal	References	Species in the World	Estimated species in the World	References	Nepal (%)
Monera						
Viruses	N/A	-	4,000	50,000–1,000,000	Groombridge and Jenkins (2002)	
			2,085	400,000	Chapman (2009)	N/A
Bacteria	N/A	-	10,000	100,000–3,000,000	Groombridge and Jenkins (2002)	
			7,643	400,000–1,000,000	Chapman (2009)	N/A
Protoctista						
Algae, Diatoms, and Protozoa	-		80,000	600,000	Groombridge and Jenkins (2002)	N/A
Algae	1001	Prasad (2013)	12,272	40,000	Chapman (2009)	8.16
Mycota (Fungi)	2,182	Kost and Adhikari (2015)		200,000–2,700,000	Groombridge and Jenkins (2002)	
	2,112	Adhikari (2012)	98,998	1,500,000	Chapman (2009)	2.13
Myxomycota	150					
Zygomycota	46					
Ascomycota (incl. Deuteromycota)	770					
Basidiomycota	1146					

Source: Chaudhary (1998); MoFSC (2014b); modified version, Shrestha et al. (2018).

Table 13.2 Plant Diversity in Nepal

Organisms (Kingdom)	Species in Nepal	References	Species in the World	Estimated species in the World	References	Nepal (%)
Plantae (Plants)						
Lichens	465*	Sharma (1995)	17,000	25,000	Chapman (2009)	5.00
Bryophytes	850	Miehe et al. (2015)				
	1,150*	Pradhan and Joshi (2008)				
	1,213	Pradhan (2016)				
	1,138	Meihe et al. (2015)	16,236	22,750	Chapman (2009)	7.01
Liverworts	350					
Mosses	780					
Hornworts	8					
Pteridophytes	534*	Thapa (2002)				
	550	Fraser-Jenkins et al. (2015)	12,000	15,000	Chapman (2009)	4.58
Gymnosperms	26*	Bista (2006)	1,021	1,050	Chapman (2009)	2.55
Angiosperms (Flowering plants)			250,000		Ziegler (2007), Byng (2014)	
	6653	Press et al. (2000), Kunwar et al. (2010)	268,600	352,000	Chapman (2009)	2.48
	6,973*	Groombridge and Jenkins (2002)	369,400	500,000	RBG Kew (2016)	

Source: *MoFSC, 2014b; modified version, Shrestha et al. (2018).

Most of the endemic species shows the distribution in the sub-alpine zone (3000–4000 m), particularly in Central Nepal (Vetaas and Grytnes, 2002). It is estimated that 63% species of flowering plants are endemic to high mountains, 38% species are from the mid-hills, and only 6% species are from lowland (Bhuju et al., 2007). According to Rajbhandari et al. (2016), the endemic species of flowering plants are categorized as follows:

- **Largest families of Endemic plants in Nepal:** Apiaceae (28 species), Asteraceae (22 species), Fabaceae (21 species), Saxifragaceae (21 species), Orchidaceae (20 species), Papaveraceae (20 species), Ranunculaceae (20 species), Poaceae (18 species), Orobanchaceae (14 species), Balsaminaceae (10 species), Gentianaceae (10 species), and Lamiaceaee (10 species).
- **Largest genera of Endemic plants in Nepal:** *Saxifraga* (21 species), *Pedicularis* (13 species), *Meconopsis* (11 species), *Aconitum* (10 species), *Impatiens* (10 species), *Astragalus* (9 species), *Corydalis* (9 species), *Acronema* (8 species), *Begonia* (7 species), *Carex* (7 species), *Saussurea* (7 species), and *Silene* (7 species).
- **Endemic trees:** *Litsea doshia* (D.Don) Kosterm. (Lauraceae), *Machilus pubescens* (Lauraceae), *Mallotus bicarpellatus* T. Kuros. (Euphorbiaceae), *Prunus himalaica* Kitam., *Prunus taplejungnica* H. Ohba & S. Akiyama, *Prunus topkegolensis* H. Ohba & S. Akiyama (Rosaceae), *Salix nepalensis* Yonek. (Salicaceae), and *Sorbus sharmae* M.F. Watson et al. (Rosaceae).

Documentation of the endemic species of non-flowering plants and allied group (fungi) is comparatively poor. It was presumed that 12 species of algae are endemic to Nepal (Baral, 1995), and 48 species of lichens as endemic to Nepal (Bhuju et al., 2007). In fungi, the estimated number of endemic species is 142 species (Kost and Adhikari, 2015). Similarly, 30 species of bryophytes are reported as endemic to Nepal (Pradhan, 2016).

13.3.2 Faunal Diversity in Nepal

The faunal species diversity in Nepal is relatively high. Nepal occupies about 0.1% of the global area but harbors over 1% of the world's known fauna (MoFSC, 2014a). Higher group of animals or the vertebrates are well documented; however, the lower groups of animals or the invertebrates are poorly documented, except some group such as butterfly, moths, and spiders (Chaudhary, 1998; Bhuju et al., 2007). A checklist of helminth parasites documented 168 species, of which 33 species are trematodes, 67 species of nematodes, and 36 species cestodes, and 32 species of plant nematodes (Gupta, 1997). Similarly, 651 species of butterflies, 785 species of moths have been documented from Nepal (Thapa, 1998). Similarly, 175 species of spiders (Thapa and Rana, 2001) are documented in Nepal. It is estimated that 5,052 species of insects occur in Nepal (Thapa, 1997).

In the animal kingdom, the invertebrates are poorly documented, whereas the vertebrates are well documented. Moreover, it has been realized that systematic documentation of fauna in Nepal is lacking. The list of invertebrate species and vertebrate species described from Nepal, and comparison with the list of described and estimated species in the world is given in Table 13.3.

Nepal comprises 187 species of fishes (Shrestha, 1995), 117 species of amphibians (frogs, toads, etc.), 78 species of reptiles (lizards, snakes, alligators, etc.) in various physiographic zones (Shah and Tiwari, 2004; Shah, 2013). Nepal has recorded 867 species of birds, among them, the resident birds are 62%, winter visitors 14%, passage migrants 12%, summer visitors 6%, residents and migrants 5%, and 1% summer and winter visitors (Fleming et al., 1976; BCN and DNPWC, 2011). Nepal is home to 35 globally threatened bird species, 19 near threatened species and 15 restricted range species. There are 208 species of mammals in Nepal (Baral and Shah, 2008; Jnawali et al., 2011), including four extinct species, representing 5.2% of the world's total. However, in the Checklist of Mammals of Nepal, 192 species are recorded (Thapa, 2014).The mid hills of Nepal have the highest number of birds (77%) and mammals (55%), whereas the Tarai-Siwalik zone has the highest number of fish species (74%) and herpetofauna (45%).

13.3.2.1 Endemic Fauna

A total of 160 animal species are endemic to Nepal which includes two species of mammal (*Apodemus gurkha*, Himalayan field mouse) which occurs in the coniferous forest of Central Nepal (2200–3600 m), and Csorba's Mouse-eared Myotis (*Myotis csorbai*) (Bhuju et al., 2007, Gotame, 2007, Jnawali et al., 2011). One species of bird *(Turdoides nipalensis*, Spiny Babbler), 6 species of fish, 14 species of Herpeto fauna, and 108 species of spiders (BPP, 1995; Bhuju et al., 2007) are reportedly endemic to Nepal (Table 13.4).

13.4 Genetic Diversity

Plant Genetic Resources (PGR), the genetic materials of plants, is an important component of biodiversity. At least 31,128 plant species are documented

Table 13.3 Species Diversity of Animals (Fauna) in Nepal

Organisms (*Animal Kingdom*)	Species in Nepal	References	Species in the World	Estimated species in the World	References	Nepal %
Invertebrates			~1,359,365	~6,755,830	Chapman (2009)	
Platyhelminthes (flatworms)	168*	Gupta (1997)	20,000	(~80,000)	Chapman (2009)	1.4
Nematoda (nematodes, roundworm)	N/A		<25,000	~500,000	Chapman (2009)	
Annelida (earthworm, etc.)	N/A		16,763	25,000–30,000	Chapman (2009)	
Mollusca (mollusks, shellfish)	192	Budha (2012), MoFSC (2014b)	~85,000	~200,000	Chapman (2009)	
Crustacea (Crabs, lobsters)	59*	Tiwari and Chhetry (2009)	47,000	150,000	Chapman (2009)	
Myriapoda (millipedes, centipedes)	N/A		16,072	~90,000	Chapman (2009)	
Rotifers	61*	Surana et al. (2005)				
Arachnida (Spiders, scorpions)	175*	Bhuju et al. (2007)	~102,248	~200,000–600,000	Chapman (2009)	0.2
Insecta (Other Insects)	5,052*	Thapa (1997)	1,015,897	~5,000,000	Groombridge and Jenkins (2002)	0.7
Lepidoptera	4,609	Bhuju et al. (2007); Haruta (2006)	174,250	300,000–500,000	Chapman (2009)	2.64
Butterflies	640	Smith (1989)				
	651*	Bhuju et al. (2007)				
Moths	3958*	Haruta (2006)				
	785	Bhuju et al. (2007)				

Organisms (*Animal Kingdom*)	Species in Nepal	References	Species in the World	Estimated species in the World	References	Nepal %
Vertebrates			64,788	~80,500*		
Pisces (Fishes)	187	Bhuju et al. (2007)	31,153	~40,000	Chapman (2009)	1.9
	230*	Rajabanshi (2013)				
Amphibia (Frogs)	59	Shah and Tiwari (2004)	6,515	~15,000	Chapman (2009)	1.8
	117*	Bhuju et al. (2007)				
Reptilia (Lizards, snakes)	118	Shah and Tiwari (2004)	10,038		Uetz (2016)	2.5
	123*	Schleich and Kastle (2002)	8,734	>10,000*	Chapman (2009)	1.9
Aves (Birds)	867*	BCN & DNPWC (2011)	9,990	>10,000*	Chapman (2009)	8.9
	878	Inskipp et al. (2016)				
Mammalia (Mammals)	185	Bhuju et al. (2007)	5,487	~5,500*	Chapman (2009)	5.2
	208*	Baral & Shah (2008); Jnawali et al. (2011)				

Source: *MoFSC (2014b), modified version.

Table 13.4 Endemic Fauna of Nepal

Categories	Number of endemic species	References
Spiders	108	BPP (1995)
Herpetofauna	14	Bhuju et al. (2007)
Fishes	6	Bhuju et al. (2007)
Birds	1	MoFSC (2014b)
	2	Groombridge and Jenkins (2002); Bhuju et al. (2007)
Mammals	1	MoFSC (2014b)
	2	Groombridge and Jenkins (2002); Jnawali et al. (2011)

as useful genetic resources. Most of the plants are used for medicinal purposes, followed by plants used in textiles and construction materials. For example, 17,810 species are documented having medicinal uses, 11,365 species are documented in textiles and construction materials, 8,140 species in agro-forestry, 5,538 species in human food, and 3,649 species are used as animal food. Furthermore, 5,338 species are possible gene sources, which are potentially useful for the genetic improvement of crops (RBG Kew, 2016).

Wild relatives of crops are recognized as the essential pool of genetic variation for the improvement of domestic crops in future. There are currently 3,546 global plant taxa recognized as crop wild relatives (Vincent et al., 2013). More than 3000 species of plants are widely used in Nepal for various domestic purposes and in trade, as the source of herbal drugs, food crops, fiber plants, etc.

Nearly 600 species of edible or food plants are reported from Nepal, of which 225 species are in cultivation, and remaining species belong to wild crops. About 15 species of cereal crops (such as rice, maize, wheat, barley, etc.) and pseudo-cereal crops (such as proso millet, foxtail millet, etc.), 18 species of grain legume crops, 23 species of fruit crops, and over 100 species of vegetable crops, spices and condiments are widely planted in various physiographic zones of Nepal. At least 83 species of the wild relative of agricultural crops have been documented so far (Shrestha and Shrestha, 1999).

Wild relative of cereal crops: Rice (*Oryza sativa*) is one of the main cereal crops or staple food in Nepal. Following species of the wild relative of rice such as *Hygrorhiza aristata, Leersia hexandra* ('Navo dhaan,' wild rice), *Oryza meyeriana, Oryza meyeriana* subsp. *granulata* (*O. granulata*), *Oryza minuta, Oryza officinalis*, and *Oryza rufipogon* (*O. nivara*) have been reported from Nepal. Similarly, several wild species of *Hordeum, Eleusine*,

Fagopyrum, and *Amaranthus* occur in Nepal, which are used as forage crops and some are weedy species (Shrestha and Shrestha, 1999; Press et al., 2000).

Wild relative of fruit crops: Several species of wild relative of fruits and vegetable crops occur as native and naturalized species in Nepal (Press et al., 2000). Some examples of wild species of fruits include *Fragaria* (*Fragaria daltoniana, F. nubicola*), *Mangifera* (*Mangifera sylvatica,* wild mango), *Morus* (*Morus macroura, M. serrata*), Musa (*Musa balbisiana,* wild banana), *Prunus* (*Prunus cerasoides, P. cornuta, P. davidiana, P. napaulensis*; as well as four endemic species *Prunus himalaica, P. jajarkotensis, P. taplejunghnica,* and *P. topkegolensis*), and *Pyrus* (*Pyrus pashia*), etc.

Wild relative of vegetable crops: The wild relative of vegetable crops include species of *Allium* (*Allium carolinianum, A. tuberosum, A. wallichii,* and one endemic species *Allium hypsistum*), *Asparagus* (*Asparagus filicinus, A. racemosus, A. tibeticus*; and one endemic species *Asparagus penicillatus*), *Dioscorea* (*Dioscorea deltoidea, D. hamiltonii, D. pentaphylla, D. pubera*), *Solanum* (*Solanum aculeatissimum, S. torvum*), *Trigonella* (*Trigonella corniculata, T. emodi*).

Other categories: Other categories of wild crop relative include species of cotton, jute, rattan, sugarcane, etc. For example, *Calamus* (*Calamus acanthospathus, C. latifolius, C. tenuis*), *Corchorus* (*Corchorus aestuans, C. olitorius*), *Gossypium* (*Gossypium arboreum, G. herbaceum*), *Saccharum* (*Saccharum arundinaceum, S. spontaneum,* and endemic species *Saccharum williamsii*).

Threats to the sustainable utilization and conservation of plant genetic resources of crop diversity, including wild crop relatives, are mainly due to degradation of habitats, shifting cultivation, encroachment, illegal harvesting, and issues of climate change. In this context, special attention should be paid to preserve the germplasm of traditional cultivars and wild relative of crops. Several national institutions such as National Agricultural Research Council (NARC), Nepal Academy of Science & Technology (NAST), Department of Plant Resources (DPR) and Tribhuvan University (TU) should take initiative *in situ* conservation in the natural habitat, *ex situ* conservation in the botanical gardens, and conserve germplasm in gene bank (Chaudhary and Bajracharya, 1995).

13.5 Threatened Plants and Animals

The 2000 IUCN Red List of Threatened Species enlisted 18,276 taxa of living organisms as threatened species in the world. This includes 816 species of Extinct species, 11,046 species of threatened species, 4,595 species of

Lower Risk categories of conservation dependent, and 1,769 infraspecific taxa of nearly threatened category (IUCN, 2013).

13.5.1   Threatened Plants in Nepal

According to IUCN Red List Categories, nearly 21% of the flowering plants of the world are currently threatened and it is estimated that one in five plants are threatened for extinction (RBG Kew, 2016). Initial assessment of IUCN Red List of Threatened species in Nepal reported nine species, of which two species were considered 'Endangered,' five species as 'Vulnerable,' and two species are 'Near Threatened' (MoFSC, 2014b). There are 18 threatened tree species found in mountains of Nepal (MoFSC, 2014a; MoFSC, 2014b). Nineteen plant species are protected in Nepal of which *Juglans regia*, *Dactylorhiza hatagirea*, and *Neopicrorhiza scroph-ulariiflora* are banned for collection, use, sale, transportation and export (MoFSC, 2014b). Moreover, Government of Nepal (Department of Plant Resources, MoFSC) has prioritized 30 species of plants as high value species for domestication and protection, of which 12 species are selected for research and development (MoFSC 2014a).

13.5.1.1   CITES List

The Convention on International Trade in Endangered species of Wild Fauna and Flora (CITES) has listed 15 species of plants, under CITES Appendices (MoFSC, 2014b). Altogether one species of fern (Appendix 1), four species of gymnosperms (Appendix II and III), eight species of flowering plants as well as all species of Orchidaceae (452 species) are placed under CITES Appendices: Appendix II and III (Table 13.5). Although information on lower group of plants with IUCN Threat categories is unknown, however, it has been estimated that 38 species of fungi are regarded as threatened species in Nepal (Kost and Adhikari, 2015).

13.5.1.2   Protected Plants of Nepal

According to National Gazette of Government of Nepal (Part 3, Sections 51 and 53), Forest Act 1993, 14 species of angiosperms, three species of gymnosperms, one species of fungi (Caterpillar fungus) and lichens (*Parmelia* species) are categorized as the protected plants of Nepal (DNPWC, 2003; IUCN, 2013; MoFSC, 2014b).

Table 13.5 CITES Listed Plants

S.N.	Plant Group	Taxa	Family	CITES Appendix
1.	Pteridophyta	*Cyathea spinulosa* Wall. ex Hook.	Cyatheaceae	II
2.	Gymnosperm	*Cycas pectinata* Griff.	Cycadaceae	II
3.	Gymnosperm	*Gnetum montanum* Markgr.	Gnetaceae	III
4.	Gymnosperm	*Podocarpus neriifolius* D. Don	Podocarpaceae	III
5.	Gymnosperm	*Taxus wallichiana* Zucc.(incl. *Taxus contorta, T. mairei*)	Taxaceae	II
6.	Angiosperm (Monocot)	*Dioscorea deltoidea* Wall. ex Griseb.	Dioscoreaceae	II
7.	Angiosperm (Monocot)	Orchids	Orchidaceae	II
8.	Angiosperm (Dicot)	*Ceropegia pubescens* Wall.	Apocynaceae (Asclepiadaceae)	II
9.	Angiosperm (Dicot)	*Meconopsis regia* G. Taylor	Papaveraceae	III
10.	Angiosperm (Dicot)	*Nardostachys grandiflora* DC.	Caprifoliaceae (Valerianaceae)	II
11.	Angiosperm (Dicot)	*Podophyllum hexandrum* Royle	Berberidaceae	II
12.	Angiosperm (Dicot)	*Rauvolfia serpentina* (L.) Benth. ex Kurz	Apocynaceae	II
13.	Angiosperm (Dicot)	*Talauma hodgsonii* Hook.f. & Thomson	Magnoliaceae	III
14.	Angiosperm (Dicot)	*Tetracentron sinense* Oliv.	Trochodendraceae (Tetracentraceae)	III

Source: DNPWC (2003), modified version, Shrestha et al. (2018).

- Category 1: Species banned for collection, use, sale, distribution, transportation and export comprises three species (*Dactylorhiza hatagirea, Juglans regia* – bark, *Neopicrorhiza scrophulariiflora*).
- Category II: Species banned for export except for processed with permission of Department of Forests, comprises five species of angiosperms namely, *Cinnamomum glaucescens* (Lauraceae), *Nardostachys grandiflora* (Caprifoliaceae), *Rauvolfia serpentina* (Apocynaceae), *Valeriana*

jatamansi, Valeriana wallichii (Caprifoliaceae), three species of gymnosperms namely, *Abies spectabilis* (Pinaceae), *Taxus contorta, Taxus wallichiana* (Taxaceae), one species of fungi namely, *Ophiocordyceps sinensis* (Ophiocordycipataceae), and lichen species (*Parmelia* spp.).

- Category III: Species banned for harvest, transportation and export for commercial purposes, comprises six species of flowering plants such as *Acacia catechu, Dalbergia latifolia* (Fabaceae), *Bombax ceiba* (Malvaceae), *Magnolia champaca* (Magnoliaceae), *Pterocarpus marsupium* (Fabaceae), and *Shorea robusta* (Dipterocarpaceae).

13.5.2 Threatened Animals in Nepal

Regarding the threatened species of animals in Nepal, as many as 26 species of birds are regarded as the threatened species (Groombridge and Jenkins, 2002). However, according to IUCN Red List of Threatened species, nine species of plants, 55 species of mammals, 149 species of birds, 15 species of herpetofauna, and 21 species of fish are considered as threatened species in Nepal (MoFSC, 2014b). As many as 130 breeding and winter species (15% of Nepal's birds) are considered as nationally threatened species, of which nine species are listed under the protected species by Government of Nepal (www.birdlifenepal.org). According to National Parks and Wildlife Conservation Act 1973 (MoFSC, 2014b), 27 species of mammals, 9 species of birds, 3 species of reptiles and 2 species of Amphibian are regarded as the protected fauna of Nepal (Table 13.6).

Three species of mammals namely, Pigmy hog (*Sus salvanius*), Indian chevrotain (*Moschiola memina*) and Cheotah (*Acinonyx jabatus*) have become extinct from Nepal. Similarly, three species of mammals namely Hispid hare (*Caprologus hispidus*), Nayan sheep (*Ovis ommon*) and Tibetan

Table 13.6 Number of Animal Species in the IUCN Red List Category

Faunal group	Mammals	Birds	Reptiles and Amphibians	Fishes
Critically Endangered (CR)	8	61	1	3
Endangered (EN)	26	38	3	1
Vulnerable (VU)	14	50	7	4
Near Threatened (NT)	7	-	4	13
Total	55	149	15	21

Source: DNPWC (2003), MoFSC (2014b), modified version.

antelope (*Pantholops hodgsonii*) are believed to be vanished, as there is no recent information (Gotame, 2007).

13.5.2.1 CITES List

The Convention on International Trade in Endangered Species of Wild Fauna and Flora (CITES) has listed 52 mammal species, 108 species of birds, 19 species of reptiles, 25 species of amphibians and three species of insects under CITES Appendices (GoN 2071 (B.S.), Jnawali et al., 2011; BCN and DNPWC, 2011; Shah, 2013). Fifty one species of animals are currently listed in Appendix I of CITES which includes 32 species of mammals, 12 species of birds, 2 species of reptiles and 5 species of amphibians. Similarly, 139 species of animals are in the Appendix II which includes 16 species of mammals, 95 species of birds, 15 species of reptiles, 10 species of amphibians and three species of insects. Altogether 7 species of mammals, birds and reptile species are in Appendix III (MoFSC, 2014a). Summary of the CITES Appendix is given in the Table 13.7.

13.5.2.2 Protected Animals of Nepal

The Government of Nepal has considered 27 species of mammals (with 14 species of Endangered species and four vulnerable species), nine species of birds (including six species of Rare species), and three species of reptiles (including one Endangered and one Vulnerable species) as the protected animals in Nepal (MoFSC, 2014a).

13.6 Threats to Biodiversity and Conservation Approach

Globally, biodiversity is under serious threat as a result of natural calamities and human activities. The six main threats to biodiversity are: (i) population

Table 13.7 CITES Listed Animals

S.N.	Animal Group	Appendix I	Appendix II	Appendix III	Total species
1.	Mammals	32	16	4	52
2.	Birds	12	95	1	108
3.	Reptiles	2	15	2	19
4	Amphibians	5	10	0	25
5.	Insects	0	3	0	3

growth and resource consumption, (ii) climate change and global warming, (iii) habitat conversion and urbanization, (iv) invasive alien species, (v) over-exploitation of natural resources, and (vi) environmental degradation. Species often become threatened or disappear when several of these factors are combined. Similarly, the environmental consequences of climate change may lead to the extinction of vulnerable species, and destruction of biodiversity values. It has been estimated that 11,000 species of plants and animals are known to be threatened, and nearly 15 million hectares of forests are deforested annually (Kluger and Dorfman, 2002). If the trend continues 15–37% of plants and animal species would face extinction by 2050. Moreover, if the trend of carbon emissions is on higher scale, the extinction rate may go up to the range of 21–52% (Thomas et al., 2004).

The world population has increased from 1.6 billion to 6.1 billion in the 20th century. The world population is expected to increase from 6.9 billion in 2010 to 9.2 billion in 2050. Nearly 100 million new people to the human population are added every year (Lovejoy, 1997). Biodiversity resources are already being lost due to the explosive growth of population resulting into the encroachment of forests and destruction of landscapes in the name of urbanization and industrialization.

The Nepal National Biodiversity Strategy and Action Plan 2014–2020 has categorized the threats into two broad groups: (i) loss and degradation of natural habitats which includes habitat loss, encroachment, expansion of cultivation, development of infrastructure, planned conversion of forest land; (ii) over exploitation and illegal exploitation of biological resources, which includes unsustainable over harvesting, uncontrolled forest fire, overgrazing, poaching and illegal wildlife trade, human-wildlife conflict, invasion by alien plant species, and stone, gravel, and sand mining; and (iii) climate change (MoFSC, 2014b).

One of the major threats to biodiversity in Nepal is the deforestation and degradation of forests, mainly due to high dependency on forest resources, unsustainable harvesting and extraction of timber, firewood and fodder, overgrazing, etc. Similarly, forest fire, encroachment for settlement and agriculture, slash and burn cultivation practice, and natural calamities such as earthquake, soil erosion and floods also resulted into massive deforestation every year (Chaudhary et al., 2016).

The annual deforestation rate in Nepal is estimated to be 1.9%. The forest area has declined from 45% in 1966 to 29% in 1994. However, the shrubland area has expanded from 4.7% in 1980's to 10.6% in mid-1990's (MoFSC, 2014b). During the last two decades Nepal has expanded its forest area substantially. Recent report revealed the growth of forest area to 40.36% and shrubland to 4.38% (DFS, 2015). Nepal government's initiative in the expansion of protected areas (23.23% of Nepal's land area) and its buffer zone, extension of community forests, leasehold forests, protected forests (including religious forests or sacred groves), and private forests have significantly contributed the forest coverage in Nepal.

The spread of invasive alien species (IAS) is transforming the habitats. IAS are non-native species that have become established in a new environment and are now recognized as one of the greatest biological threats to the planet's environment and economic well-being (Mooney et al., 2004). At least 13,168 species of vascular plants have been naturalized, outside their native habitat; and a total of 4,979 species are documented as invasive species (RBG Kew, 2016). Invasive species are one of the major drivers for biodiversity loss. Invasive species are the introduced or exotic species in origin, not a native or indigenous species, which spreads profusely in any type of environment and may damage the forests, agriculture lands, marginal land, and near settlements. There are at least 219 alien species of flowering plants naturalized in Nepal (Tiwari et al., 2005). Tiwari et al. (2005) documented 21 naturalized plant species to be invasive in Nepal. Shrestha (2016) revised the list and finalized a number of 25 invasive alien species in Nepal. Invasive species are highly centered on the southern half of the country running east to west with tropical to subtropical climate.

- **Wetland/marshy species:** *Eichhornia crassipes, Monochoria vaginalis* (Pontederiaceae), *Leersia hexandra* (Poaceae), *Myriophyllum aquaticum* (Haloragaceae), *Pistia stratiotes* (Araceae), *Polygonum hydropiper* (Polygonaceae), and *Trapa, quadrispinosa* (Lythraceae).
- **Agroecosystem species:** *Ageratum conyzoides, Ageratum houstonianum, Galinsoga parviflora* (Asteraceae), *Alternanthera sessilis, Amaranthus viridis* (Amaranthaceae), *Cynodon dactylon, Echinocloa crussgalli, Setaria glauca* (Poaceae), and *Trifolium repens* (Fabaceae).
- **Forest and shrubland species:** *Ageratina adenophora, Chromolaena odorata, Mikania micrantha* (Asteraceae), *Cuscuta reflexa* (Convolvulaceae), and *Lantana camara* (Verbenaceae).
- **Marginal lands:** *Amaranthus spinosus* (Amarathaceae), *Ipomoea fistulosa* (Convolvulaceae), *Parthenium hysterophorus,* and *Xanthium strumarium* (Asteraceae).

Climate change has become a major environmental and policy issue in national and international arenas. Climate change impacts the ecological functioning and hence management of Protected Areas and pervades the discourse on Protected Areas. It has been reported that Alpine warming will further threaten endangered animals like snow leopards or blue sheep (Salick and Byg, 2007) and therefore climate change directly threatens biodiversity. In general, increase of temperature by 2°C may lead to extinction of 15–40% of currently existing species.

13.6.1 Conservation Approaches

The maintenance of viable populations of species in their natural habitat is identified as a fundamental requirement for the conservation of biological diversity by the Convention on Biological Diversity (CBD). Nepal tried various conservation approaches to conserve the biodiversity. Among these, protected area management and community forestry are two successful approaches in biodiversity conservation in Nepal.

13.6.1.1 Protected Area Network and World Heritage Sites

It has been estimated that 25,000 species, nearly 10% of flowering plants, on Earth are classified as extinct, endangered or threatened plants (Prance, 1990). Extinction of nearly 25% plants by 2050 has been predicted due to the consequences of human activities and different anthropogenic factors. To overcome the loss of biodiversity, minimize the impact of habitat loss fragmentation, and to increase the population of threatened species, more protected areas have been designated in last few decades. The establishment of protected areas has been fundamental to the conservation of biodiversity, and today well-managed protected areas not only support healthy ecosystems but also provide multiple benefits to people (Bertzky et al., 2012). The governments in the international forums have adopted the Strategic Plan for Biodiversity 2010–2020 and its 20 Aichi Biodiversity Targets, and the 2030 Agenda for Sustainable Development and its Sustainable Development Goals. These constitute two of the most important environment and sustainable development commitments ever made by governments, and both recognize the important role of protected areas as a key strategy for biodiversity conservation and sustainable development in the targets they contain, for example, Aichi Biodiversity Target 11, SDG goals 14 and 15) (UNEP-WCMC and IUCN, 2016).

Establishment of biosphere reserves, national parks, conservation areas, wildlife reserves and sanctuaries, and Ramsar wetland sites are some significant conservation efforts for the *in-situ* conservation of rare, threatened and endemic species. There are 202,467 terrestrial and inland water protected areas recorded in the World Database on Protected Areas (WDPA), covering 14.7% (19.8 million km^2) of the world's extent of these ecosystems (excluding Antarctica) (UNEP-WCMC and IUCN, 2016). The Yellowstone National Park in the USA, established in 1872, is the first national park of the world. Now there are 6,555 national parks, of which Northeast Greenland National Park, established in 1974, is the largest national park of the world, covering an area of 972,000 km^2 (Jeffries, 2005). There are approximately 10,900 protected areas in the Asia Region, covering 13.9% of terrestrial and inland water areas with an area of 2.9 million km^2, and 1.4% of marine and coastal areas (Juffe-Bignoli et al., 2014).

13.6.1.2 Protected Areas in Nepal

Nepal's rich biodiversity is under threat from various factors. However, in the last four decades Nepal has made significant efforts and tangible achievements in developing a network of Protected Areas for the long-term protection of biodiversity. A formal protection initiative started after the approval of the 1973 National Parks and Wildlife Conservation Act which was a major commitment to the development of the Protected Area (PA) System in Nepal. Chitwan National Park (previously Royal Chitwan National Park) was established as the first Protected Area in 1973, and its establishment was followed in 1976 by National Parks in Langtang, Sagarmatha and Rara, and Royal Karnali Wildlife Reserve (later expanded to Bardiya National Park in 1988). The 1973 Act also laid the foundation for the conservation of all 208 known mammal species (Jnawali et al., 2011), 871 bird species (Baral et al., 2012) and other species and ecosystems.

At present, Nepal has a network of 20 protected areas consisting of twelve national parks, one wildlife reserve, six conservation areas and one hunting reserve; and thirteen buffer zone areas, covering an area of 34,419.75 km2 (23.39% of total land area of the country) (DNPWC, 2017). The percentage of Nepal's land area which has protected status is one of the highest in Asia, compares well with the global figure of 12.7% and comfortably meet the Convention on Biological Diversity's 2010 Biodiversity Target to conserve at least 17% of national territories by 2020 (Bertzky et. al., 2012). The Protected Area system has made a significant contribution to the conservation of biodiversity in Nepal, and at least 85% of mammal species, 95% of bird species, 67% of herpeto-fauna and 75% of fish are found within Protected Areas. Among the protected areas, Chitwan National Park and Sagarmatha National Park have been listed as World Heritage Sites and are managed to protect their natural features for which they are of outstanding international significance. Similarly, 9 wetlands of Nepal are considered to be of international importance as a waterfowl habitat and have been listed as Ramsar sites, and 6 of them (Koshi Tappu, Beeshazar Tal, Gokyo Tso, Gosain Gunda and associated lakes, Rara Lake, and Phoksundo Lake) lie within the protected area system (Table 13.8). The management models of these protected areas consist of various mixes of IUCN categories based upon different socio-economic settings, with National Parks and the Conservation Areas covering more than three quarters of the total protected areas (Bajracharya et al., 2015).

The protected area management system in Nepal made gradual shift and gave priority to a people-centered approach. The third amendment to the National Park and Wildlife Conservation Act in 1989 gave legal recognition of Conservation Area as a new category of protected area in Nepal which is a new approach to biodiversity protection (Bajracharya et al., 2007). The Annapurna Conservation Area is the first Conservation Area of Nepal and this approach paved the way for community-based conservation in Nepal. Similarly, the people-centered approach in management was applied to existing protected areas by establishing buffer zones around national parks and wildlife reserves, and this was formalized by the Buffer Zone Management Regulation in 1996. At present, buffer zones have been established in 13 national parks and wildlife reserves covering 5,708.41 km^2 area (DNPWC, 2015). The national parks, conservation areas, wildlife reserves, hunting reserve, and buffer zones in Nepal represent all physiographic zones, majority of ecosystems, vegetation, and forest types from tropical to nival zones.

The Shey-Phoksundo National park is the largest national park of Nepal, covering an area of 3,555 km^2. In terms of area, the Annapurna Conservation Area is the largest protected area of Nepal with an area of 7,629 km^2, and Rara National Park is the smallest protected areas in Nepal (Figure 13.1, Table 13.9). Recognizing the importance of the protected areas, the Chitwan National Park and Sagarmatha National Park have been declared as UN World Heritage Sites in 1979 and 1984, respectively.

13.6.1.3 World Heritage Sites

World Heritage Sites are cultural and/or natural sites considered to be of 'Outstanding Universal Value,' which have been inscribed on the World Heritage List by the World Heritage Committee. These places or buildings are thought to have special importance for everyone, and represent unique, or the most significant or best examples of the world's cultural and/or natural

Table 13.8 Ramsar Sites in Nepal

S. No.	Ramsar sites	Listed	Area (ha)	Average Depth (m)	Elevation (m)	Physiographic zone	District	Province
1	Bishazari and associated lakes	2003	3,200		150	Lowland Tarai	Chitwan	Central Nepal (Province 3)
2	Ghodaghodi Tal	2003	2,563	4	205	Lowland Tarai	Kailali	Mid-Western Nepal (Province 7)
3	Gokyo and associated lakes	2007	7,770		4700–5000	High hill	Solukhumbu	Eastern Nepal (Province 1)
4	Gosaikunda and associated lakes	2007	1,030		4380	High hill	Rasuwa	Central Nepal (Province 3)
5	Jagdishpur Reservoir	2003	225	2.0–7.0	197	Lowland Tarai	Kapilvastu	Western Nepal (Province 5)
6	Koshi Tappu	1987	17,500	5	75–81	Lowland Tarai	Sunsari	Eastern Nepal (Province 1)
7	Mai Pokhari	2008	90		2100	Mid hill	Ilam	Eastern Nepal (Province 1)
8	Phoksundo Lake	2007	494	145	3611	High hill	Dolpa	Mid-Western Nepal (Province 6)
9	Rara Lake	2007	1,583	167	2990	Mid hill/high hill	Mugu	Farwestern Nepal (Province 6)
10.	Lake cluster of Pokhara Valley	2016	26,106	23	742	Mid hill	Kaski	Western Nepal (Province 4)

Source: Chaudhary (1998), Bhuju et al. (2007); Karki (2008), MoFSC (2014b), DNPWC (2015), modified version.

Figure 13.1   Protected Areas of Nepal (Courtesy: National Trust for Nature
Conservation).

heritage. Among the World Heritage sites of Nepal, Chitwan National Park
and Sagarmatha National Park were inscribed as World Heritage Sites.

The Sagarmatha National Park (SNP) was inscribed as the 120th World
Heritage site by the 3rd session of the World Heritage Committee in 1979.
It was inscribed with the reason that the Park is an exceptional area with
dramatic mountains, glaciers, and deep valleys dominated by Mount Everest, the highest peak in the world (8,848 m). Several rare species, such as
the snow leopard and the lesser panda, are found in the Park. The presence
of the Sherpas, with their unique culture, adds further interest to this site.
Similarly, Chitwan National Park was inscribed as the 284th World Heritage
Site by the WHC's 8th session in 1984. The park was inscribed because it is
one of the few undisturbed areas of the Tarai region with its very rich flora
and fauna including single-horned Asiatic rhinoceros and the Bengal tiger.

13.6.1.4 Community Forestry Program

Public awareness and peoples participation is essential in the conservation
of the forests and biodiversity resources. Of the several approaches adopted,

Table 13.9 General Features of the Protected Areas in Nepal

S.N	Protected areas	Establishment	Area (km²)	Buffer zone (km²)	Elevation (m)	Physiographic zone	Districts	Region/Province
	National Parks							
1	Banke National Park	2010	550	343	360–480	Lowland Tarai/ Churiya	Banke, Dang, Salyan	Mid-western Nepal (Province 5,6)
2	Bardia National Park	1976	968	507	152–1441	Lowland Tarai/ Churiya	Bardiya	Mid-western Nepal (Province 5)
3	Chitwan National Park	1973	932	750	100–815	Lowland Tarai/ Churiya	Chitwan, Makwanpur, Nawalparasi, Parsa	Central Nepal (Province 2,3,4)
4	Khaptad National Park	1984	225	216	1000–3276	Midhill/High hill	Bajhang, Bajura, Accham, Doti	Far-western Nepal (Province 7)
5	Langtang National Park	1976	1,710	420	792–7245	Midhill/High hill	Rasuwa, Nuwakot, Sindhupalchok	Central Nepal (Province 3)
6	Makalu Barun National Park	1991	1,500	830	435–8463	Midhill/High hill	Sankhuwasabha, Solukhumbu	Eastern Nepal (Province 1)
7	Rara National Park	1976	106	198	2800–4039	Midhill/High hill	**Mugu,** Jumla	Far-western Nepal (Province 6)
8	Sagarmatha National Park	1976	1,148	275	2945–8848	Midhill/High hill	Solukhumbu	Eastern Nepal (Province 1)

Table 13.9 (Continued)

S.N	Protected areas	Establishment	Area (km²)	Buffer zone (km²)	Elevation (m)	Physiographic zone	Districts	Region/Province
9	Shey-Phoksundo National Park	1984	3,555	1,349	2130–6885	Midhill/High hill	**Dolpa**, Mugu	Mid-western Nepal (Province 6)
10	Shivapuri Nagarjun National Park	2002	159	118.61	1400–2732	Midhill	**Kathmandu**, Nuwakot, Sindhupalchok	Central Nepal (Province 3)
11	Shuklaphanta National Park	1976	305	243.5	174–1386	Lowland Tarai/ Churiya	Kanchanpur	Far-western Nepal (Province 7)
	Hunting Reserve							
12	Dhorpatan Hunting Reserve	1987	1,325	N/A	2850–5500	Midhill/High hill	**Baglung**, Myagdi, Rukum	Western Nepal (Province 4,5)
	Wildlife Reserve							
13	Koshi Tappu Wildlife Reserve	1976	175	173	75–81	Lowland Tarai/ Churiya	**Sunsari**, Saptari, Udaypur	Eastern Nepal (Province 1, 2)
14	Parsa Wildlife Reserve	1984	627.39	285.3	435–950	Lowland Tarai/ Churiya	**Parsa**, Bara, Makwanpur	Central Nepal (Province 2, 3)
	Conservation Area							
15	Annapurna Conservation Area	1992	7,629	N/A	790–8091	Midhill/High hill	Manang, Mustang, Lamjung, Kaski, Myagdi	Central Nepal (Province 4)

S.N	Protected areas	Establishment	Area (km²)	Buffer zone (km²)	Elevation (m)	Physiographic zone	Districts	Region/Province
16	Api Nampa Conservation Area	2010	1,903	N/A	518–7132	Midhill/High hill	Darchula	Far-West Nepal (Province 7)
17	Blackbuck Conservation Area	2009	16.95	N/A	120–230	Lowland Tarai/ Churiya	Bardiya	Mid-western Region (Province 5)
18	Gaurishankar Conservation Area	2010	2,179	N/A	1000–7134	Midhill/High hill	**Dolakha,** Ramechhap, Sindhupalchok	Central Nepal (Province 3)
19	Kangchenjunga Conservation Area	1997	2,035	N/A	1200–8586	Midhill/High hill	Taplejung	Eastern Nepal (Province 1)
20	Manaslu Conservation Area	1998	1,663	N/A	1400–8156	Midhill/High hill	Gorkha	Central Nepal (Province 4)

one of the most popular and successful approaches is the involvement of local communities in biodiversity conservation. A large group of communities have been involved in the protection, management and sustainable utilization of the forest products to improve their livelihoods. Community forestry program has been highly successful in the rural areas of Asia, particularly in China, India, Philippines, and in Nepal.

To complement the government's campaign in biodiversity conservation, the community forestry program was introduced in Nepal during late 1970s with the aim to protect the degraded hills of Nepal. The community forestry intended to improve forest quality by meeting the basic forest resources requirements by the local communities through their active participation in forest development and management. Giving priority to the community forestry, Nepal Government has endorsed the Forest sector Masterplan in 1987 to legalize it. In the early 1990s, the Community Forest User Groups were legally recognized as an independent and self-governing entity with the right to utilize and manage the forests nearby the settlements.

Forests account for approximately 40% of the total national land area in Nepal (nearly 5.5 million hectares). It is one of the major productive resources and contributes around 10% to Nepal's Gross Domestic Product (GDP). Community forestry occupies nearly 23% of national total forest (over 1.22 million hectares). About 75% of community forests lie in the middle hills, 16% in the high mountains, and 9% in the lowland Tarai. There are 17,685 Community Forest User Groups (CFUGs), with over 1.6 million households as members (equivalent to about 33% of the total rural population), throughout the country (MoFSC, 2013).

Community Forest User Groups (CFUGs) and other Community Based Forest Management Groups (such as leasehold forestry groups, religious forestry groups, buffer zone and traditional forest management groups) are well established in all districts of Nepal. The Federation of Community Forestry Users Nepal (FECOFUN), established in 1995, is a formal network organization of CFUGs Community-based forest management groups, spreading all over the country (www.fecofun.org.np). Except the government managed national parks and other categories of protected areas, more government forests are likely to be managed by the local communities, adding more forest areas under the community forests in the future (Figure 13.1).

13.6.1.4.1   Landscape-Level Conservation Initiatives in Nepal

The Nepal Biodiversity Strategy (MoFSC, 2002), and Nepal Biodiversity Strategy and Action Plan (MoFSC, 2014b) identified and advocated for the landscape level biodiversity conservation in Nepal. This approach aims to conserve biodiversity encompassing a large area with diverse ecosystems, comprising forests, agricultural land, rangelands, and wetlands. The National Biodiversity Strategy and Action Plan 2014–2020 (MoFSC, 2014b) has taken landscape conservation as its strategic approach (Bajracharya et al., 2015).

1. **Tarai Arc Landscape (TAL):** It was initiated in 2000 by the Ministry of Forests & Soil Conservation (MoFSC), with the support of WWF Nepal Program. This trans-boundary landscape covers an area of 23,199 km^2 in 14 Tarai and Midland districts of Nepal. It extends from Bagmati River westwards along the Indo-Nepal border to Raja Ji National Park (India).

2. **Sacred Himalayan Landscape (SHL):** It was initiated in 2006 by MoFSC and WWF Nepal Program. It has an area of 28,681 km^2, spreading into the northern region of 18 mountain and Midland districts of Nepal. It extends from Langtang National Park (C. Nepal) through Sikkim and Darjeeling to Toorsa Strict Nature Reserve in Western Bhutan.

3. **Chitwan Annapurna Landscape (CHAL):** The Ministry of Forests and Soil Conservation initiated an in-country landscape, the Chitwan Annapurna Landscape (CHAL) in 2011. The landscape covers an area of 32,067 km^2, of the Gandaki Basin extending from lowland Tarai to the Trans-Himalayan region. It encompasses part of 19 districts of Central Nepal.

4. **Kailash Sacred Landscape (KSL):** The MoFSC in collaboration with International Centre for Integrated Mountain Development (ICIMOD) initiated Kailash Sacred Landscape Conservation Initiative in 2013. It is a trans-boundary landscape, which covers SW Tibet Autonomous Region of Nepal, Northwest part of India and Nepal. In Nepal, it covers an area of 13,289 km^2, encompassing four districts of far-west Nepal (Baitadi, Bajhang, Darchula, and Humla).

5. **Kangchenjunga Landscape (KL):** The Kangchenjunga Landscape was initiated by MoFSC and ICIMOD in 2014. It extends from the southern part of Mt. Kangchenjunga in Eastern Nepal and India (Sikkim) to Bhutan. In Nepal, it covers an area of 5,190 km^2, comprising parts of four districts: Taplejung, Panchthar, Ilam, and Jhapa.

13.6.1.4.2   Protected Forest Systems in Nepal

The Ministry of Forests & Soil Conservation declared following eight protected forests, with an area of 133,754.8 hectares under the Forest Act 1993.

Table 13.10 Protected Forests in Nepal

Name of Protected Forests	Districts	Area (ha)	Year of Establishment
Barandabhar	Chitwan	10,466	2011
Basanta	Kailali	69,001.2	2011
Dhanushadham	Dhanusha	430	2012
Kankrebihar	Surkhet	175.5	2002
Khata	Bardia	4503.7	2011
Laljhadi-Mohana	Kailali and Kanchanpur	29,641.7	2011
Madane	Gulmi	13,761	2010
Panchase	Kaski, Shyanga and Parbat	5,775.7	2011

These protected forests were declared during 2002–2012 due to rich in biodiversity and important wildlife corridors (MoFSC 2014b), as shown in Table 13.10.

The main objective of the protected forests is to conserve biodiversity, biological corridors, promote income generation activities of local communities by sustainable utilization of biological resources, and initiation of innovative activities to promote ecotourism and educational programs.

13.7 Concluding Remarks

Understanding biodiversity, sustainable utilization of biological resources, and conservation of rare and threatened species and biodiversity hotspots, including Important Plant Areas (IPAs) and Important Bird Areas (IBAs) are some key issues globally. Serious commitments from the government and non-government organizations and active participation of the local communities are essential to conserve the forests and biodiversity resources for the future generations. Due to the explosive growth of human population since last few decades, many virgin lands, forests, and landscapes have been encroached and destroyed. Obviously, the human beings are mainly responsible for the irreversible loss of biodiversity and deteriorating the ecosystems. If the present trend of human population increase and encroachment of natural habitats continue, no doubt there will be a high risk of extinction of many valuable species and their habitats within next couple of decades. It has been realized that efforts should be made for the *in situ* and *ex situ* conservation of commercially important species, and expansion of protected areas and community forests is essential to preserve the pristine global wealth.

Nepal's effort to conserve biodiversity through an extensive network of protected areas system and community forestry program has been considered exemplary and successful. A gradual increase in the population of some of the mega fauna like the tiger, rhino, and growing engagement of local communities in the protected area management can be taken as vivid indicators. A significant forest area has been handed over to the local communities as community forests. Besides effective engagement of local communities in the forest management, the approach has also helped to increase the forest coverage in Nepal.

Several researches are still on-going throughout the world for more than two centuries in exploring and documenting global biodiversity. Systematic explorations to document the biodiversity, especially in the less known areas, and a relatively unknown group of organisms should be the primary goal of taxonomists and biodiversity experts. The documentation of biological diversity, especially the species diversity, in Nepal is not satisfactory. Except for the vertebrates in faunal diversity and flowering plants in plant diversity, very little attention has been paid to document invertebrates and non-flowering plants of Nepal. To fulfill Nepal Government's commitment to United Nation's Convention on Biological Diversity (CBD) and Sustainable Development Goals (SDGs), it is high time to consolidate our effort for the systematic documentation of biological resources of Nepal.

Similarly, research should be continued to understand the goods and services provided by the ecosystems to the human being and their intricate relationship within the ecosystem. Human activities and climate change are threatening biodiversity and ecosystem services directly or indirectly to many parts of the world. The increase of carbon concentration in the atmosphere means the loss of biodiversity and other ecosystem services. It is quite evident that the biodiversity loss is irreversible, and we should be prepared to deal with dire sequences and worst environment that can happen in future.

Keywords

- ecosystem
- endemism
- fauna
- flora
- forests
- protected areas
- threatened species
- vegetation

References

Adhikari, M. K., (2012). *Researches on the Nepalese Mycoflora–2: Checklist of Macrofungi (Mushrooms)*. K. S. Adhikari, Kathmandu.

Bajracharya, S. B., Chaudhary, R. P., & Basnet, G., (2015). Biodiversity conservation and Protected Area Management in Nepal, In: *Nepal: An Introduction to the Natural History, Ecology and Human Environment of the Himalayas,* Miehe, G., & Pendry, C., (eds.). A companion volume to the flora of Nepal. Royal Botanical Garden Edinburgh. Edinburgh.

Bajracharya, S. B., Gurung, G. B., & Basnet, K., (2007). Learning from community participation in conservation area management. *J. Forest and Livelihood*, *6*(2), 54–66.

Baral, H. S., Regmi, U. B., Poudyal, L. P., & Acharya, R., (2012). Status and conservation of birds in Nepal. In: *Biodiversity Conservation in Nepal: A Success Story,* Acharya, K. P., & Dhakal, M., (eds.) Kathmandu: DNPWC, pp. 61–90.

Baral, H. S., & Shah, K. B., (2008). *Wild Mammals of Nepal.* Himalayan Nature, Kathmandu.

Baral, S. R., (1995). *Enumeration of the Algae of Nepal.* Biodiversity Profile Project. Publication No. 11. Kathmandu: GoN Department of National Parks and Wildlife Conservation.

BCN and DNPWC, (2011). *The State of Nepal's Birds 2010.* BCN and DNPWC, Kathmandu.

Bertzky, B., Corrigan, C., Kemsey, J., Kenney, S., Ravillious, C., Besancon, C., & Burges, N., (2012). *Protected Area Planet Report 2012: Tracking Progress Towards Global Targets for Protected Areas.* IUCN and UNEP-WCMC, Gland and Cambridge.

Bhuju, U., Shakya, P. R., Basnet, T. B., & Shrestha, S., (2007). *Biodiversity Resource Book: Protected Areas, Ramsar Sites, and World Heritage Sites.* ICIMOD, and Ministry of Science & Technology, GoN, Kathmandu, Nepal.

Bista, M., (2006). Gymnosperms of Nepal. In: *Environment and Plants: Glimpses of Research in South Asia,* Jha, P. K., Chaudhary, R. P., Karmacharya, S. B., & Prasad, V., (eds.), Ecological Society, Kathmandu.

Bista, M. S., Adhikari, M. K., & Rajbhandari, K. R., (2001). *Flowering Plants of Nepal (Phanerogams).* Department of Plant Resources, Kathmandu, Nepal.

BPP, (1995). An assessment of representation of the terrestrial ecosystems in the protected areas system of Nepal. In: *Biodiversity Profiles Project.* Kathmandu: HMG/N.

Budha, P. B., (2012). *Review of Freshwater and Terrestrial Molluscan Studies in Nepal: Existing Problems and Future Research Priorities.* Paper presented at the Entomological Review Workshop organized by Nepal Agriculture Research Council, Khumaltar.

Byng, J. W., (2014). *The Flowering Plants Handbook.* A practical guide to families and genera of the World. Plant Gateway Ltd. Hertford, United Kingdom.

Chapman, A. D., (2009). *Numbers of Living Species in Australia and the World.* 2nd Edition. Department of the Environment, Water, Heritage and the Arts. Australian Government.

Chaudhary, R. P., (1998). *Biodiversity in Nepal–Status and Conservation.* Devi, S., Saharanpur, India & Tecpress, Bangkok, Thailand.

Chaudhary, R. P., & Bajracharya, D., (1995). Role of institutions in conservation and use of plant genetic resources: their structure and function. In: *Plant Genetic Resources Nepalese Perspectives*, Upadhyay, M. P., Saiju, H. K., Baniya, B, K., & Bista, M. S., (eds.). Nepal Agricultural Research Council, Knumaltar, Lalitpur and International Plant Genetic Resources Institute, APO, Singapore. pp. 137–146.

Chaudhary, R. P., Uprety, Y., & Rimal, S. K., (2016). Deforestation in Nepal: Causes, consequences and responses. In: *Biological and Environmental Hazards, Risks, and Disasters,*

Shroder, J. F., & Sivanpalli, R., (eds.). Elsevier, Inc., Netherlands, UK and USA, pp. 331–372.

DFRS, (2015). State of Nepal's Forests. *Forest Resource Assessment (FRA) Project*, Kathmandu. Department of Forest Research & Survey, Kathmandu, Nepal.

DNPWC, (2003). *Convention on International Trade in Endangered Species of Wild Fauna and Flora (CITES)*. Department of National Park and Wildlife Conservation, Babarmahal, Kathmandu.

DNPWC, (2017). *Protected Areas of Nepal*. Department of National Parks and Wildlife Conservation. Kathmandu, Nepal.

Fleming, R. L., Fleming, R. L., & Bangdel, L. S., (1976). *Birds of Nepal, With Reference to Kashmir and Sikkim*. Sue Lowell Natural History & Travel Books, London.

Fraser-Jenkins, C. R., Kandel, D. R., & Pariyar, S., (2015). *Ferns and Fern Allies of Nepal, vol. 1*. Department of Plant Resources, MFSC, National Herbarium & Plant Laboratories, Godawari, Lalitpur, Nepal.

GoN, (2017). *Flora and Fauna of Nepal in CITES Annexes*, MoFSC, Department of National Park and Wildlife Reserve, Kathmandu, Nepal.

Gotame, B., (2007). Protected Mammals of Nepal. *The Initiation*, 112–116.

Groombridge, B., & Jenkins, M. D., (2002). *World Atlas of Biodiversity: Earth's Living Resources in the 21st Century*. University of California Press, Berkeley.

Gupta, R., (1997). *Diversity of Parasitic Helminth Fauna in Nepal*. Research Division, Tribhuvan University Kirtipur, Kathmandu.

Hara, H., Chater, A. O., & Williams, L. H. J., (1982). *An Enumeration of the Flowering Plants of Nepal, vol. III,* The British Museum (Natural History), London.

Hara, H., Stearn, W. T., & Williams, L. H. J., (1978). *An Enumeration of the Flowering Plants of Nepal, vol. I*. The British Museum (Natural History), London.

Hara, H., & Williams, L. H. J., (1979). *An Enumeration of the Flowering Plants of Nepal, vol. II*. The British Museum (Natural History), London.

Haruta, T., (2006). *Moths of Nepal, vol. 1–16*. TINEA, Japan Heterocerist's Society, Tokyo.

Hobohm, C., & Tucker, C. M., (2014). The increasing importance of endemism: Responsibility, the media and education. In: *Endemism in Vascular Plants*, Hobohm, C., (eds.). Springer, 3–9.

Inskipp, C., Baral, H. S., Phuyal, S., Bhatt, T. R., Khatiwada, M., Inskipp, T., Khatiwada, A., Gurung. S., Singh, P. B., Murray, L., Poudyal, L., & Amin, R., (2016). *The Status of Nepal's Birds*: *The National Red List Series*. Zoological Society of London, UK.

IUCN, (2013). *IUCN Red List of Threatened Species* (www.iucnredlist.org).

Jeffries, M. J., (2005). *Biodiversity and Conservation*, Second Edition. Routledge Taylor& Francis Group London and New York.

Jnawali, S. R., Baral, H. S., Lee, S., Acharya, K. P., Upadhyay, G. P., Pandey, M., Shrestha, R., Joshi, D., Lamichhane, B R., Griffiths, J., Khatiwada, A., Subedi, N., & Amin, R., (2011). *The Status of Nepal Mammals: The National Red List Series*. DNPWC, Kathmandu.

Juffe-Bignoli, D., Bhatt, S., Park, S., Eassom, A., Belle, E. M. S., Murti, R., Buyck, C., Raza Rizvi, A., Rao, M., Lewis, E., MacSharry, B., & Kingston, N., (2014). *Asia Protected Planet 2014*. UNEP-WCMC: Cambridge, UK.

Karki, J. B., (2008). Koshi Tappu Ramsar site: Updates on ramsar information sheet on wetlands. *The Initiation* (Suffrec), *8*, 10–16.

Kluger, I., & Dorfman, A., (2002). The challenges we face. Special report: How to save the Earth. *Time, 160*, 6–12.

Kost, G., & Adhikari, M. K., (2015). Mycota. In: *Nepal: An Introduction to the Natural History, Ecology and Human Environment of the Himalayas,* Meihe, G., Pendry, C. A., & Chaudhary, R., (eds.). Royal BotaniC Garden Edinburgh, 203–210.

Kunwar, R. M., Shrestha, K., Dhungana, S. K., Shrestha, P. R., & Shrestha, K. K., (2010). Floral Biodiversity of Nepal: An Update. *J. Nat. Hist. Mus., 25*, 295–311.

Lovejoy, T. E., (1997). Biodiversity: Why is it? In: *Biodiversity II: Understanding and Protecting Our Biological Resources,* Reaka-Kudla, M. L., Wilson, D. E., & Wilson, O. E. (eds.). Joseph Henry Press, Washington D.C., pp. 7–14.

LRMP, (1986). *Land Utilization Report:* Appendix I. Land Resource Mapping Project. HMG/N Government of Canada/ Keating Earth Sciences, Kathmandu, Nepal.

Miehe, G., Pendry, C. A., Chaudhary, R., (2015). *Nepal: An Introduction to the Natural History, Ecology and Human Environment of the Himalayas*. Edinburgh: Royal Botanic Gardens Edinburgh.

MoFSC., (2002). *Nepal Biodiversity Strategy*. Ministry of Forests & Soil Conservation, Kathmandu, Nepal.

MoFSC, (2009). Nepal Fourth National Report to the Convention on Biological Diversity, Government of Nepal, Ministry of Forests and Soil Conservation, Kathmandu, Nepal.

MoFSC, (2013). *Persistence and Change: Review of 30 Years of Community Forestry in Nepal*, Ministry of Forests and Soil Conservation, Multi Stakeholder Forestry Program/ SSU.

MoFSC, (2014a). *Nepal Fifth National Report on Convention on Biological Diversity*. Government of Nepal, MoFSC, Kathmandu, Nepal.

MoFSC, (2014b). *Nepal Biodiversity Strategy and Action Plan: 2014–2020*. Ministry of Forest & Soil Conservation, Government of Nepal, Kathmandu, Nepal.

Mooney, H. A., McNeely, J. A., Neville, L. E., Schi, P. J., & Waage, J. K., (2004). *Invasive Alien Species: Searching for Solution*. Island Press, Washington, D.C.

Olson, D. M., Dinerstein, E., Wikramanayake, E. D., Burgess, N. D., Powell, G. V. N., Underwood E. C., D'Amico, J. A., Itoua, I., Strand, H. E., Morrison, J. C., Loucks, C. J., Allnutt, T. F., Ricketts, T. H., Kura, Y., Lamoreux, J. F., Wettengel, W. W., Hedao, P., Kassem. K. R., (2001). Terrestrial ecoregions of the world: New map of life on earth. *Bioscience, 51*(11), 933–938.

Pradhan, N., (2016). Bryophytes of Nepal. In: *Frontiers of Botany,* Jha, P. K., Siwakoti, M., & Rajbhandary, S. (eds.). Central Department of Botany, Tribhuvan University, pp. 100–123.

Pradhan, N., & Joshi, S. D., (2008). A diversity account of Bryaceae (Bryophyte: Musci) of Nepal. *J. Nat. Hist. Mus., 23*, 19–26.

Prance, G. T., (1990). Flora. In: *The Earth as Transformed by Human Action: Global and Regional Changes in the Biosphere Over the Past 300 Years,* Turner. B. L., (eds.). Cambridge Press, NY. pp. 387–391.

Prasad, V., (2013). Biodiversity: Algae. In: *Biological Diversity and Conservation,* Jha, P. K., Neupane, F. P., Shrestha, M. L., & Khanal, I. P., (eds.). Nepal Academy of Science & Technology, Lalitpur., pp. 97–103.

Press, J. R., & Shrestha, K. K., (1999). Collections of Flowering Plants by Francis Buchanan-Hamilton from Nepal, 1802–1803. *Bull. Br. Mus. Nat. Hist. (Bot.), 30*(2), 101–130.

Press, J. R., Shrestha, K. K., & Sutton, D. A., (2000). *Annotated Checklist of the Flowering Plants of Nepal*. The History Museum, London.

Rajbanshi, K. G., (2013). *Biodiversity and Distribution of Freshwater Fishes of Central Himalaya Regional Nepal Fisheries Society*, Kathmandu.

Rajbhandari, K. R., (2015). *A Handbook of the Orchids of Nepal*. Department of Plant Resources, MoFSC Kathmandu.

Rajbhandari, K. R., (2016). History of botanical explorations in Nepal: 1802–2015. In: *Frontiers of Botany*, Jha, P. K., Siwakoti, M., & Rajbhandary, S., (eds.). 1–99.

Rajbhandari, K. R., Rai, S. K., & Bhatt, G. B., (2016). Endemic flowering plants of Nepal: An update. *Bul. Dept. Pl. Res., No. 38*, 106–144.

RBG, Kew, (2016). *The State of the World's Plants Report–2016*. Royal botanic gardens, Kew.

Salick, J., & Byg, A., (2007). *Indigenous Peoples and Climate Change*. Oxford: Tyndall Centre for Climate Research.

Schleich, H, H., & Kastle, W., (2002). *Amphibians and Reptiles of Nepal*. A. R. G. Ganter Verlag, K. G. Germany.

Shah, K. B., (2013). Biodiversity: Amphibians and reptiles. In: *Biological Diversity and Conservation*, Jha, P. K., Neupane, F. P., Shrestha, M. L., & Khanal, I. P., (eds.), NAST, Lalitpur.

Shah, K. B., & Tiwari, S., (2004). *Herpetofauna of Nepal. A Conservation Companion*. IUCN–The World Conservation Union, Nepal.

Sharma, L. R., (1995). *Enumeration of the Lichens of Nepal*, Biodiversity Profile Project, DNPWC/ MoFSC, Kathmandu, Nepal.

Shrestha, B. B., (2016). Invasive alien plant species in Nepal. In: *Frontiers of Botany*, Jha, P. K., Siwakoti, M., & Rajbhandary, S., (eds.). 269–284.

Shrestha, G. L., & Shrestha, B., (1999). An overview of wild relatives of cultivated plants in Nepal. In: *Wild Relatives of Cultivated Plants in Nepal*, Shrestha, R., & Shrestha, B., (eds.). The green energy mission/ Nepal, Anam Nagar, Kathmandu, Nepal, 19–23.

Shrestha, J., (1995). Enumeration of the fishes of Nepal. In: *Biodiversity Profile Project Publication No. 10. HMG/N Department of National Parks & Wildlife Conservation*, Ministry of Forest & Soil Conservation, Kathmandu.

Shrestha, K. K., (2015). Flora of Nepal: Issues, progress update and way forward. In: *Taxonomic Tools and Flora Writing*, Siwakoti, M., & Rajbjandary, S., (eds.). Department of Plant Resources, MoFSC and Central Department of Botany, TU. pp. 25–52.

Shrestha, K. K., (2016). Global biodiversity and taxonomy initiatives in Nepal. In: *Frontiers of Botany*, Jha, P. K., Siwakoti, M., & Rajbhandary, S., (eds.). Central Department of Botany, Tribhuvan University, pp. 177–223.

Shrestha, K. K., Bhattarai, S. & Bhandari, P., (2018). *Handbook of Flowering Plants of Nepal* (Volume 1. Gymnosperms and Angiosperms: Cycadaceae - Betulaceae). Scientific Publishers, Jodhpur, India.

Shrestha, T. B., & Joshi, R. M., (1996). *Rare Endemic and Endangered Plants of Nepal*. WWF Nepal Program, Kathmandu, Nepal.

Smith, C., (1989). Butterflies of Nepal. In: *Wild is Beautiful*. Majpuria Publication, Craftsman Press Bangkok, Thailand, pp. 352.

Stainton, J. D. A., (1972). *Forests of Nepal*. John Murray, London.

Stearn, W. T., (1960). *Allium* and *Milula* in the Central and Eastern Himalaya. *Bull. Br. Mus. Nat. Hist. (Bot.)*, *2*, 159–191.

Surana, R., Subba, B. R., & Limbu, K. P., (2005). Community structure of zooplanktonic group of Chimdi Lake, Sunsari, Nepal. *Our Nature*, *3*, 81–82.

Thapa, N., (2002). *Pteridophytes of Nepal, Bull. Dept. Pl. Resources No. 19*. Department of Plant Resources Thapathali, Kathmandu, Nepal.

Thapa, R. B., & Rana, H. B., (2001). Spider fauna occurring in rice field in Chitwan, Nepal. *Natural History Society of Nepal (NAHSON) Bulletin, vol. 1*.

Thapa, S., (2014). A checklist of mammals of Nepal. *J. Threatened Taxa.*, *6*(8), 6061–6072. http: /dx.doi.org/10: 11609/JOTT.o3511.

Thapa, V. K., (1997). *An Inventory of Nepal's Insects, (Protura–Odonta), vol. I*. IUCN Nepal Biodiversity Publication Series 1, IUCN Nepal, Kathmandu.

Thapa, V. K., (1998). *An Inventory of Nepal's Insects, (Lepidoptera), vol. II*. IUCN Nepal Biodiversity Publication Series 3, IUCN Nepal, Kathmandu.

Thomas, C. D., Cameroon, A., Green, R. E., Bakkenes, M., Collingham, Y. C., Erasmus, B. F., De Siqueria, M. F., Grainger, A., Hannah, L., Hughes, L., Huntley, B., Van Jaarsveld, A. S., Midglwy, G. F., Miles, L., Ortega-Huerta, M. A., Peterson, A. T., Phillips, O. L., & Williams, S. E., (2004). Extinction risk from climate change. *Nature*, *427*(6970), 145–148.

Tiwari, R. B., & Chhetry, P., (2009). Diversity of zooplankton in Betna Wetlands, Belbari, Morang. *Our Nature*, *7*, 236–237.

Tiwari, S., Adhikari, B., Siwakoti, M., & Subedi, K., (2005). *An Inventory and Assessment of Invasive Alien Plant Species of Nepal*. IUCN Nepal, Kathmandu.

Uetz, P., (2016). *The Reptile Database*. (www.reptile-database.org): Last Updated October.

UNEP-WCMC and IUCN, (2016), Protected Planet Report 2016. UNEP-WCMC and IUCN: Cambridge, UK and Gland, Switzerland.

Vetaas, O. R., & Grytnes, J. A., (2002). Distribution of vascular plant species richness and endemic richness along the Himalayan elevation gradient in Nepal. *Global Ecol Biogeogr.*, *11*, 291–301.

Vincent, H., Wiersema, J., Kell, S., Fielder, H., Dobbie, S., Castaneda-Alvarez, N. P., Guarino, L., Eastwood, R., Leon, B., & Maxted, N., (2013). A prioritized crop wild relative inventory to help underpin global food security. *Biol. Conserv.*, *167*, 265–275.

Watson, M. F., Akiyama, S., Ikeda, H., Pendry, C., Rajbhandari, K. R., & Shrestha, K. K., (2011). *Flora of Nepal, (Magnoliaceae–Rosaceae), vol. 3*. Royal Botanic Gardens, Edinburgh, U.K.

Wikramanayake, E. D., Dinerstein, E., Loucks, C. J., Olson, D. M., Morrison, J., Lamoreux, J., McKnight, M., & Hedao, P., (2002). *Terrestrial Ecoregions of the Indo-Pacific: A Conservation Assessment*. Island Press, Washington DC.

Zeigler, D., (2007). *Understanding Biodiversity*. Praeger Publishers, West Port, Connecticut, London.

Alpine scrub: Far-West Nepal

Tropical Sal forest (Chitwan National Park)

Blue pine forest: Mid-West Nepal

Rice cultivation (Dhading, Central Nepal)

Mai Pokhari: Ilam (East Nepal)

Mikania micrantha (Chitwan NP)

Plate 13.1

Pinus wallichiana (Pinaceae)

Caltha palustris (Ranunculaceae)

Coelogyne nitida (Orchidaceae)

Nelumbo nucifera (Nelumbonaceae)

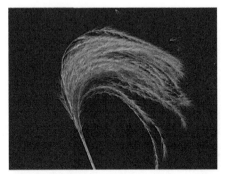

Miscanthus nepalensis (Poaceae)

Cotoneaster microphyllus (Rosaceae)

Plate 13.2

Boehmeria rugulosa (Urticaceae)

Haldina cordifolia (Rubiaceae)

Schima wallichii (Theaceae)

Saussurea gossypiphora (Asteraceae)

Rhododendron cinnabarinum (Ericaceae)

Wild mushroom (Annapurna CA)

Plate 13.3

Rhinoceros (Chitwan NP)

Swamp Deer (Shuklaphanta NP)

Snow Leopard (Sagarmatha NP)

Egyptian Vulture (Syangja, C. Nepal)

Musk Deer (Manaslu CA)

Tibetan Snowcock (Langtang NP)

Plate 13.4

CHAPTER 14

Biodiversity in Vietnam

P. K. LOC,[1] M. D. YEN,[2] and LEONID AVERYANOV[3]

[1]Department of Botany, Faculty of Biology, Hanoi University of Science, Vietnam,
E-mail: pkeloc@yahoo.com
[2]Department of Zoology, Faculty of Biology, Hanoi University of Science, Vietnam
[3]Komarov Botanical Institute, Russian Academy of Sciences, Russia

14.1   Introduction

Vietnam comprising the largest area in the eastern part of the Indo-Chinese Peninsula is situated along the Southeastern margin of the Peninsula. Vietnam extends from 8°34' to 23°23'N latitude, a distance of more than 1600 km from its northern border with the P.R. of China to the Southern-most point at Cap Camau. The East-West extend of the country is variable and not large, about 600 km in the North and only a little more than 40 km at its narrowest part within Quang Binh province. The total land area is 331.212 km² (2007) and the coastline is 3451 km.

The Northeastern and Southwestern frontiers of Vietnam are formed by the waters of the Bac Bo (Tonkin) and Thailand gulfs. A large number of islands mainly are continental origin. Two oceanic archipelagos of islands with coral atolls origin in the center of East Sea are Hoang Sa (Paracels) and Truong Sa (Spratly). The Truong Son (Annamite) mountain chains form the natural western Vietnamese boundary with Laos and Cambodia. The Northern border with China lies within the mountain systems of Southeastern Yunnan and Guangxi. The contiguous island border of Vietnam with China, Laos, and Cambodia extends nearly 4639 km.

About three-quarters (3/4) of Vietnam is either hilly or mountainous. The largest mountain formation is Hoang Lien Son with the highest peak is Fan Si Pan (3143 m a.s.l.). It is composed mainly of granite, gneiss, shale, schist, and sandstones. Ancient highly metamorphosed and solid highly karst eroded limestone is very typical and widely distributed in the north portion of the country. Towards the East, the mountains gradually give way to vast alluvial plains, often at minimum elevations. Southwards the mountainous regions are associated with relatively level and dissected plateau, where average elevations are from 500 to 1500 m a.s.l. The Southern portion of the country is the vast partially swamped lowland plain of Mekong River delta.

The hydrographic network within Vietnam is well developed within addition to numerous small and medium-sized streams and rivers. Vietnam has 9 rivers with their watershed over 10,000 km². Two big and important river systems of Vietnam are Red river in the North and Mekong river in the South. Vietnam has two big deltas: Red River delta (15.000 km²) in the North and Mekong River delta in the South (60.000 km²).

There are several notable geological features found in Vietnam but the most striking by far is the Karst formations. Karst consists of irregular limestone in which erosion has produced fissures, sink holes, caves and underground rivers. The Northern part of Vietnam has a spectacular assemblage of these formations notably around Halong Bay, Bai Tu Long Bay, and Tam Coc. In the South, there is a less impressive collection around the Ha Tien area in the Mekong Delta. The marble mountain near Danang in Central Vietnam is yet another example. Not all of Vietnam's mountains are limestone. The coastal ranges near Nha Trang and those at Hai Van Pass (Da Nang) are composed of granite. The Western part of the Central Highland (near Buon Me Thuot and Pleiku) is known for its red volcanic soils. The Mekong river has produced one of the world's great deltas, composed of the silt which has washed downstream for millions of years. The Mekong delta continues to grow at a rate of about 100 m per year.

Winter and summer monsoons are the main factors determining the climate formation within Southeast-Asia and in Indochina Peninsula including Vietnam. Winds from North and Northeast dominate in the region from October to March. Basically, these winds are cold and dry. Winds from south and southeast dominate in Southeast Asia from April to September. These winds come from the Indian and Pacific oceans and are formed during summer. Basically, summer monsoon winds are warm and wet. They bring hot and rainy weather. At the same time, the complicated orography of the region and specific configuration of the coastline of Vietnam locally shift the direction of monsoon winds, change the times of the dry and rainy seasons and influence the amount of precipitation. Inland regions south to the Mekong Delta experience smaller seasonal fluctuations in temperature than areas to the north with summer rains and a dry season of zero to five months. On the high plateau of the central region, temperatures are lower and conditions wetter with dry seasons lasting for only three months. Coastal areas experience a rainy season in the Autumn and Winter (September to January) followed by a dry period of up to seven months. Further south in the Mekong Delta, temperatures are quite warm and stable year round. Rains fall in the summers from

May to October, with the heaviest rains occurring in July and August. The dry season varies from two to six months. The hottest period is from March to May with high humidity in the latter month. Climate also affects marine areas with cold streams flowing along the coast from northeast to southwest and warm currents flowing from the southwest to the northeast. South of the Hai Van Pass waters remains above 20°C year round. In contrast, the northern water temperatures during the winter can drop to 13°C (Lap, 1978; Thao, 1997).

14.2   Ecosystems Diversity Vegetation Ecoregions

14.2.1   Ecosystems Diversity

As above descriptions, the country of Vietnam has a high variation in topography, in geology, in climate, and in landscape. Therefore, Vietnam has many types of Ecosystems. Each ecosystem type has its own biological community, ecological processes, and ecological services.

For Terrestrial Ecosystems, Vietnam has the following types:

• Subalpine, Montane, Hill, Deltaic, Coastal, Cave, and Island.

For Aquatic Ecosystems, Vietnam has the following types:

• For Freshwater – lotic: Pond, Lake, Ricefield, Dam Reservoirs, Wetland.
• For Freshwater – lentic: River, Stream, Estuary.
• For groundwater: groundwater.
• For Brackish water: lagoon, Muddy/Sandy/Rocky Coastal Beach, Mangrove, Sea grasses.
• For marine water: Coral reef, Continental shelf, Oceanic (Yen, 1994; Yen and Sung, 1998).

Through a long history of land use, there are a variety of human-generated ecosystems, i.e., artificial ecosystems. Vietnam is a country with a high density of human population, therefore, Vietnam's natural ecosystems have been severely degraded.

Each ecosystem type has its own ecological services such as Supporting, Provisioning, Regulating, and Cultural. Normally, the Natural Ecosystem in Vietnam in comparison with the same type in the other countries (not in tropical region) has higher in biological productivity. Many researchers in Vietnam on Biomass and Biological productivity in Forest ecosystems, Mangrove ecosystems, and Freshwater lakes have confirmed these conclusions.

14.2.2 Vegetations

The Monsoon tropical forests are the most typical vegetation of the eastern Indo-Chinese Peninsula and Vietnam. Wet closed evergreen broad-leaved, mixed or conifers forests are present in mountainous regions. In some hilly lowland and valley areas with more arid conditions these forests give way to open semi-deciduous and deciduous broad-leaved forest, woodlands and semi-savannas. Along semi-desert coastal plains and on seasonally drier sea slopes of southern Vietnam, such woodlands develop scrubby derivatives, which after appear as more or less dense xerophytic scrub. Swampy grass-sedge communities and mangrove thickets are very common on flat coastal plains and are especially well developed within the deltas of large rivers. At present all types of native primary vegetation are severely affected by human activities and most are completely converted over most of the country. Secondary forest, scrubs and grassland, bamboo thickets, tree plantations, pastures, crop and vegetable fields and various weed communities usually lacking all native plants now replace native vegetation over vast areas of the territory (Averyanov et al., 2003; Vidal, 1979). A simplified vegetation map of Vietnam is represented with eight major types of plant communities roughly outlined (Figure 14.1).

1. Evergreen broad-leaved forests on alkaline soils.
2. Evergreen and semi-deciduous broad-leaved, mixed and coniferous limestone mountain forests.
3. Evergreen lowland forests on silicate rocks at 0–1,000 m a.s.l.
4. Evergreen montane and highland forests on silicate rocks at 1,000–3,143 m a.s.l.
5. Semi-deciduous dry lowland forests.
6. Deciduous dry lowland forests and savanna-like woodlands.
7. Coastal vegetation, lowland wetlands, and mangrove thickets.
8. Secondary, weed and agricultural plant communities, timber and industrial plantations.

As of almost developing countries, the forest cover of Vietnam was decreased regularly mainly since 1943 to now, for example, the Figure 14.2 illustrated the forest cover change in Vietnam mainland in period 1983–2004 (World Bank in Vietnam, Ministry of Natural Resources and Environment, Embassy of Sweden, 2005).

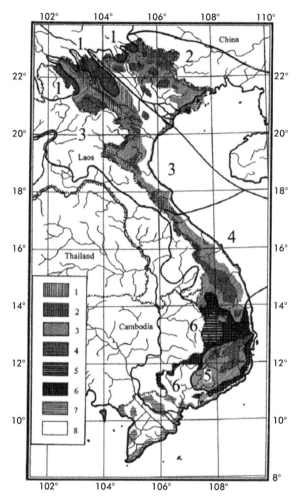

Figure 14.1 Eight major types of vegetation and six floristic regions of Vietnam. **1–8 Vegetation types:** 1. Evergreen broad - leaved, mixed and coniferous limestone mountain forests. 3. Evergreen lowland forests on silicate rocks at 0–1000 m a.s.l. 4. Evergreen montane and highland forests on silicate rocks at 1000–3000 m a.s.l. 5. Semi-deciduous dry lowland forests. 6. Deciduous dry lowland forests and savanna-like woodlands. 7. Coastal vegetation, lowland wetlands and mangrove thickets. 8. Secondary, weed and agricultural plant communities, timber and industrial plantations. **1–6. Floristic regions:** 1. Sikang–Yunnan Province. 2. South Chinese Province. 3. North Indo-Chinese Province. 4. Central Truong Son (Annamese) Province. 5. South Truong Son (Annamese) Province. 6. South Indo-Chinese Province. (Source: Averyanov, 2003)

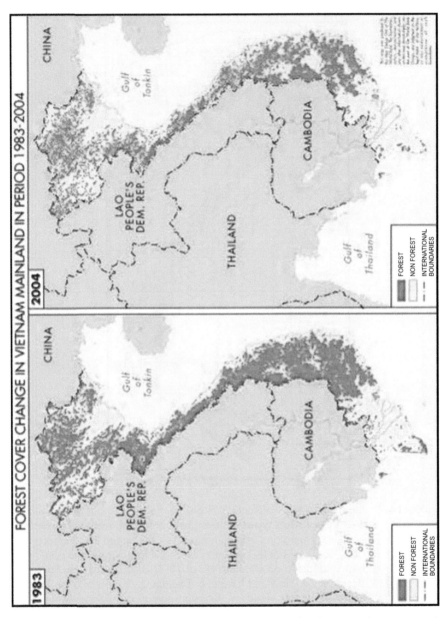

Figure 14.2 The forest cover change in Vietnam mainland in period 1983–2004 (Source: World Bank, 2005).

14.2.3 Ecoregions

Within Vietnam, the geographical distribution of biodiversity across the country is uneven. The varied pattern of biodiversity across the country can be illustrated in a number of ways, depending on the number of factors taken into account. The definition of ecoregions/or biodiversity regions is a useful way of describing areas of land or water containing a characteristic set of natural communities that share a majority of their species, dynamics and environmental conditions. Figure 14.3 identifies 10 distinct terrestrial and 9 coastal/marine ecoregions (World Bank in Vietnam, Ministry of Natural Resources and Environment, Embassy of Sweden, 2005).

According to many Biodiversity Surveys in terrestrial ecosystems, there are in Vietnam many regions which have the highest biodiversity importance in mainland (see Figure 14.4) (Vietnam Socialist Republic Government, GEF VIE/91/G31, 1995, World Bank in Vietnam, Ministry of Natural Resources and Environment, Embassy of Sweden, 2005).

14.3 Species Diversity Including Plants, Animals, and Microorganisms

Vietnam is assessed as one among the 25 highest level of Biodiversity countries of the world. By general estimation, Vietnam flora contains around 20,000 of plant species, 3,000 of Fishes, more 1,000 of Birds, and more 300 of Mammals. Vietnam biodiversity both for Flora and Fauna is considered as typical of the Oriental region in Biogeographical analysis. There are now in Vietnam so enough Botanists and Zoologists who carried out the researches on different taxonomic groups of plants, animals, and microorganisms.

14.3.1 Species Diversity of Plants

The following botanists are the persons who carried out many researches on Flora of Vietnam: Lecomte (1907–1951), Ho (1991–1993), Thin (1997), Averyanov et al. (2003), Tien (1996–1997). By compilation of the number of plant species of different taxonomic groups of Vietnam in the different Inventories, we have the Table 14.1.

There are some remarks on Species Diversity of Plants of Vietnam:

- For almost taxonomic groups of plant diversity, there are more or less 10% number species are endemic.

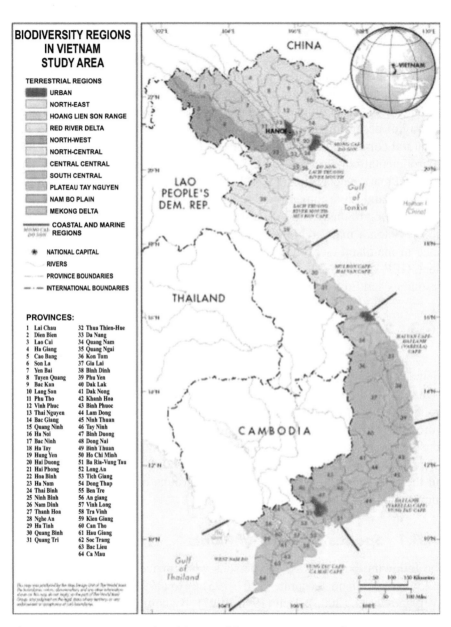

Figure 14.3 10 terrestrial and 9 coastal/marine ecoregions of Vietnam (Source: World Bank, 2005).

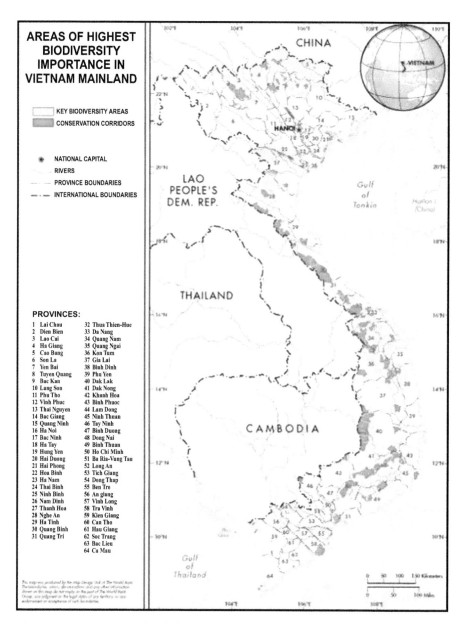

Figure 14.4   Geographical distribution of regions with highest biodiversity importance (Source: World Bank, 2005).

- The diversity of Plant species in Vietnam could be grouping into six centers such as Mountain Northeast, Hoang Lien Son mountain, calcareous mountain North – Central – Cuc Phuong, Karst – mountain Phong Nha – Ke Bang, center of Truong Son mountain, and Lang Bian mountain (Vietnam Socialist Republic Government – Project GEF VIE/91/G31, 1995)
- The 10 largest families of vascular plants in the Flora of Vietnam are Orchidaceae, Fabaceae, Poaceae, Euphorbiaceae, Rubiaceae, Cyperaceae, Asteraceae, Lauraceae, Fagaceae, Acanthaceae. The first largest one is Orchidaceae comprising 1212 listed species, the 10 largest one is Acanthaceae which has 177 species.
- In Flora geographical analysis, Vietnam could be divided into six floristic provinces such as Sikang Yunnan, South Chinese, North Indo-Chinese, Central Truong Son, South Truong Son, and South Indo-Chinese (Figure 14.1) (Averyanov et al., 2003).
- Vietnam is a developing agricultural country. The flora both wild and cultivated plays a significant role in socio-economical development (Yen and Hamel, 2002).
- The diversity of Algae both for Freshwater and Marine water is very high, for Freshwater there are 1,000 species and for Marine water, there are more than 500 species.

14.3.2   Orchid Diversity in Vietnam

The orchids (Orchidaceae) are most significant family among so-called key groups of flowering plants having a very important role in national culture and economy of Vietnam. They also play an outstanding role as a precise marker in vectors of present nature transformation and represent a classic model for fundamental studies of evolution, florogenesis, and understanding of whole plant diversity.

Similarly, to other tropical countries with the humid climate, the orchids form the largest family in the flora of Vietnam. According to present scientifically documented data, the orchid family in Vietnam includes at least 1212 species from 174 genera, which represent from 10 to 12% of all vascular plant diversity of the country (Averyanov et al., 2015a,b, 2016a,b). According to these data, the orchid flora of Vietnam appears as the richest among all other regional floras of mainland Asia. It is noteworthy, that orchid flora of Vietnam represents almost all taxonomic groups of the family with the particular well presentation of most primitive genera like *Aphyllorchis*

Table 14.1   Number of Plant Species of Different Taxonomic Groups of Vietnam
(Source: Compiled by Authors, 2016)

Taxonomic groups	Number of Species
1. Whole country	
Fungi (Macro fungi)	826
Algae	#1000
Bryophyta	793
Psilotophyta	2
Lycopodiophyta	57
Equisetophyta	2
Polypodiophyta	669
Gymnospermae	63
Angiospermae	9812
2. Freshwater algae	
Cyanophyta_Cyanobacteria	344
Pyrrophyta	30
Chrysophyta	14
Bacillariophyta	388
Rhodophyta	4
Euglenophyta	78
Chlorophyta	530
Charophyta	3
3. Marine water	
Algae	662
Rhodophyta	309
Phaeophyta	124
Chlorophyta	152
Cyanophyta	77
Sea grasses	15
Mangrove forest's plants	94

Source: Compiled by authors, 2016.

Blume, *Apostasia* Blume, *Corymborkis* Thouars, *Cyrtosia* Blume, *Eryth-rorchis* Blume, *Galeola* Lour., *Lecanorchis* Blume, *Neuwiedia* Blume, *Paphiopedilum* Pfitz., *Tropidia* Lindl. and *Vanilla* Sw. (Apostasioideae, Cypripedioideae, primitive tribes of Spiranthoideae and Epidendroideae). Other Vietnamese orchids often illustrate highest specialization in numerous

well advances phylogenetic evolutional traits (many genera of evolution-
arily advanced tribes of Spiranthoideae, Orchidoideae, Epidendroideae, and
Vandoideae). Largest genera of orchids in the flora of Vietnam are – *Dendrobium* Sw. (118 sp.), *Bulbophyllum* Thouars (117 sp.), *Eria* Lindl. (53 sp.),
Liparis Rich. (52 sp.), *Habenaria* Willd. (37 sp.), *Oberonia* Lindl. (32 sp.),
Cleisostoma Blume (30 sp.), *Coelogyne* Lindl. (28 sp.), *Cymbidium* Sw. (27
sp.), *Calanthe* Lindl. and *Paphiopedilum* Pfitzer (each with 20 sp.). Geographically Vietnam is also important diversity center for such genera of
high horticultural value as *Anoectochilus* Blume, *Ascocentrum* Schltr., *Ascocentropsis* Senghas & Schildh., *Christensonia* Haager, *Coelogyne* Lindl.,
Cymbidium Sw., *Dendrobium* Sw., *Holcoglossum* Schltr., *Paphiopedilum*
Pfitzer, *Phalaenopsis* Blume, *Pleione* D.Don, *Renanthera* Lour. and *Vanda*
R.Br.

The orchid flora of Vietnam is heterogenic in its fluorogenic composition
and geographic elements. Its complicated nature is the result of a long geological history of the Indochina territory, various extensive migrations and
biological isolation of numerous mountain systems. The modern orchid species formation and their present high diversity are well supported by diverse
natural conditions observed presently in the country. In their geography,
orchid species in Vietnam form several main groups having similar distribution (Averyanov et al., 2003a,b). These groups are listed in Table 14.2 in the
order of decreasing of their distribution square.

As seen from Table 14.1 the largest part of Vietnamese orchids are also
occur in the tropical Himalayas (43%), much less are connected in their
distribution with Malesia (11%), or have broad distribution in SE. Asia
(12.5%). Many orchids of Vietnam are endemics of Indo-Chinese Peninsula
(11%) and strict endemics of Vietnam (19.2%). Very few Vietnamese species has broad distribution in the tropical zone of the world.

Table 14.2 Orchid Species in Vietnam Having Similar Distribution

Distributional type	Approximate portion in orchid flora of Vietnam
Pantropical, Paleotropical, Asian – Australian	3.1%
Himalayan – Malesian	12.5%
Himalayan	43.2%
Malesian	11.0%
Indo-Chinese endemics	11.0%
Vietnam endemics	19.2%

(Compiled by Averyanov et al. (2003b)

The level of proper orchid endemism of Vietnamese flora is very high (19.2%). The orchid flora of the country comprises at least 230 endemic species. Most of them belong mainly to largest genera such as *Bulbophyllum*, *Eria*, *Liparis*, *Habenaria*, *Oberonia*, *Cleisostoma*, *Cymbidium*, and *Paphiopedilum*. Ten orchid genera are endemic in Vietnam. They are – *Ascocentropsis* Senghas & Schildh. (1 sp.), *Bidoupia* Aver. (2 sp.), *Christensinia* Haager (1 sp.), *Cleisostomopsis* Seidenf. (1 sp.), *Deceptor* Seidenf. (1 sp.), *Eparmatostigma* Garay (1 sp.), *Lockia* Aver. (1 sp.), *Theana* Aver. (1 sp.), *Vietorchis* Aver. (2 sp.) and *Zeuxinella* Aver. (1 sp.). All they include 1 or 2 species with very limited distribution.

The orchids are well presented in any zonal and azonal primary plant communities in Vietnam. In Vietnam they reach highest diversity in primary evergreen broad-leaved, mixed and coniferous submontane forests at elevation 700–1500 m a.s.l. Here may be found 650–750 species or 55–65% of all species occurring in Vietnam, depending of mother rocks (limestone orchid floras usually richer than habitats of non-limestone mountains). Broad-leaved and particularly coniferous montane forests at elevation 1500–2500 m a.s.l. are also very rich in orchids. They give home to about 500 orchid species (40–45% of all orchids). Highland cloud mountain forests and ericaceous thickets higher 2000 m exhibit less orchid diversity with maximum 200–300 species not exceeding 20% of orchid flora. Similar figures of orchid diversity are observed in lowland evergreen, semideciduous and deciduous forests, broad-leaved and coniferous woodlands and savanna-like plant communities. Few orchids inhabit lowland coastal, wetland, mangrove habitats, as well as secondary forests, secondary scrub, timber plantations and grasslands of different genesis. A portion of orchids occurring here do not exceed 10–15%, but usually much less. Very few orchid species can survive in agricultural and urbanized territories. Among them are a few semi-weedy species of such genera as *Arundina* Blume, *Geodorum* Andrews, *Spathoglottis* Blume, *Spiranthes* Rich., and *Zeuxine* Lindl.

Photographs of some of the most primitive and most specialized orchids in the flora of Vietnam are given in Figure 14.5(a–i) while representatives of genera endemic in the flora of Vietnam are given in Figure 14.6(a–i).

14.3.3 Species Diversity of Animals

The following Zoologists are the persons who carried out the researches on different groups taxonomic of animals (Thanh et al., 2001, 2002; Than, 2003; Thung and Sarti, 2005); Hoc (2012) on insects; Yen (1978, 1992, 1995) on fishes; Sang et al. (2009) on Herpetofauna; Delacour and Jabouille

Figure 14.5   Most primitive and most specialized orchids in the flora of
Vietnam (Subfam. Apostasioideae: a – *Apostasia odorata* Blume,
b – *Neuwiedia zollingeri* Rchb.f.; subfam. Cypripedioideae: c
– *Paphiopedilum coccineum* Perner & R.Herrm.; primitive tribes
of subfam. Dendrobioideae: d – *Tropidia curculigoides* Lindl.,
e – *Miguelia cruenta* Aver. & Vuong; highly specialized spcies
of subfam. Spiranthoideae: f – *Anoectochilus roxburghii* Lindl.;
subfam. Orchidoideae: g – *Corybas annamensis* Aver.; subfam.
Dendrobioideae: h – *Dendrobium unicum* Seidenf.; subfam.
Vandoideae: i – *Schoenorchis scolopendria* Aver.).

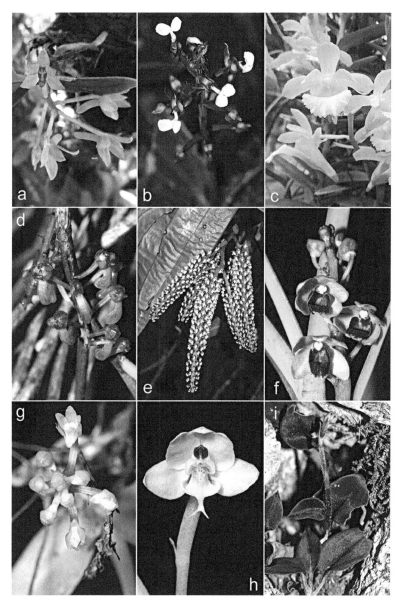

Figure 14.6 Representatives of genera endemic in the flora of Vietnam (a – *Ascocentropsis pusilla* (Aver.) Senghas & Schildh., b – *Bidoupia khangii* Aver., c – *Christensinia vietnamica* Haager, d – *Cleisostomopsis eberhardtii* (Finet) Seidenf., e – *Deceptor bidoupensis* (Tixier & Guillaumin) Seidenf., f – *Lockia sonii* Aver., g – *Theana vietnamica* Aver., h – *Vietorchis furcata* Aver. & Nuraliev, i – *Zeuxinella vietnamica* (Aver.) Aver.).

(1931) and Quy (1975, 1981), Quy and Cu (1999) on birds; Tien (1985), Can (1985), Can et al. (2008), Sterling et al. (2006) on mammals.

By compilation of the number of animal species of different taxonomic groups of Vietnam in the different inventories, we have the Table 14.3.

There are some remarks on Species Diversity of Animals of Vietnam:

Table 14.3 Number of Animal Species of Different Taxonomic Groups of Vietnam

Taxonomic groups	Numbers of species
1. Whole country	
Protozoa	Many hundreds
Invertebrates (excluding Insects)	>4000
Insects	>7000
Fishes marine	2527
Fishes freshwater	544
Amphibia	197
Reptiles	447
Birds	1026
Mammals	304
2. Freshwater	
Invertebrates	#1000
Crustaceae	183
Insectalar larvae	102
Rotatoria	100
Polychaeta	30
Oligochaeta	47
Hirudinae	9
Mollusca	141
Fishes	544
3. Marine water	#5000
Zooplankton	659
Corals	346
Mollusca	#1500
Fishes	2527
Reptiles	20
Birds	43
Mammals	12

Source: Complied by authors, 2016.

- There are many animal species which are endemic to Vietnam, particularly for freshwater fishes (around 50 species), freshwater invertebrates (around 50 species), small mammals (around 50 species and subspecies).
- In the Zoogeographical analysis, almost all Zoologists agreed that the territory of Vietnam could be divided into five units at the level of province, such as North–West, North–East, North–Center, Central Highland, and South Vietnam.

The main taxonomic groups of vertebrate animals at the level of orders with high diversity of species are:

- Freshwater fishes: Cypriniformes, Siluriformes
- Marine fishes: Perciformes, Clupeiformes
- Amphibia: Anura
- Reptile: Lacertilia, Serpentes, Testudinae
- Birds: Passeriformes, Piciformes, Charadriformes, Gruiformes, Galliformes, Coraciformes
- Mammals: Rodentia, Chiroptera, Fissipedia, Artiodactyla, Primates

Now, in Vietnam, there are many wild animals which were domesticated (aquatic animals, fishes, mammals).

14.3.4   Species Diversity of Microorganisms

The researches on species (varieties, strains) diversity of microorganisms in Vietnam are beginning. The main groups of microorganisms are bacteria, yeasts, microscopic fungi, virus, all of which are of great signification on socio-economical development and public health. Many species are pathogenic to plants, to livestock, to human health. Some others are useful for agriculture, animal husbandry, aquaculture, food processing, and for protection environment.

Now many Research Institutes on Microbiology have begun to keep in their Laboratories their own collection of strains of microorganisms as in the Table 14.4. There are three groups of strains: common for all Laboratories, natives, i.e., only of Vietnam, have been analyzed in DNA (Hop, 2015).

14.4   Genetic Diversity with Emphasis on Crop Plants/Cultivated Plants and Cultured Animals

Vietnam has 54 ethnic groups living in a territory as long with 15 range of latitude in the tropical region and around of surface area 330,000 km^2.

Table 14.4 Numbers of Different Strains of Microorganism Conserving in the Collections of Different Institutes of Vietnam (2015)

Name of Institute	Number of strains	
1. Institute of Microbiology and Biological Engineering (Hanoi National University)	National standard VTCC	8976
2. Institute of Research of Food Engineering	Food engineering microorganisms	1160
3. Institute of Agro-chemistry and Pedology	Agricultural microorganisms	890
4. Institute of Veterinary Medicine	Veterinary microorganisms	2720
5. Hanoi Medicine University	Medicine microorganisms	82
6. Institute of Epidemiology	Medicine microorganisms	82
7. National Center of Science and Technology	Strains *Bacillus thuringiensis*	7364
8. University of Pedagogy Hanoi	Fungi of Mangrove's forest	278

Source: D.V. Hop. Institute of Microbiology and Biological Engineering-Hanoi National University (Personal communication), 2015.

Vietnam has a long history. Recent archaeological finds indicate that the earliest human habitation of northern Vietnam goes back about 500,000 years. Mesolithic and Neolithic cultures existed in northern Vietnam 10,000 years ago, these groups may have engaged in primitive agriculture as early as 7,000 B.C. Vietnam was considered as one among 12 centers of origin of Cultivated plants and Domestic animals. This center is called as Indo-Chinese – Indonesian center (the tropical Asian center of origin). The analysis in genetic diversity of cultivated plants and domestic animals, we normally divided them into four groups:

- Domestication/acclimatization/in situ by local people very long ago.
- Introduction from other countries long ago (exotic species).
- Introduction from other countries recently (exotic species), most of these species are appreciated in the worldwide by high productivity, for exportation, cheap, in production, OGM varieties.
- Domestication/acclimatization in site by local people recently (Yen and Hamel, 2002).

Vietnam is now considered as a high genetic diversity center for crop plants/cultivated plants and domestic animals.

14.4.1 Genetic Diversity on Crop Plants/Cultivated Plants

Normally, in Vietnam now, we divided the cultivated plants into the following groups:

- Cereal plant/food plants
- Vegetables plants
- Fruits plants
- Exciting/stimulus plants
- Industry plants
- Medicine plants
- Ornamental plants
- Green manure plants
- Animal forage plants
- Soil cover plants
- Wood plants
- Street/town plants
- Divine plants

All the above cultivated plants, with preliminary inventory we have some following data (Yen and Hamel, 2002):

- In Hanoi city center (10 districts/wards) there are 411 plant species with 148 plant species in two sides of the streets, 75 species of fruits (Chuyen, 1999).
- The total number of cultivated plants species for crops: 134 species; for vegetables and exciting/stimulus: 109 species; for industrial plants: 28 species; medicine plants: 179 species; fruit plants 130 species; ornamental plants: 150 species.
- There are many varieties in important/economical cultivated plants, for examples, with Rice: 156, Maize: 47, Sweet potato: 9, Potato: 8, Soybean: 22, Peanut: 14, Tomato: 14, Rubber: 14, Coffee: 14, Green bean: 7, Cotton: 9, Mango: 5, Durian 5 (Dat, 2008).
- The Indonesian, Indo-Chinese center of origin in Cultivated plants/ Domestic animals is considered as important region with for crops such as bamboos, tropical fruits trees, ginger plants, *Cocos nucifera, Colocasia esculenta, Dioscorea* spp., *Musa* spp., wild and weedy *Oryza* spp., *Piper* spp., and *Saccharum officinarum*.
- The number of wild plant species now have been cultured increasing quickly.

14.4.2 Genetic Diversity on Domestic Animals (Livestock)

As in the other countries in tropical regions, Vietnam has the following domestic animals (excluding aquaculture, ornamental animals):

a) Mammals
1. Pig (*Sus scrofa*)
2. Water buffalo (*Bubalus bubalis*)
3. Cow (*Bos* spp.)
4. Horse (*Equus* spp.)
5. Goat (*Capra* spp.)
6. Sheep (*Ovis* spp.)
7. Dog (*Canis* spp.)
8. Cat (*Felis* spp.)
9. Rabbit (*Oryctolagus cunniculus*)
10. Deer (*Cervus nippon*)
11. Rusa (*Rusa unicolor*)

b) Birds
12. Chicken (*Gallus gallus*)
13. Duck (*Anas platyrhynchus*)
14. Goose (*Anser anser*)
15. Perching duck (White winged duck)
16. Star chicken (*Polyplectron germaini*)
17. Japan quail (*Coturnix japonica*)
18. Dove (Pigeon) (*Columba livia*)
19. Pheasant (*Phasianus colchicus*)
20. Ostrich (*Struthio colchicus*)

c) Reptiles
21. Chinese softshell turtle (*Trionyx sinensis*)
22. Crocodyle (*Crocodylus siamensis*)

d) Amphibia
23. Common frog (*Rana rugulosa*)

Some remarks on the domestic animals of Vietnam:

- All of above domestic animals are for food, most of them are self-provisioned to local people.
- There are many varieties of some common domestic animals (livestock, poultry, cattle) such as for chicken: 30 varieties, duck: 12 varieties, goose: 3 varieties, pig: 24 varieties, cow: 4 varieties, water buffalo:

3 varieties, horse: 2 varieties, rabbit: 2 varieties, dog: many varieties, cat: many varieties (Tieu et al., 2010).

- There are three species in domestic animals which are domesticated long ago by local people living in Vietnam territories such as chicken (*Gallus gallus*), pig (*Sus scrofa*), and water buffalo (*Bubalus bubalis*).
- The following domestic animals are very common in the rural human settlements everywhere in Vietnam: chicken, duck, pig, water buffalo, cow, dog, cat.
- The following domestic animals are recently domesticated by local people: Deer, Rusa, Pheasant, Crocodyle, Common frog.
- Some domestic animals are exotic origin: Japan quail, Ostrich.
- Now, there are many wild animals recently domesticated, for example: for birds: *Anas platyrhynchos, Phalacrocorax carbo, Pavo muticus,* swiftlet (*Aerodramus fuciphagus*); for mammals: *Rhizomus prunosus, Vivera indica, Atherurus macrourus, Elephas maximus, Canis dingo*; for reptiles *Varanus salvator, Trionyx sinensis, Naja hannah, Gecko gecko*, for amphibia *Rana rugulosa*.

14.5   Threats to Biodiversity: Endangered Plants and Animals, Protected Areas

14.5.1   Threats to Biodiversity in Vietnam

The biodiversity of Vietnam now has been decreased quickly. The main reasons are:

- Land conversion without a proper scientific base, surface areas of natural forests are reduced quickly.
- Infrastructure developments, building up many dams, roads, new urban and rural human settlements.
- Introduction of many alien invasive species (plants, animals).
- Overexploitation of Natural resource/Illegal exploitations in Fishing, Hunting, Forestry.
- Environmental pollutions (air, water, soil).
- Climate change, natural disaster, extreme weather events.
- Pressure from population growth and free migrations.
- Fire to forests.
- Deteriorations of many kinds of natural ecosystems.
- Replacement of exotic crop plants and domestic animals.

14.5.2   Endangered Plants and Animals

As above mentioned, the biodiversity of Vietnam has decreased quickly in all three components: Biodiversity in species, Biodiversity in ecosystems, and Biodiversity genetic.

The Vietnam Red list (2007) (new version) identified 882 species (418 animals and 464 plants) as threatened. This represents as an increase of 161 species considered as threatened from the first assessment (1992–96 edition of the Vietnam list). Also, between first and second assessment were 10 species that moved from being classified as "Endangered (EN) to Extinct in the wild (EW). The number of species in the Red list 2007 by taxon of Vietnam is illustrated in the Table 14.5 (Ministry of Science and Technology, Vietnam Academy of Science and Technology, 2007).

Table 14.5   Number of Endangered Species in the Red Books (2007) of Vietnam (Source: Ministry of Science and Technology, 2007)

	EX	EW	CR	EN	VU	LR	DD
Flora	1		37	178	200	4	
Magnoliophyta							
Dicots			29	96	147		
Monocots	1		4	69	34	3	
Pinophyta			4	4	18	1	
Pteridophyta				1	1		
Lycopodiophyta					1		
Rhodophyta				5	2		
Phaeophyta							
Mycophyta				3			
Fauna	4	5	48	113	189	17	30
Mammals	4	1	12	30	30	5	8
Birds			11	17	25	11	9
Reptiles_Amphibia		1	11	22	19		
Fishes		3	4	28	31		3
Invertebrates			10	16	64	1	10

EX: Extinct, EW: Extinct in the wild, CR: Critically endangered, EN: Endangered, VU: Vulnerable, LR: Low risk, DD: Data deficient.

Source: Ministry of Science and Technology, Academy of Science and Technology, 2007.

The Decree of Vietnam Government N° 160/2013/ND-CP gives a List of Endangered, Precious and Rare species prioritized for the protection of 17 plants species and 82 animal species (48 mammals, 22 birds, 11 reptiles) Annex 14.1 (Ministry of Natural Resources and Environment, 2015).

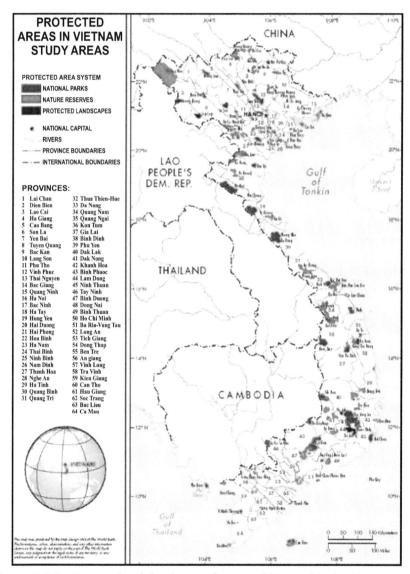

Figure 14.7 Geographical distribution of system of National Parks, Nature Reserves, and Landscape Protected Areas (Source: World Bank, 2005).

14.5.3 Protected Areas

Vietnam Government has built up many types of Protected Areas. (Ministry of Natural Resources and Environment, 2015).

1. In terrestrial habitats: National Parks (Annex 14.2), Nature Reserves, Species protected areas, Landscape protected areas.
2. In freshwater habitats: Inland water protected, Wetland protected areas.
3. In marine habitats: Marine protected areas.

The total surface of Protected Areas in the mainland is around 2 millions hectares, in seawater around 200,000 hectares.

Besides the system of Protected Areas as above, Vietnam has also some other types of Biodiversity conservation such as Biosphere Reserves, Green Corridors, Zooparks, Botanic gardens, etc.

The geographical distribution of the system of Nationals Parks, Nature Reserves, and Landscape Protected areas is illustrated on Figure 14.7.

Keywords

• freshwater habits
• marine habits

• protected areas
• terrestrial habits

References

Averyanov, L. V., Duy, N. V., Vinh, T. T., Hoi, Q. V., & Cong, V. K., (2015b). Four new species of orchids (Orchidaceae) in eastern Vietnam. *Phytotaxa*, *238*(2), 136–148.

Averyanov, L. V., Loc, P. K., Hiep, N. H., & Harder, D. K., (2003b). *Phytogeographic Review of Vietnam and Adjacent Areas of Eastern Indochina, vol. 3.* Komarovia, pp. 1–83.

Averyanov, L. V., Nguyen, K. S., Tich, N. T., Nguyen, P. T., Nong, V. D., Nguyen, V. C., & Xuan, C. C., (2015a). New orchids in the flora of Vietnam. *Wulfenia*, *22*, 137–188.

Averyanov, L. V., Nong, V. D., Nguyen, K. S., Maisak, T. V., Nguyen, V. C., Phan, K. T., Nguyen, T., Nguyen, T. T., & Truong, B. V., (2016b). New species of orchids (Orchidaceae) in the flora of Vietnam, *Taiwania*, *61*(4), 319–354.

Averyanov, L. V., Ormerod, P. A., Duy, N. V., Tien, T. V., Chen, T., & Zhang, D. X., (2016a). *Bidoupia phongii*, New orchid genus and species (Orchidaceae, Orchidoideae, Goodyerinae) from southern Vietnam. *Phytotaxa*, *266*(4), 289–294.

Averyanov, L., Cribb, P., Loc, P. K., & Hiep, N. T., (2003a). *Slipper Orchids of Vietnam With an Introduction to the Flora of Vietnam.* Royal Botanic Gardens, Kew. Compass Press Limited.

Can, D. N., Endo, H., Son, N. T., Oshida, T., Can, L. X., Phuong, D. H., Lunde, D. P., Kawada, S. I., Hayashida, A., & Sasaki, M., (2008). *Checklist of Wild Mammal Species of Vietnam.* Shoukadoh Book sellers, Kyoto Japan.

Chuyen, V. V., (1999). *List of Plants in Hanoi Capital.* Annex 1 Encyclopedia of Hanoi T. 6. Polytechnic Dictionary Publishers Hanoi (in Vietnamese).

Dat, D. H., (2008). Biodiversity on agricultural species of Vietnam-Data based for sustainable development. *Proceedings on Biodiversity Conservation of Truong Son Mountain Range.* Hanoi (in Vietnamese).

Delacour, J., & Jabouille, D., (1931). *Birds of French Indochina, 1–4.* Paris.

Ho, P. H., (1991–1993). *An Illustrated Flora of Vietnam.* Tom 1, 2, 3 Mekong Publishers, Santa Ana/Montreal.

Hoc, T. Q., (2012). *Vietnam Nature Environment and Sustainable Development.* Sciences and Technology Publishers (in Vietnamese).

Hop, D. V., (2015). *Summary on Conservation and Management of Biodiversity on Microorganism of Vietnam* (in Vietnamese, personal communication).

Lap, V. T., (1978). *Natural Geography of Vietnam. Tomes I, II, III.* Educational Publishers Hanoi (in Vietnamese).

Lecomte, H., (1907–1951). *Flora of Indochina. Tomes 1–7.* Paris.

Ministry of Natural resources and Environment, (2009–2013). *Vietnam-Fifth National Report to the United Nations.* Convention on biological diversity. Reporting period: 2015.

Ministry of Science and Technology, (2007). Vietnamese Academy of Science and Technology. *Vietnam Red List.* Natural Science and Technology Publishers, Hanoi.

Quy, V., & Cu, N., (1999). *Checklist of the Birds of Vietnam.* Hanoi Center for Natural Resources and Environment studies/Agricultural Publishers Hanoi (in Vietnamese).

Quy, V., (1975). *Birds of Vietnam Tome 1.* Science and Technology Publishers Hanoi (in Vietnamese).

Quy, V., (1981). *Birds of Vietnam Tome 2.* Science and Technology Publishers Hanoi (in Vietnamese).

Sang, N. V., Cuc, H. T., & Truong, N. Q., (2009). *Herpetofauna of Vietnam.* Edition Chimaira Frankfurt am Main.

Sterling, E. J., Hurley, M. M., & Minh, L. D., (2006). *Vietnam a Natural History.* Yale Univ. Press, New Haven and London.

Thanh, D. N., (2003). *East Sea IV Marine Biodiversity and Marine Ecology.* Hanoi National Univ. Publishers Hanoi (in Vietnamese).

Thanh, D. N., et al., (2002). *Hydrobiology of Freshwater bodies of Vietnam.* Science and Technology Publishers Hanoi (in Vietnamese).

Thanh, D. N., Hai, H. T., Tien, D. D., & Yen, M. D., (2001). Biodiversity of aquatic species composition in the island freshwater of Vietnam. *Proceedings of Resources on Ecology and Biological Resource 1996–2000.* Agricultural Publishers (in Vietnamese).

Thao, L. B., (1997). *Vietnam: The Country and Its Geographical Regions.* Hanoi The Gioi Publishers (in Vietnamese).

Thin, N. N., (1997). *Manual on Research of Biodiversity.* Agricultural Publishers Hanoi (in Vietnamese).

Thung, D. C., & Sarti, M., (2005). Marine biodiversity and its reduction – A challenge for Vietnam Marine resources and environment. *First National Workshop on Ecology and Biological Resources* (in Vietnamese).

Tien, D. D., (1996–1997). *Freshwater Algae of Vietnam.* Agricultural Publishers Hanoi (in Vietnamese).

Tien, D. V., (1985). *Surveys on Mammals in the North Vietnam.* Science and Technology Publishers Hanoi (in Vietnamese).

Tieu, H. V., et al., (2010). *Vietnam Livestock Genetic Resources Conservation and Exploration (VLGC&E) from 1990 to 2007.* National Institute of Animal Husbandry Research Paper (in Vietnamese).

Vidal, J. E., (1979). Outline of the ecology and vegetation on the Indo-Chinese peninsula. In: *Tropical Botany.* Larsen, K., & Holm–Nielson, L. B., (eds.), London Academic Press. pp. 109–124.

Vietnam Socialist Republic Government, (1995). GEF VIE/91/G31, *Biodiversity Action Plan of Vietnam.* Hanoi.

World Bank in Vietnam, (2005). Ministry of Natural Resources and Environment, Embassy of Sweden. Vietnam Environment monitor. *Biodiversity.*

Yen, M. D., & Hamel, C., (2002). *National Training on the Conservation of Biological Diversity.* Manual textbook given to the students of University of Montréal, Univ. Nationale de Hanoi, University of Science of Hue and Univ. Nationale of HochiMinh ville.

Yen, M. D., & Sung, C. V., (1998). Vietnam's Ecosystems. In: *Environment and Bioresources of Vietnam.* The Gioi Publishers, Hanoi, pp. 7–22.

Yen, M. D., (1978). *Identification of Freshwater Fishes of the North Vietnam.* Science and Technology Publishers, Hanoi (in Vietnamese).

Yen, M. D., (1992). *Identification of Freshwater Fishes of the South Vietnam.* Science and Technology Publishers Hanoi (in Vietnamese).

Yen, M. D., (1994). Contributions to the study on the ecosystems of Vietnam. *Proceedings of the VI International Congress of Ecology,* Manchester, U.K.

Yen, M. D., (1995). The biodiversity of freshwater fishes and different measures applied for its conservation in Vietnam Rep. *Suwa hydrobiol., 9,* 13–18.

Appendices

Appendix 14.1 Lists of Endangered, Precious and Rare Species Prioritized for Protection (Decree N° 160/2013/ND-CP dated Nov. 12, 2013) in Vietnam (*Source*: Ministry of Natural Resources and Environment)

A.	**Flora**	12.	*Rhinopithecus avunculus*
1.	*Taiwania cryptomerioides*	13.	*Trachypithecus (phayrei) barbei*
2.	*Cunninghamia konishii*	14.	*Nomascus gabriellae*
3.	*Glyptostrobus pensilis*	15.	*Nomascus leucogenys*
4.	*Xanthocyparis vietnamensis*	16.	*Nomascus nasutus*
5.	*Cupressus tonkinensis*	17.	*Nomascus concolor*
6.	*Keteleeria davidiana*	18.	*Cuon alpinus*
7.	*Abies delavayi* subsp. *fansipanensis*	19.	*Helarctos malayanus*
8.	*Shorea falcata*	20.	*Ursus thibetanus*
9.	*Hopea pierrei*	21.	*Lutra sumatrana*
10.	*Hopea cordata*	22.	*Lutrogale perspicillata*
11.	*Hopea reticulata*	23.	*Lutra lutra*
12.	*Berberis* spp.	24.	*Aonyx cinerea*
13.	*Coptis quinquesecta*	25.	*Arctictis binturong*
14.	*Coptis chinensis*	26.	*Neofelis nebulosa*
15.	*Panax bipinnatifidus*	27.	*Panthera pardus*
16.	*Panax stipuleanatus*	28.	*Catopuma temminckii*
17.	*Panax vietnamensis*	29.	*Panthera tigris*
B.	**Fauna Mammalia**	30.	*Prionailurus viverrinus*
1.	*Cynocephalus variegatus*	31.	*Pardofelis marmorata*
2.	*Nycticebus bengalensis*	32.	*Elephas maximus*
3.	*Nycticebus pygmaeus*	33.	*Rhinoceros sondaicus*
4.	*Trachypithecus villosus*	34.	*Axis porcinus*
5.	*Trachypithecus poliocephalus*	35.	*Moschus berezovskii*
6.	*Pygathrix nigripes*	36.	*Muntiacus vuquangensis*
7.	*Pygathrix nemaeus*	37.	*Muntiacus truongsonensis*
8.	*Pygathrix cinerea*	38.	*Rucervus eldii*
9.	*Trachypithecus hatinhensis*	39.	*Bos javanicus*
10.	*Trachypithecus francoisi*	40.	*Bos gaurus*
11.	*Trachypithecus delacouri*	41.	*Bos sauveli*

Appendix 14.1   (Continued).

42.	*Pseudoryx nghetinhensis*	63.	*Polyplectron germaini*
43.	*Naemorhedus sumatraensis*	64.	*Polyplectron bicalcaratum*
44.	*Bubalus arnee*	65.	*Grus antigone*
45.	*Manis javanica*	66.	*Houbaropsis bengalensis*
46.	*Manis pentadactyla*	67.	*Anorrhinus tickelli*
47.	*Nesolagus timminsi*	68.	*Aceros nipalensis*
48.	*Sousa chinensis*	69.	*Aceros undulatus*
49.	*Dugong dugon*	70.	*Buceros bocornis*
C.	**Aves**	71.	*Garrulax ngoclinhensis*
50.	*Pelecanus philippensis*	**D.**	**Reptilia**
51.	*Anhinga melanogaster*	72.	*Ophiophagus hannah*
52.	*Egretta eulophotes*	73.	*Dermochelys coriacea*
53.	*Gorsachius magnificus*	74.	*Eretmochelys imbricata*
54.	*Leptoptilos javanicus*	75.	*Lepidochelys olivacea*
55.	*Ciconia episcopus*	76.	*Caretta caretta*
56.	*Platalea minor*	77.	*Chelonia mydas*
57.	*Pseudibis davisoni*	78.	*Cuora trifasciata*
58.	*Pseudibis gigantea*	79.	*Cuora galbinifrons*
59.	*Cairina scutulata*	80.	*Mauremys annamensis*
60.	*Arborophila davidi*	81.	*Rafetus swinhoei*
61.	*Lophura edwarsi*	82.	*Pelochelys cantorii*
62.	*Tragopan temminckii*		

Appendix 14.2   System of National Parks of Vietnam (*Source*: Ministry of Natural Resources and Environment, 2015)

1.	Cuc Phuong National Park
2.	Tam Dao National Park
3.	Ba Vi National Park
4.	Bach Ma National Park
5.	Yok Don National Park
6.	Cat Tien National Park
7.	Hoang Lien National Park
8.	Ba Be National Park
9.	Xuan Son National Park
10.	Bai Tu Long National Park
11.	Cat Ba National Park
12.	Xuan Thuy National Park
13.	Ben En National Park
14.	Pu Mat National Park
15.	Vu Quang National Park
16.	Phong Nha_Ke Bang National Park
17.	Chu Mom Ray National Park
18.	Kon Ka Kinh National Park
19.	Chu Yang Sin National Park
20.	Bu Gia Map National Park
21.	Bidoup Nui ba National Park
22.	Nui Chua National Park
23.	Phuoc Binh National Park
24.	Lo Go Xa Mat National Park
25.	U Minh Thuong National Park
26.	U Minh Ha National Park
27.	Tram Chim National Park
28.	Con Dao National Park
29.	Phu Quoc National Park
30.	Mui Ca Mau National Park

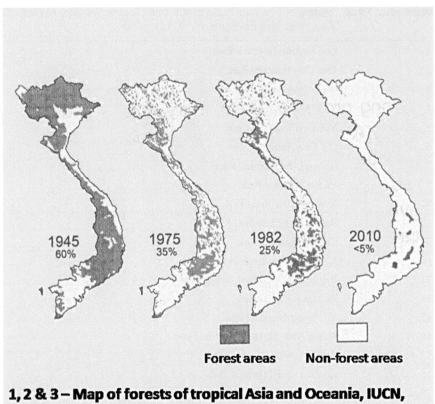

1, 2 & 3 – Map of forests of tropical Asia and Oceania, IUCN, 1991; 4 – Unpublished data of Dung V. V.

Plate 14.1

Index

Printed and bound by CPI Group (UK) Ltd, Croydon, CR0 4YY

23/10/2024

01777704-0016